W9-DCJ-720

Emerging Wireless Technologies and the Future Mobile Internet

This book provides a preview of emerging wireless technologies and their architectural impact on the future mobile Internet. The reader will find an overview of architectural considerations for the mobile Internet, along with more detailed technical discussion of new protocol concepts currently being considered at the research stage.

The first chapter starts with a discussion of anticipated mobile/wireless usage scenarios, leading to an identification of new protocol features for the future Internet. This is followed by several chapters that provide in-depth coverage on next-generation wireless standards, ad hoc and mesh network protocols, opportunistic delivery and delay-tolerant networks, sensor network architectures and protocols, cognitive radio networks, vehicular networks, security and privacy, and experimental systems for future Internet research. Each of these contributed chapters includes a discussion of new networking requirements for the wireless scenario under consideration, architectural concepts, and specific protocol designs, many still at the research stage.

Dipankar Raychaudhuri is Professor-II, Electrical and Computer Engineering and Director, Wireless Information Network Lab (WINLAB) at Rutgers University. WINLAB's research scope includes topics such as RF (Radio Frequency)/sensor devices, cognitive radio, dynamic spectrum access, 4G systems, ad hoc mesh networks, wireless security, future Internet architecture, and pervasive computing. Raychaudhuri is widely recognized as a leader in the future Internet research field and has lectured extensively on the topic at both national and international forums. During 2005–2007, he organized and co-hosted the NSF (National Science Foundation) "Wireless Mobile Planning Group" (WMPG) workshops that inspired and set the stage for much of the content in this book.

Mario Gerla is a Professor in the Computer Science Department at the University of California, Los Angeles. He has led the ONR (Office of Naval Research) MINUTEMAN (Multimedia Intelligent Network of Unattended Mobile Agents) project, designing the next-generation scalable airborne Internet for tactical and homeland defense scenarios and two advanced wireless network projects under U.S. Army and IBM funding. Dr. Gerla is an active participant in future Internet research activities in the United States, co-hosting the NSF WMPG workshops from 2005 to 2007. His research group is an active contributor to the emerging field of vehicular networking and is credited with the "CarTorrent" protocol for peer-to-peer file transfer between vehicles.

Emerging Wireless Technologies and the Future Mobile Internet

Edited by

DIPANKAR RAYCHAUDHURI

WINLAB, Rutgers University

MARIO GERLA

University of California, Los Angeles

CAMBRIDGE
UNIVERSITY PRESS

CAMBRIDGE UNIVERSITY PRESS
Cambridge, New York, Melbourne, Madrid, Cape Town,
Singapore, São Paulo, Delhi, Tokyo, Mexico City

Cambridge University Press
32 Avenue of the Americas, New York, NY 10013-2473, USA

www.cambridge.org
Information on this title: www.cambridge.org/9780521116466

First published 2011

Printed in the United States of America

A catalog record for this publication is available from the British Library.

Library of Congress Cataloging in Publication Data

Raychaudhuri, Dipankar, 1955–
Emerging Wireless Technologies and the Future Mobile Internet /
Dipankar Raychaudhuri, Mario Gerla.
p.   cm.
ISBN 978-0-521-11646-6
1. Wireless Internet.   I. Gerla, Mario, 1943–   II. Title.
TK5103.4885.R39   2011
384.3–dc22   2010049729

ISBN 978-0-521-11646-6 Hardback

CONTENTS

Contents vii

CONTRIBUTORS

Arati Baliga Security R & D Laboratory

Deborah Estrin Department of Computer Science, UCLA

Sachin Ganu Aruba Networks

Mario Gerla Department of Computer Science, UCLA

Omprakash Gnawali Department of Computer Science, Stanford University

Marco Gruteser WINLAB, Rutgers University

Shweta Jain York College, City University of New York

Hisashi Kobayashi Princeton University

Hang Liu InterDigital

George Nychis Department of Electrical & Computer Engineering, Carnegie Mellon University

Max Ott NICTA (National Information and Communications Technology Australia)

Sanjoy Paul InfoSys Technologies Limited

Radha Poovendran College of Engineering, University of Washington

Dipankar Raychaudhuri WINLAB, Rutgers University

Sasank Reddy Department of Computer Science, UCLA

Srinivasan Seshan School of Computer Science, Carnegie Mellon University

Ivan Seskar WINLAB, Rutgers University

Mani Srivastava Department of Computer Science, UCLA

Peter Steenkiste Departments of Computer Science and Electrical & Computer Engineering, Carnegie Mellon University

Wade Trappe WINLAB, Rutgers University

Matt Welsh School of Engineering and Applied Sciences, Harvard University

Suli Zhao Qualcomm

FOREWORD

The current Internet is an outgrowth of the ARPANET (Advanced Research Projects Agency Network) that was initiated four decades ago. The TCP/IP (Transmission Control Protocol/Internet Protocol) designed by Vinton Cerf and Robert Kahn in 1973 did not anticipate, quite understandably, such extensive use of wireless channels and mobile terminals as we are witnessing today. The packet-switching technology for the ARPANET was not intended to support real-time applications that are sensitive to delay jitter. Furthermore, the TCP/IP designers assumed that its end users – researchers at national laboratories and universities in the United States, who would exchange their programs, data, and email – would be trustworthy; thus, security was not their concern, although reliability was one of the key considerations in the design and operation of the network.

It is amazing, therefore, that given the age of the TCP/IP, the Internet has successfully continued to grow by supporting the ever increasing numbers of end users and new applications, with a series of ad hoc modifications and extensions made to the original protocol. In recent years, however, many in the Internet research community began to wonder how long they could continue to do "patch work" to accommodate new applications and their requirements. New research initiatives have been launched within the past several years, aimed at a grand design of "a future Internet." Such efforts include the NSF's FIND (Future Internet Design) and GENI (Global Environment for Network Innovations), the European Community's FP 7 (Frame-network Program, Year 7), Germany's G-Lab, and Japan's NWGN (New Generation Network).

It is therefore extremely timely that Drs. Raychaudhuri and Gerla are publishing this book at this juncture, because better understanding of rapidly evolving wireless technologies and emerging new applications will be crucial in deciding the right architecture for the future Internet. It is not clear at this point which approach among several alternatives proposed or being pursued – ranging from so-called clean-slate architectures to continuous enhancements of the current IP network – will eventually prevail, but there is no question that the *future Internet* architecture must be built with wireless technologies as its major components, and mobility of end users/terminals and security of applications and services must be adequately supported.

The conventional architecture of treating a wireless network as an L-2 level *access* network connected to the core network (i.e., L-3 layer) through a gateway is becoming outdated. As pervasive computing in smart devices and wireless sensors/actuators attached to numerous things are expected to become predominant end users/devices in a future network, a novel network architecture and protocols with end-to-end control and routing, including heterogeneous wireless subnetworks as an integrated part of the entire network, will be called for to provide mobility services with satisfactory performance, security, and scalability. Up to now, wireless technologies have been largely treated as synonymous with wireless *communication* links, where a wireless channel serves merely as an interface between the end mobile user and the core network. In the future network, however, we anticipate that in-network *computing* (or *processing*) of data from sensors and *storing* (or *caching*) of data based on its content ought to be performed.

The introductory chapter of this book presents a variety of emerging wireless networking scenarios and identifies requirements for a new architecture and protocol for each of the mobile networking scenarios. These requirements are then aggregated into a number of key protocol features. Technical issues associated with implementing these wireless/mobility requirements into a unified comprehensive future Internet architecture protocol are then discussed. In the concluding chapter, Drs. Raychaudhuri and Gerla review the overall challenge of evolving the current Internet to meet these mobile networking needs and provide a roadmap for the future.

Hisashi Kobayashi
The Sherman Fairchild University Professor Emeritus,
Princeton University, Princeton, New Jersey

ACKNOWLEDGMENTS

The editors of this book gratefully acknowledge support from the National Science Foundation (NSF) in the form of a planning grant entitled "Planning Grant: New Architectures and Disruptive Technologies for the Future Internet – A Wireless & Mobile Network Community Perspective," CNS-0536545, 2005–07. This grant provided the seed funding for a series of future mobile Internet workshops that led to the publication of the "Wireless Mobile Planning Group (WMPG)" report that motivated this book. We wish to thank our NSF program officer at the time, Dr. Guru Parulkar, for providing the original vision and inspiration behind the future Internet research program in the United States. We are also grateful to Darleen Fisher, Jie Wu, Ty Znati, and Suzanne Iacono of NSF CISE (Computer and Information Science and Engineering) and Chip Elliott of BBN Technologies for ongoing technical discussions, support for community workshops, and guidance of future Internet research under the FIND and GENI programs.

Dr. Raychaudhuri would like to acknowledge sabbatical leave support from his home institution, Rutgers University, during the academic year 2008–2009, when much of this book was planned and organized. Thanks are also due to the Clean Slate Program at Stanford University (led by Prof. Nick McKeown) for providing him with office space and other resources during his sabbatical visit in 2008. Finally, he would like to express his gratitude to his wife, Arundhati Raychaudhuri, for her encouragement and support throughout the course of this project.

Dr. Gerla would like to acknowledge the support of the NSF grant "The Health Guardian – A Gateway to Networked Wellness" and of the NSF-GENI grant "Campus Vehicular Testbed" that helped him focus on mobile and vehicular communications. Also, the International Technical Alliance project (led by IBM) supported some of the time dedicated to the research that went into this book. Finally, Dr. Gerla wishes to express his gratitude to his doctoral student, Eun Kyu Lee, for his outstanding editorial work during the final and very critical phase of integrating all the chapters to a consistent manuscript.

1

Introduction

Dipankar Raychaudhuri and Mario Gerla

1.1 Background

Over the next ten-to-fifteen years, it is anticipated that significant qualitative changes to the Internet will be driven by the rapid proliferation of mobile and wireless computing devices. Wireless devices on the Internet will include laptop computers, personal digital assistants, cell phones (more than 3.5 billion in use as of 2009 and growing!), portable media players, and so on, along with embedded sensors used to sense and control real-world objects and events (see Figure 1.1). As mobile computing devices and wireless sensors are deployed in large numbers, the Internet will increasingly serve as the interface between people moving around and the physical world that surrounds them. Emerging capabilities for opportunistic collaboration with other people nearby or for interacting with physical-world objects and machines via the Internet will result in new applications that will influence the way people live and work. The potential impact of the future wireless Internet is very significant because the network combines the power of cloud computation, search engines, and databases in the background with the immediacy of information from mobile users and sensors in the foreground. The data flows and interactions between mobile users, sensors, and their computing support infrastructure are clearly very different from that of today's popular applications such as email, instant messaging, or the World Wide Web.

As a result, one of the broad architectural challenges facing the network research community is that of evolving or redesigning the Internet architecture to incorporate emerging wireless technologies – efficiently, and at scale.[1] The Internet's current TCP/IP protocol architecture was designed for static hosts and routers connected by wired links. Protocol extensions such as mobile IP have been useful for first-generation cellular mobile services involving single-hop radio links from mobile devices to base stations or access points.[2] However, incremental solutions based on IP are inadequate for dealing with the

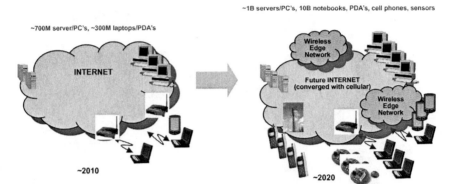

Figure 1.1. Migration of Internet usage from fixed PCs and servers to mobile devices and sensors.

requirements of fast-growing wireless usage scenarios such as multihop mesh,[3] peer-to-peer,[4] disruption-tolerant networks (DTN),[5] sensor systems,[6] and vehicular applications.[7] These emerging wireless scenarios motivate us to consider "clean-slate" network architectures and protocols capable of meeting the needs of these and other emerging wireless scenarios. In the next section (1.2), we present an overview of these emerging wireless networking scenarios, identifying new architecture and protocol requirements for each of these usage cases. These mobile network architecture requirements will then be aggregated into a number of key protocol features in Section 1.3 that follows. Technical challenges associated with implementing these new wireless/mobility requirements into a unified comprehensive future Internet architecture protocol will then be discussed briefly in Section 1.4. Each of the emerging wireless technology scenarios identified in this introductory chapter will then be discussed in greater depth in each of the chapters that follows. In the concluding chapter, we will review the overall challenge of evolving the current Internet to meet these mobile networking needs, and provide a brief view of the road ahead.

1.2 Wireless Technology Roadmap

Wireless and mobile networks represent an active research and new technology development area. The rapid evolution of core radio technologies, wireless networks/protocols, and application scenarios is summarized for reference in the technology roadmap given in Figure 1.2. It can be seen from the chart that in addition to 2.5G/3G cellular data and WLAN systems developed during the 1990s, emerging wireless scenarios include personal-area networks, wireless peer-to-peer (P2P), ad hoc mesh networks, cognitive radio networks, sensor networks, RFID systems, and pervasive computing.

Each of the previously mentioned wireless technologies or usage scenarios is associated with unique network architecture and service requirements that

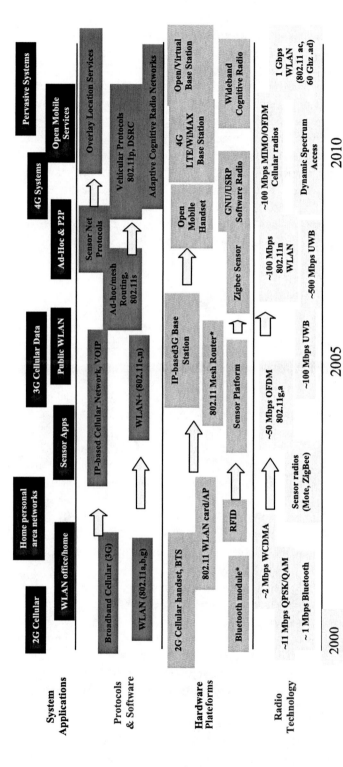

Figure 1.2. Wireless technology roadmap.

3

affect both the access and infrastructure portions. The default approach adopted
by most of the research community is to treat the wireless access portion as a
"layer 2" local area network connecting to the Internet (i.e., layer 3 IP) through a
gateway. This approach is pragmatic, but it precludes uniform dissemination of
control and routing information through the entire network and creates a potential processing bottleneck at the gateway. A more integrated end-to-end control
and routing architecture is important for optimizing mobile/wireless service features such as location management, dynamic handoff, quality-of-service (QoS)
or cross-layer transport. Also, a local-area wireless network may contain one
or more routing elements, which can create inconsistencies in protocol layering
and addressing. If compatibility with the current IP network is not viewed as an
essential constraint, it may be possible to develop a clean-slate network architecture that can accommodate emerging wireless networks in a single unified
protocol structure.

1.3 Wireless Networking Scenarios

The most important wireless technology in use today is the cellular network that
provides mobile phone and data services on handheld devices. Cellular networks
are ubiquitous in all parts of the world, with almost 4 billion cell phones in use
worldwide at the time of writing of this book. Cellular networks have evolved
from first-generation analog systems (such as the AMPS system used in the
United States prior to 1990) to second-generation digital systems (such as GSM
and CDMA[8] used in most parts of the world between 1990 and 2005), and then
to third-generation, or 3G, systems such as CDMA2000 and UMTS/WCDMA
in use since about 2005. Second-generation cellular systems such as GSM are
capable of supporting packet data services at bit-rates of $\sim$100 Kbps, whereas
3G systems such as UMTS or CDMA2000 can deliver between $\sim$300 Kbps –
2 Mbps, depending on signal quality. Further evolution from 3G to 4G cellular
systems with the goal of supporting service bit-rates in the range of $\sim$10–100
Mbps is planned by the industry over the next three to five years. Examples of
4G systems are LTE and WiMAX/IEEE 802.16.

   From a network architecture point of view, cellular has always been built
as a separate custom network with its own set of protocols for key interfaces,
such as mobile terminal to base station and base station to mobility service
gateways such as the MSC and GGSN. These networks were initially built for
integration with the telephone network that was based on a set of signaling
protocols defined by the ITU. More recently, 3G networks have been migrating
toward integration with the IP network using voice-over-IP (VoIP) protocols such
as SIP[9] for signaling and mobility protocols such as mobile IPv6.[2] As data services for mobile devices continue to grow, this may be expected to lead to a
gradual migration of mainstream cellular services to the Internet. However, gradual migration of cellular networks to the Internet involves the use of overlays

and gateways for interfacing between mobile network features such as authentication, addressing, and mobility – an approach that has scalability and performance limitations.

In addition to cellular, a number of short-range wireless data technologies such as WiFi, Bluetooth, and Zigbee have started to penetrate the market for enterprise and home networks starting in the late 1990s. Of these technologies, WiFi (based on the IEEE 802.11 standard) is the most ubiquitous as an Internet access link, with more than 500 million devices in use today, with the number expected to grow to a billion by 2012.

Most of these WiFi devices are used as wireless local area networks (WLAN) that connect to the Internet as "layer 2" networks similar to the widely used Ethernet LANs. When WiFi is used as a home or office LAN, it is the last hop for Internet access, but does not provide mobility or global roaming features associated with the cellular network. As we will see in later chapters, 802.11 WLAN technology is also being used in the ad hoc mode to build new kinds of networks such as peer-to-peer (P2P), vehicular networks (V2V and V2I), and mesh networks. In addition to 802.11 radios, there are several short-range radio standards such as Bluetooth and Zigbee that are used to provide short-range access to devices such as wireless speakers and sensors/actuators. Power and size limitations on the sensor devices imply the need for a more general wireless network architecture that provides connectivity to a range of heterogeneous radios with different transmission ranges. In contrast to the cellular network, the emerging wireless network will incorporate multiple radio technologies operating under a decentralized control framework.

This is illustrated in Figure 1.3, which shows that the overall network architecture is evolving from the separate special-purpose cellular and WiFi networks toward a more general, heterogeneous wireless access network with multiple radio technologies, opportunistic ad hoc association, self-organization, multihop routing, and so on. The long-term architectural goal would be to evolve the Internet architecture to seamlessly meet all the requirements associated with the general wireless "network of networks" shown in the right-hand side of the figure.

Next, let us consider some of the key wireless networking scenarios of importance to the future Internet architecture. The first and most well-understood emerging wireless service scenario is that of anytime, anywhere access to the Internet from personal mobile devices. As shown in Figure 1.4, this scenario implies the need for a network addressing and routing scheme capable of handling roaming and continuous mobility across multiple points of attachment.

User mobility of this sort is handled quite effectively in today's cellular network using the concepts of a "home network" and "visited network." In particular, users of the network have a permanent address to which all communication is initially addressed, and a forwarding (or visiting) address used to temporarily forward connections during mobility outside the home area. A modified form of this approach has been used in the mobile IP specification

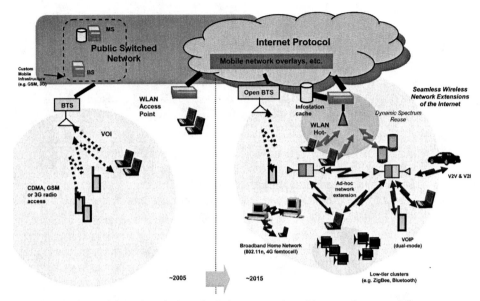

Figure 1.3. Anticipated evolution of wireless network architecture from special-purpose networks to heterogeneous "wireless network of networks."

that is part of IPv6, but is not widely implemented in the Internet today. For connection-oriented traffic, an additional requirement that is also met by today's cellular networks is that of dynamic handoff by which an existing connection can be smoothly migrated from one point of radio attachment to another without setting up a new connection. Clearly, end-user roaming and dynamic mobility support is a key requirement for the future Internet given the rapid increase in mobile data devices. Although mobile IPv6 does provide a solution for this requirement, it may be appropriate to consider alternative approaches toward achieving this functionality in the future network. Mobility also involves security

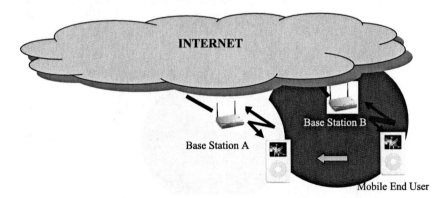

Figure 1.4. Mobile data service scenario.

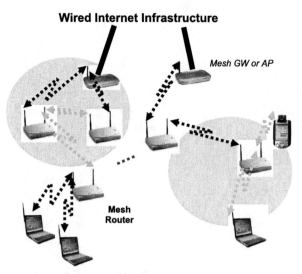

Figure 1.5. Wireless mesh network with multihop routing of data packets between radio nodes.

considerations such as user authentication, which will need to be an integral part of any solution.

A second emerging wireless usage scenario is that of an ad hoc or mesh network in which multiple wireless devices with short-range radios form a multihop network with increased coverage and connectivity. Ad hoc networks were first proposed to support tactical communications between small groups of mobile radio nodes. More recently, multihop mesh architectures (illustrated in Figure 1.5) have been used to extend wireless access network coverage in both urban and rural areas using low-cost short-range radios such as WiFi. In these ad hoc and mesh scenarios, each radio node serves as a router with the capability of forwarding packets to their destination across multiple wireless hops.

Traffic to or from the Internet must pass through one or more gateways or access points that are designed to have both wired and wireless network interfaces. Specialized ad hoc network routing protocols (such as the MANET specification from IETF[10]) have been devised for this purpose, and there is a considerable body of research on this class of routing protocols. Routing in mesh and ad hoc networks generally requires an awareness of cross-layer parameters from the radio links that make up a potential path. Given the growing importance of multihop wireless routing, it may be useful for the future Internet protocol to provide seamless routing across both wired and wireless portions of the network. As for the mobile data service scenario in Figure 1.3, the network needs to support end-user roaming and dynamic mobility as part of the basic transport service.

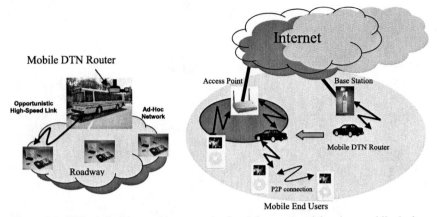

Figure 1.6. P2P wireless networking scenarios involving sensor pickup by a mobile device, or opportunistic content delivery to a passing vehicle.

The third scenario of current significance is the P2P network model in which short-range radios associate with each other opportunistically for content delivery or some type of machine-to-machine (M2M) interaction. This scenario is also sometimes referred to as delay-tolerant networking (DTN), because intermittent opportunistic connectivity implies the need for delay-tolerant applications designed to wait for transmit/receive opportunities. Figure 1.6 shows two kinds of P2P or DTN applications, one in which a bus is picking up data from sensors in the roadway and storing this data for later delivery to the wired network core (perhaps using WiFi or other short-range radios once parked inside its regular garage). The second part of the figure shows the P2P and "Infostations" service models in which users associate opportunistically with each other to exchange content, or when users associate for short periods with wireless data caches (or Infostation) to download popular or personal content. Both these scenarios are important because of the fact that opportunistic short-range radio access is fundamentally faster and more efficient than continuous cellular-type connectivity. Moreover, continuous long-range wireless access may not be feasible for small, low-power sensor devices such as those shown in Figure 1.6. It is noted that the TCP/IP protocol stack used in the Internet today was not designed to support discontinuous or opportunistic connectivity of this type, indicating the need to consider this requirement further when designing future Internet protocols.

Another emerging wireless scenario of importance is that of vehicular networking, involving both V2V (vehicle-to-vehicle) and V2I (vehicle-to-infrastructure) modes. In vehicular networking, cars on the highway may exchange safety information with those in proximity, or might download content (such as navigational maps or audio/video files) from infrastructure access points placed along the highway. The vehicular scenario shares some common elements with the ad hoc and P2P cases considered earlier, but have the additional property of location or geographic awareness. Referring to Figure 1.7, it is observed that

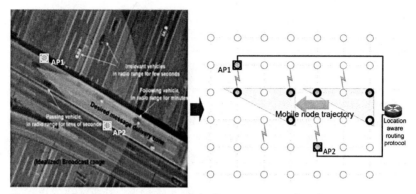

Figure 1.7. Vehicular networking scenario (figure courtesy of Prof. Marco Gruteser).

a typical transmission in a V2V situation is a "geographic multicast" in which a message is propagated to all receivers along a certain section of roadway, but not to those outside that region. This requirement motivates a new service, called geocasting, in which a message is forwarded to all radio nodes within a defined geographic area. This type of network routing is very different from device-address-based routing currently used in the Internet. Given the fact that there are approximately 500 million vehicles worldwide and growing, it would be desirable to consider this geographic routing capability as a requirement when designing future Internet protocols.

Another important wireless scenario is that of sensor networks and pervasive computing (see Figure 1.8). The sensor network scenario generally involves a hierarchical network structure with clusters of low-power sensors connected as

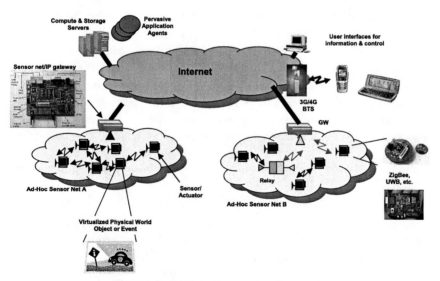

Figure 1.8. Wireless sensor network scenario.

ad hoc multihop networks at the lowest tier. The function of a sensor (or actuator) is to provide a virtualized representation of a physical-world object or event, thus making it possible to design "pervasive computing" applications that allow us to observe and interact with the physical world. The sensor network clusters connect into the Internet cloud through multiple gateways that convert from the localized sensor network protocol to the global Internet protocol.

Within the sensor network cluster, there may also be a tiering of nodes including low-power sensors, relays, forwarding nodes, and gateways. Of course, ad hoc routing considerations similar to those discussed earlier for the ad hoc/mesh case continue to apply. However, there is an additional requirement of energy efficiency because of severe power constraints at each sensor, and there may also be unique data aggregation requirements involving processing and aggregation of data at each transit node. Sensor network applications involve computing and storage servers in the network cloud as shown in Figure 1.8, and there are many related issues of how to architect the computing and networking system given the greater importance of content and location over the physical address itself. Applications will also have an end-user interface, typically a mobile device such as a cellular handset or PDA. Currently sensor systems are built as special-purpose networks with gateways to the Internet, but a long-term goal is to improve scalability and performance by using a single unified protocol across both sensor and Internet clouds.

In concluding this subsection, it is noted that core radio technology itself is going through a fundamental change, moving from hardware radios to cognitive software-defined radios. Examples of early cognitive radio prototypes are the USRP (Universal Software Radio Prototype), WARP from Rice University, the Microsoft Research Software Radio, and the WINLAB WiNC2R software radio platform. Cognitive radios are motivated by the need to use radio spectrum more efficiently to accommodate rapidly increasing wireless traffic. The use of cognitive radios as network elements will enable dynamic spectrum sharing and adaptive networking methods that are inherently more flexible than the radio access technology standards in use today. This implies the need for extensions to control and resource management protocols in the access network, providing for features such as dynamic spectrum coordination, cross-layer awareness, and the ability to set and control radio parameters based on networking requirements.

1.4 Classifying Wireless Networking Scenarios

The NSF Wireless Mobile Planning Group report[1] written in 2005 provides a useful classification for the full range of future wireless networking scenarios, some of which were individually discussed earlier, in Section 1.3. In that report, three distinct clusters of usage scenarios are identified as summarized below.

1.4.1 Scenario A – Individual Wireless Devices Interfacing with the Internet ("Mobile Computing")

The simplest scenario involves a single wireless device that interfaces with the broader Internet. The mobile device may be a cellular phone, a PDA, a media player, a digital camera, or some type of combination consumer device. Mobile computing devices may connect through a wireless local area network, a mesh-style wireless network, or a wide-area wireless technology (such as cellular 3G or WiMAX). Service models to be considered include mobile services, hot-spot services with limited mobility, as well as cached content delivery via opportunistic wireless links. High mobility, the potential for intermittent connectivity, and heterogeneity of radio access are key characteristics of this scenario.

A typical example of this mode of operation is that of a mobile customer downloading a real-time video stream (e.g., a live sporting event) to a portable media player from the Internet. Seamless connectivity should be maintained as the customer moves from a shopping mall (WiFi coverage) to outdoors (2.5G or 3G cellular connectivity), and then to the car (Bluetooth within the car, WiMAX radio to the Internet). At each step, the wireless media player needs to be aware of available connectivity options and then select the best service. The multimedia server must also be aware of current connectivity constraints so that it can deliver a stream with parameters (data rate, format, etc.) consistent with the configuration. The same mobile customer should be efficiently tracked by the network and reachable by VoIP calls if he/she so chooses. Location- or context-aware queries (such as "where is the nearest pharmacy?") and delay-tolerant services (e.g., seamless suspension and resumption of a large file transfer when the user walks or drives through areas without coverage) should be supported. Caching of files for rapid downloading within a hot spot may also be useful in this scenario.

1.4.2 Scenario B – Constellations of Wireless Devices ("Ad hoc Nets")

The second type of wireless scenario is motivated by a variety of settings in which multiple radio devices may be in close physical proximity and can collaborate by forming an ad hoc network. For example, wireless devices in an office or home environment can set up an ad hoc network between themselves to improve coverage and communications quality. Another popular application involving constellations is that of community mesh networks formed by rooftop radios for the purpose of shared broadband access. In the important emerging application of vehicular communication, clusters of cars on the highway may participate in an ad hoc network for the purpose of collision avoidance and traffic flow management. Constellations may include heterogeneous radio and computing devices with different capabilities and resource levels. Emerging cognitive radio technologies also offer the capability of highly adaptive wireless ad hoc networks with physical layer negotiation between nodes, scavenging unused spectrum at low cost to support a private ad hoc network. Opportunistic association, changing network topologies, varying link quality, and potentially large scale (in terms of number of nodes) are some of the characteristics of this scenario.

A simple example of opportunistic constellations is the formation of an ad hoc network between several user laptops in a meeting room with limited Internet access coverage. The ad hoc network enables high bandwidth communication between participants at the meeting and allows them to use a favorably positioned (e.g., with good cellular network throughput) node as a forwarding relay to the Internet. Another example is the cooperative downloading of popular files from the Internet by drivers on a highway, when hot-spot "Infostations" with WiFi service are spaced by several miles on the highway, and a car traveling at 60 miles per hour may not be able to download an entire file through short V2I (vehicle-to-infrastructure) mode access. If several drivers are interested in the same file, it is possible for the cars to collaborate and exchange segments in a P2P opportunistic networking arrangement similar to that used in Bit Torrent (see Chapter 7). This allows the download to be completed without requiring a car to stop at a hot spot, saving time for the end-user and avoiding traffic congestion problems. The same ad hoc networking capability can also be used by cars to exchange control information necessary for traffic flow management or collision avoidance.

Ad hoc radio constellations also apply to civilian disaster recovery and in tactical defense environments. These applications usually involve communications between a number of first responders or soldiers who work within close proximity of each other. The response team may need to exchange text messages, streaming media (e.g., voice or video), and use collaborative computing to address a shared task such as target recognition or identification of a spectral jammer. Individual nodes may also need to access the Internet for command-and-control purposes or for information retrieval. This application has similarities with the ad hoc mesh network for suburban or rural broadband access mentioned earlier.

1.4.3 Scenario C – Pervasive Systems and Sensor Networks ("Sensor Nets")

Sensor nets refer to a broad class of systems involving embedded wireless devices connected to the Internet. The first generation of sensor networks involves collecting and aggregating measured data from large numbers of sensors in a specified geographic area. In the near future, sensor net applications will also include closed-loop sensor/actuator systems for real-time control of physical world objects. Current sensor net applications are in science (ecology, seismology, ocean and atmospheric studies, etc.) and engineering (water quality monitoring, precision agriculture, livestock tracking, structural monitoring), as well as consumer-oriented applications (home security and energy management, hobbyist and sports enthusiast applications of distributed imaging, eldercare, pet monitoring, etc.). Sensor networks share several characteristics of ad hoc scenarios but are differentiated by the fact that tiny sensor devices have more stringent processing power, memory, and energy constraints. These constraints generally imply the need for a hierarchical ad hoc network structure in which low-tier sensor nodes connect to the Internet via one or more levels of repeating wireless gateways. Other important characteristics of this scenario are the data-centric nature of applications, potential for large scale (in terms of numbers of sensors), and geographic locality.

Traditionally, large "sensor fabrics" such as those installed to monitor the environment have been designed as vertically optimized systems, with an ad hoc network designed to meet specific energy and processing constraints and optimized to support specialized queries dictated by the application at hand. The interface to the Internet has been via edge nodes that isolate the Internet stack from the sensor fabric architecture. However, more recent trends indicate an increased need for sensor networks that provide open access via the Internet, in a more extensive and capillary way that can be supported via edge nodes. For instance, scientists interested in the correlation between data found in different data bases (e.g., soil characteristics, pollutants carried in the local water supplies, productivity of local vineyards, production and sale of local wines) can be permitted to access specific regions within a sensor fabric directly from the Internet to extract the required data rather than overburdening the access gateways. Moreover, new types of sensor networks based on "mobile" sensor platforms are becoming available – for example, vehicles in the urban grid or firefighters in a disaster recovery operation equipped with a variety of sensors (video, chemical, radiation, acoustic, etc.). These sensor platforms have practically unlimited storage, energy, and processing resources. The vehicle grid then becomes a sensor network that can be accessed from the Internet to monitor vehicle traffic congestion and to help investigate accidents, chemical spills, and possible terrorist attacks. Likewise, firefighters carry cameras and several other sensors, allowing the commander to be aware of the conditions in the field and to direct the operations to maximize the use of the forces while preserving the life of his responders. These latter examples also show that the gap between sensor networks and ad hoc networks tends to diminish in mobile sensor systems at least in terms of communications capabilities and Internet access. In the longer term, pervasive systems involving personal mobile devices, smart offices/homes, and densely deployed multimodal sensors/actuators will serve as a platform for development of various new applications ranging from tracking and inventory control to personal productivity, public safety, and resource management.

1.5 Future Network Requirements

Before moving to more detailed discussions of future wireless scenarios and their networking protocols in the following chapters, let us briefly consider the general future network design requirements that arise from the scenarios introduced in this chapter.

Considering the wide range of future wireless network usage scenarios (4G cellular/mobile, WLAN, mesh, P2P, DTN, sensor networks, vehicular networks, sensor/pervasive systems), it is important to extract a set of common requirements general enough to meet these needs, as well as those of future applications that cannot easily be predicted today. We suggest an approach for decomposing these requirements into two major categories, the first reflecting the intrinsic properties of the radio medium and the second reflecting the needs of future mobility and pervasive services. It is important to note that these requirements should apply to future access networks and the Internet protocol stack as a whole in view of the increasingly predominant role of wireless end-user devices. The current approach of designing specialized networking solutions for cellular systems, ad hoc nets, sensor applications, and so on leads to undesirable fragmentation (and hence poor scalability, lack of interoperability, inefficiencies in application development, etc.) among different parts of the network, and needs to be replaced by a unified end-to-end protocol architecture that supports emerging requirements of both wired and wireless networks.

To elaborate further, basic transport services of future Internet protocols should reflect intrinsic radio properties such as spectrum use, mobility, varying link quality, heterogeneous PHY, diversity/MIMO, multihop, multicast, and so on, and the capabilities of emerging radio technologies such as LTE, next-generation WLAN, Bluetooth, Zigbee, vehicular standards such as 802.11p, and of course, cognitive software-defined radio (SDR). In addition, Internet protocol service capabilities should be designed to serve emerging uses of wireless technology, not only for conventional mobile communications, but also for content delivery, cloud computing, sensing, M2M control, and various other pervasive system applications.

Here, we briefly identify some of the key requirements for a future network designed to support the range of wireless usage scenarios discussed in Sections 1.3 and 1.4. Of course, it might not be feasible to achieve the full set of requirements in a single networking architecture, but it is still instructive to understand all the needs in a top-down manner before considering implementation issues. Examples of specific mobile network protocol features that may be useful are:

1. *Dynamic spectrum coordination capability*: Historically, network protocols have been designed to support resource management in terms of wired network concepts such as link bandwidth and buffer storage. As radios

become an increasingly important part of the network, it will be useful to be able to specify and control radio resources within the networking protocol itself. For example, control protocols should be able to support dynamic assignment of spectrum to avoid conflicts between multiple radio devices within the network. Just as current IP networks incorporate protocols such as dynamic host control protocol (DHCP) for address assignment, future networks could incorporate a distributed repository of spectrum usage information that could then be used to assign nonconflicting spectrum to radio devices when they join the network.

2. *Dynamic mobility for end-users and routers*: As more and more end-user devices become wireless, networks will need to be designed to support mobility as a normal mode of operation rather than as a special case. This means that end-user devices should be able to attach to any point in the network (i.e., global roaming), with the network providing for fast authentication and address assignment at a very large scale. Currently available mechanisms such as DHCP and mobile IP represent a first in this direction, but a more general solution could involve a clean separation of naming and addressing where each device would have a unique name, but would only be assigned a routable address local to the network with which it is currently associated (and this routable address may be as general as a geographic location, i.e., geo-address). The main challenge is to provide a distributed global name resolution and address assignment service that scales to the level of billions of mobile devices. Because wireless devices may also serve as routers in some of the ad hoc environments discussed earlier, the network should be able to support dynamic migration of subnetworks. In addition, dynamic handoff of traffic from one point of attachment to another may be required for certain connection-oriented services.

3. *Fast discovery and ad hoc routing*: Because several wireless usage scenarios involve ad hoc associations and continuously changing network topology, it is important for the network to support fast discovery of neighboring network elements. Discovery protocols for ad hoc networks should support efficient topology formation in multihop wireless environments taking into account both connectivity requirements and radio resources. Multihop wireless scenarios further require efficient ad hoc routing between network elements with dynamically changing topologies and radio link quality. The ad hoc routing protocol used in wireless access networks should seamlessly integrate with the global routing protocol used for end-to-end connectivity.

4. *Cross-layer protocol stack for adaptive networks*: Routing in multihop wireless networks requires a greater awareness of radio link layer parameters to achieve high network throughput and low delay. This means that the network's control plane should include information about radio link parameters to be used for algorithms that support topology discovery and routing. A

key architectural issue is that of determining the appropriate granularity and degree of aggregation with which this cross-layer information is exchanged across different parts of the network (i.e., access, regional, core, etc.).

5. *Incentive mechanisms for cooperation*: Ad hoc mobile networking scenarios typically involve cooperation among independent wireless devices. It will be important for future Internet protocols to include protocols that enable such cooperation, first by advertising resources and capabilities to neighboring radios and second by providing mechanisms for exchange of credits or barter of resources in return for services such as relaying or multihop forwarding.

6. *Routing protocols for intermittent disconnection*: Today's Internet routing and transport protocols are designed under the assumption of continuous connectivity. However, this assumption is no longer valid for mobile devices that frequently experience disconnection due to radio signal fading and/or service unavailability. Future protocols should be designed for robustness in presence of occasional disconnection. In order to achieve this, the network generally needs to be able to store in-transit data during periods of disruption, while forwarding messages opportunistically when a path becomes available.

7. *Transport protocols for time-varying link quality*: Reliable delivery of data on the Internet is currently accomplished using transport control protocol (TCP) for end-to-end flow control and error control. TCP is known to perform poorly in wireless access networks that are characterized by higher packet error rates than wired links, along with time-varying bandwidth caused by variations in radio channel quality and medium access control (MAC) layer contention. Future transport layer protocols should be designed to work efficiently in presence of packet errors and varying end-to-end bandwidth – this will require the ability to distinguish between congestion in the network and channel quality variations.

8. *Efficient multicasting and multipath routing*: The wireless channel has inherent multicast capabilities, that is, a single packet sent by a radio is simultaneously received by all receivers within the transmission range. This property can be exploited to improve network performance in various scenarios, but the routing and transport protocols have to be enhanced to support multicast operation as a core capability. Radio multicast also opens up the possibility of multipath routing in which multiple independent paths are used for routing a single packet to improve end-to-end reliability and delay.

9. *Location awareness and geographic routing*: As discussed earlier, emerging pervasive computing applications (i.e., vehicular, sensor, M2M) often require the ability to delivery packets to an entire geographic region rather than to a specific IP address. Also, for mobility services, knowledge of the current geographic location is central to providing various new services such as navigation and geographic search. This means that future networks should provide location information as a basic control plane capability. In

addition, it would be desirable to optionally offer geographic multicast and routing modes by which packets can be delivered to a specified geographic region.

10. *Content- and context-awareness*: A number of future network service scenarios involve content addressability or content routing. For example, an M2M application might involve a query for a particular functionality (such as "printer"), and it would thus be useful if the network protocol can resolve a content query to one or more specific network addresses. Another network capability to be considered is that of content routing by which network routers forward traffic based on content attributes of the data being carried in the packet rather than the IP address in the header.

11. *In-network storage for content caching*: A network with content addressability capabilities can also be enhanced to provide in-network storage and caching services in an integrated manner. Caching of popular or personal content can provide significant improvements in both end-user application performance and network throughput. Although these capabilities can be provided above the network as an "overlay," it is worth considering whether content caching should be fully integrated with the network layer protocol to minimize control overheads and delay.

12. *Programming model for in-network processing*: Emerging sensor and pervasive applications may involve in-network computation for functions such as data aggregation, data-dependent routing or local content search. Whereas these functions are typically implemented above the networking layer as overlays, it is worth considering basic computing features for a future mobile network in which an increasing proportion of applications would benefit from in-network computation. A key issue is the design of a programming model by which to specify optional computational functions at each network element.

13. *Enhanced security and privacy for radio medium*: Because the wireless channel is open to eavesdroppers and potential denial-of-service attackers, it is important to consider enhanced security and privacy features for emerging mobile networks. User mobility implies the need for stronger authentication features as a baseline for any device joining the network, while the open radio medium means that transmissions should generally use strong encryption. In addition, if the network has information about location or content, it would be important to build in privacy guarantees that prevent tracking of users or their content.

1.6 Discussion

In Section 1.5, we have used a top-down approach to identify a number of new network protocol capabilities that would be desirable for the future mobile Internet. Clearly, it is very difficult to incorporate all or most of these features into a single network architecture even if we start from a so-called clean slate. Moreover, clean-slate design of an existing network as large and complex as the Internet is not really a practical option, and any practical attempt to upgrade functionality must eventually consider factors such as backward compatibility, evolutionary upgrade of equipment, equipment cost, software complexity, and so on. However, the top-down clean-slate design methodology described in this book is expected to be beneficial because it exposes key requirements and design issues without being constrained by current practices. Although a single new Internet protocol is unlikely to emerge from this methodology, it may be expected that researchers will design and validate several of the key network capabilities outlined in Section 1.5, and eventually some of these ideas will migrate into the mainstream Internet protocol. In the chapters that follow, we will explore the details of protocol design for each of the emerging wireless service scenarios outlined in this introductory chapter. In the concluding chapter, we will briefly discuss a roadmap to the future, including some strategies for how to put all these ideas together into a unified network architecture.

References

[1] Raychaudhuri, D. and Gerla, M. 2005. New Architectures and Disruptive Tech-
     nologies for the Future Internet: The Wireless, Mobile and Sensor Network Per-
     spective. *Report of NSF Wireless Mobile Planning Group (WMPG) Workshop.*
     http://www.winlab.rutgers.edu/WMPG
[2] Perkins, C. 1998. Mobile IP. *IEEE Communications Magazine,* 35(5), 84–86, 91–99.
[3] Akyildiz, I., Wang, X., and Wang, W. 2005. Wireless Mesh Networks: A Survey,
     Computer Networks. *Computer Networks and ISDN Systems,* March.
[4] Lamming, M. and Bohm, D. 2003. SPECs: Another Approach to Human Context and
     Activity Sensing Research, Using Tiny Peer-to-Peer Wireless Computers. *Lecture
     Notes in Computer Science,* October.
[5] Fall, K. 2003. A Delay Tolerant Network Architecture for Challenged Internets. *ACM
     SigComm.*
[6] Culler, D., Estrin, D., and Srivastava, M. 2004. Guest Editors' Introduction: Overview
     of Sensor Networks. *IEEE Computer Magazine,* August.
[7] Zanella, A., Fasolo, E., Padova, C., Chiasserini, F., Meo, M., Torino, M., Frances-
     chinis, M., and Spirito, A. 2006. Inter-vehicular communication networks. *Second
     Internal NEWCOM Workshop.*
[8] Rappaport, T. 2002. *Wireless Communications: Principles and Practices.* Second ed.
     Prentice Hall.
[9] Rosenberg, J. et al. 2002. SIP: Session Initiation Protocol. *Internet Engineering Task
     Force RFC 3261.*
[10] Perkins, C., and Belding-Royer, E. 2003. Ad hoc On-Demand Distance Vector
     (AODV) Routing. *Internet Engineering Task Force, RFC 3561.*

2

Next-Generation Wireless Standards and Their Integration with the Internet

Hang Liu

Abstract

Standards provide the foundation for developing innovative technologies and enabling them to be widely adopted in market. Several major international standard bodies are developing next-generation wireless standards, including the Institute of Electrical and Electronics Engineers (IEEE), the Internet Engineering Task Force (IETF), the International Telecommunication Union Radiocommunication Sector (ITU-R), the European Telecommunications Standards Institute (ETSI), and the Third Generation Partnership Project (3GPP). The standardization activities of IEEE 802 committee mainly focus on physical (PHY) and media access control (MAC) layers, that is, layers 1 and 2 of the network protocol stack, including WLAN, WMAN, and WPAN network interfaces. IETF standards deal with layer 3 and above, in particular with standards of the TCP/IP and Internet protocol suite, including mobile IP and mobile ad hoc networks (MANET) related protocols. ITU-R is one of the three sectors of the ITU and is responsible for radio communications. It plays a vital role in the global management of the radio-frequency spectrum and satellite orbits, and developing standards for radio communications systems to assure the necessary performance and quality and the effective use of the spectrum. ETSI is a European standards organization for producing globally applicable standards for information and communications technologies (ICT), including fixed, mobile, broadcast, and Internet technologies. ETSI inspired the creation of, and is a partner of, 3GPP – a collaboration project between groups of telecommunications associations worldwide. 3GPP's original scope was to produce technical specifications and technical reports for a globally applicable 3G cellular mobile system based on evolved Global System for Mobile communications (GSM) core networks and radio access technologies, as well as maintain and develop GSM technical specifications and reports. It is currently developing 4G mobile network system. 3GPP standardization

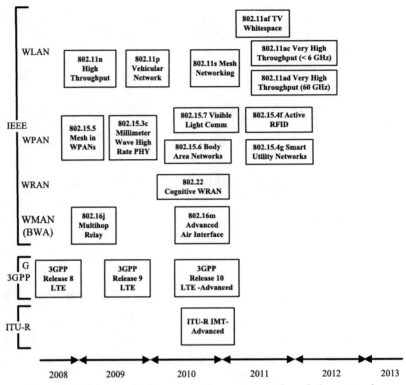

Figure 2.1. Major standards processes for next-generation wireless networks.

encompasses radio access, core network, and service architecture. Figure 2.1 illustrates major standards processes for next-generation wireless networks in IEEE, 3GPP, and ITU-R.

2.1 Technology and Service Trends of Emerging Wireless Standards

The standardization efforts for future wireless networks focus on both new radio access interfaces and improved network architectures. The standardization work on new radio interfaces aims at increasing network capacity to match or shorten the gap with wireline broadband access, and improving bandwidth efficiency and coverage range by employing advanced physical and MAC layer techniques such as multiple-input and multiple-output (MIMO), orthogonal frequency-division multiple access (OFDMA), and space-division multiple access (SDMA), as well as extending battery life and reducing latency for real-time communications. As shown in Table 2.1, future WLAN and WPAN standards will support up to 1 Gbps data rate, and future WMAN and cellular standards can support a peak

Table 2.1. *Emerging Wireless Interfaces*

Standard	Maximum PHY Rate	PHY Technology	MAC Technology	Operating Frequency
802.11n WLAN	600 Mbps (4 × 4 MIMO, 4 spatial streams, 40 MHz bandwidth); 200 Mbps (3 × 3 MIMO, 3 spatial streams, 20 MHz bandwidth)	MIMO and OFDM	EDCA and HCCA	<6 GHz, typical 2.4 GHz and 5 GHz
802.11ac WLAN	>1 Gbps for multi-station; >500 Mbps for a single link	MU-MIMO and OFDM	SDMA	<6 GHz, typical 2.4 GHz and 5 GHz
802.11ad WLAN	>1 Gbps	TBD	TBD	60 GHz
802.15.3c high rate WPAN	5 Gbps on 2 GHz bandwidth	Single carrier and OFDM	TDMA and CSMA-CA	60 GHz
802.15.4/4a low rate WPAN	250 kbps with 802.15.4; 27 Mbps with 802.15.4a UWB PHY; 1 Mbps with 802.15.4a spread spectrum PHY	Spread spectrum and UWB	TDMA and CSMA-CA	Spread spectrum PHY: typical 2.4 GHz, 915 MHz, 868 MHz; UWB PHY: 3 GHz to 5 GHz, 6 GHz to 10 GHz, and less than 1 GHz
802.16m WMAN	300 Mbps for downlink (4 × 4 MIMO, 20 MHz bandwidth); 135 Mbps for uplink (2 × 4 MIMO, 20 MHz bandwidth)	MU-MIMO and OFDM	OFDMA in downlink and uplink	<6 GHz
3GPP LTE E-UTRAN	300 Mbps downlink (4 × 4 MIMO, 20 MHz bandwidth); 75 Mbps uplink for a user (SC-FDMA, 20 MHz bandwidth)	MU-MIMO and OFDM	OFDMA in downlink and SC-FDMA in uplink	<6 GHz

downlink rate of several hundred Mbps and a peak uplink rate of ~100 Mbps under high mobility.

It is critical to utilize the spectrum efficiently and ensure the coexistence of different wireless systems. Cognitive and dynamic spectrum access schemes provide a promising solution. In addition, new FCC regulations for unlicensed devices to operate in the TV whitespace requires that the secondary whitespace devices have cognitive radio and dynamic spectrum access capabilities and shall not interfere the operation of primary users. Several standard working groups and committees such as IEEE 802.22, IEEE SCC41, IEEE 802.19, and IEEE 802.11 are developing or plan to develop the standards for radio systems to operate in TV whitespace using cognitive radio technology.

The standardization work on the mobile network architecture aims at optimizing network performance, improving cost efficiency, facilitating the fixed-mobile convergence and mass-market IP-based services with seamless mobility and global roaming capability, as well as enhanced network QoS and security. New network architecture to integrate various radio access technologies under IP is defined in 3GPP to support seamless global roaming, interworking, and vertical handover between different access systems. In addition, IEEE 802.21 also defines a layer 2 solution to support mobility and media independent handover.

Multihop wireless networks are emerging as a promising architecture to extend wireless coverage in a flexible and cost-effective way. They have broad applications in Internet access, emergency networks, public safety, and so forth. Technical solutions for multihop wireless networks are being specified in IEEE 802.11s, 802.16j, 802.16m, 802.15.5, and 3GPP LTE-advanced. IETF has also defined routing protocols for mobile ad hoc networks.

2.2 Radio Technologies in Next-Generation Wireless Standards

2.2.1 Emerging IEEE WLAN Standards

The throughput of wireless LANs[1] keeps increasing with advances in radio technologies. The new IEEE 802.11n standard[2] is able to achieve up to 600 Mbps data rate when operating on 40 MHz bandwidth by using advanced physical layer techniques including MIMO and channel bonding. 802.11n supports backward compatibility with 54 Mbps 802.11a/g radios. At the MAC layer, it is still based on carrier-sensing multiple access with collision avoidance (CSMA/CA) contention-based media access, called enhanced distributed channel access (EDCA) and polling-based content-free media access, called hybrid coordination function controlled channel access (HCCA). To take advantages of high physical layer data rate and reduce protocol overhead, 802.11n defines two levels of aggregation at MAC layer. MAC Service Data Unit (MSDU) aggregation is processed at the top of MAC by packing multiple MSDUs into an aggregated MSDU, and MAC Protocol Data Unit (MPDU) aggregation is processed at the bottom of the MAC by packing multiple MPDUs into an aggregated MPDU. Block acknowledgment mechanism defined in 802.11e is also enhanced in 802.11n for better performance. These MAC features reduce the overhead, thus increasing the user-level data rate.

As wireless usage grows, there exists an increasing need for additional capacity. To provide comparable throughput as gigabit per second wired LAN products, a new task group (TG), 802.11ac[3] Very High Throughput for Operation in Bands below 6GHz, was formed in September 2008 to develop the enhancements to both the 802.11 PHY and MAC that enable modes of operation capable of supporting a maximum multistation (STA) throughput of at least 1 Gbps and a maximum single-link throughput of at least 500 Mbps while ensuring backward compatibility and coexistence with legacy IEEE 802.11 devices in the 5 GHz unlicensed band. 802.11ac will also provide enhancements over 802.11n on a set of other interdependent performance indicators including range of operation, spectrum efficiency, and power consumption.

In order to provide higher throughput than IEEE 802.11n, Space-Division Multiple Access (SDMA) has been proposed in the 802.11ac TG to handle multiple simultaneous communications between an access point and its associated stations. In general, SDMA employs multiuser MIMO (MU-MIMO) as a channel access method and allows a station to transmit (or receive) signal to (or from) multiple other stations in the same band simultaneously. Compared to point-to-point MIMO or single-user MIMO used in 802.11n, MU-MIMO leverage the availability of multiple independent stations and their diverse channel conditions to create parallel spatial channels using beam forming for superior communications performance in radio multiple access systems. Other techniques proposed to 802.11ac include backward compatibility and coexistence with 802.11n and other WiFi systems, support of more than 40 MHz channel bonding, and more than 4 MIMO antenna elements. The projected timeline for this task group is to have an initial draft by November 2010 and the approved standard in 2012.

For wireless access in vehicular environments (WAVE), IEEE 802.11 TGp[6] is specifying amendments to 802.11 to support Intelligent Transportation Systems (ITS) applications, which include data exchange between high-speed vehicles and between the vehicles and the roadside infrastructure in the licensed ITS band of 5.9 GHz. It specifies the functions and services that allow WAVE-conformant 802.11 stations to operate in a rapidly varying environment and to establish communications quickly each other. IEEE 1609 Family of Standards for Wireless Access in Vehicular Environments is a higher layer standard on which IEEE 802.11p is based.

IEEE 802.11ad[4] is developing technology to enable WLAN operation in the 60 GHz frequency band (typically 57–66 GHz). Due to high available bandwidth at 60 GHz band, multi-gigabit per second throughput can be achieved to support high throughput applications such as simultaneous streaming of multiple HDTV video streams or less compressed/uncompressed video streams, very-high-speed Internet access, wireless data bus for cable replacement, and so forth. It is expected that future mobile devices can be equipped with multiband WLAN access capabilities, short-range multi-Gbps throughput using 60 GHz band, and middle-range Gbps throughput operating at 5GHz band with seamless session transfer. 802.11ad is investigating the fast session transfer techniques between 60 GHz and 2.4/5 GHz. It is also studying the mechanisms that enable coexistence with other systems in the band, including IEEE 802.15.3c[14] systems.

2.2.2 Emerging IEEE WPAN Standards

Unlike wireless LANs, WPANs are used to convey information over relatively short distances, generally up to 10 meters, among a relatively few participants via power efficient and inexpensive networks. WPAN involves little or no infrastructure. IEEE 802.15 Task Group 3c (TG3c) is developing a millimeter-wave-based high-rate WPAN. The 802.15.3c WPAN will operate in the 60 GHz unlicensed band. The standard defines three PHY modes with different modulation and channel coding techniques, which can achieve a data rate up to 5 Gbps on the 2.16 GHz channel bandwidth.

802.15.3c MAC is based on 802.15.3 piconet with enhancements. A piconet is an ad hoc network that allows a number of devices to communicate with each other. One device acts as a piconet coordinator (PNC) that provides the basic timing for the piconet with beacons, and manages the QoS requirements, power save modes, and access control to the piconet. A piconet is formed without preplanning and as long as the piconet is needed.

Timing in the 802.15.3 piconet is based on the superframe composed of beacon, contention access period (CAP), and channel time allocation period (CTAP). The beacon is used to set the timing allocations and to communicate management information for the piconet. The CAP uses CSMA/CA as the medium access mechanism for commands and asynchronous data. CTAP is composed of channel time allocations (CTAs) that can be used for commands, isochronous streams, and asynchronous data connections.

Sensor networks will become part of Internet to provide various types of information. The IEEE 802.15 TG4 has defined the PHY and MAC specifications for low data rate, low complexity, and low power consumption WPANs for inexpensive devices. The 802.15 TG4 and its later enhancements TG4a, TG4c, and TG4d have defined various physical layer modes. These PHYs use different techniques such as spread spectrum or ultra-wideband (UWB), support different data rates from 20 Kbps to 27.24 Mbps, operates at different frequency band to meet different country's regulations – for example, 2.4 GHz ISM band, 915 MHz, 3 GHz to 5 GHz, and the like – and targets different applications such as sensors, interactive toys, smart badges, remote controls, and automation. The IEEE 802.15.4 standard is the basis for the ZigBee, WirelessHART, and MiWi specifications, each of which further offers a complete networking solution by developing the upper layers not covered by 802.15.4.

Depending on the application requirements, an IEEE 802.15.4 low-rate WPAN (LR-WPAN) may operate in either the star topology or the peer-to-peer (P2P) topology. It can be formed automatically. At the MAC layer, 802.15.4 LR-WPAN can use unslotted CSMA-CA or a superframe structure. A superframe contains contention free period (CFP) with guaranteed time slot for low-latency applications or applications requiring specific data bandwidth, as well as CAP with slotted CSMA-CA. The standard was developed with limited power supply availability of the devices in mind. A device may spend most of its operational life in a sleep state, only periodically listening to the channel in order to determine whether a message is pending.

Moreover, TG4f[10] is currently defining the new PHY layer and enhancements to the 802.15.4 MAC layer for active radio-frequency identification (RFID) systems. TG4g[11] is defining an amendment to 802.15.4 to facilitate very large scale process control applications such as the utility smart-grid networks, capable of supporting large, geographically diverse networks with minimal infrastructure, and potentially millions of nodes. The IEEE 802.15 TG6[12] is developing a standard for body area networks, and the IEEE 802.15 TG7[13] is defining a PHY and MAC standard for visible light communications (VLC). The low-power and low-cost sensor networks are expected to connect to the Internet in certain ways to provide various types of information.

2.2.3 Emerging 3GPP and IEEE Mobile Broadband Access Standards

Regarding cellular networks, the ITU-R has commenced the process of developing the International Mobile Telecommunications-Advanced (IMT-Advanced) systems standards[26,27,28] for next-generation (4G) mobile networks. The first invitation for the submission of proposals for candidate radio interface technologies (RITs) or a set of RITs (SRITs) for the IMT-Advanced was issued in March 2008. Under the current schedule, the deadline for submission of candidate RIT and SRIT proposals was October 2009, and it is anticipated that the development of radio interface specification recommendations will be completed in 2011.

According to ITU-R requirements, IMT-Advanced provides enhanced data rates to support advanced services and applications (100 Mbps for high mobility and 1 Gbps for low mobility were established as target peak downstream rates), as well as improved spectrum efficiency and battery life. It will be fully IP-based system with voice carried by VoIP, which is different from hybrid circuit-switching and packet-switching IMT-2000 (3G) mobile communications systems. IMT-Advanced also has capabilities for supporting high-quality multimedia applications in a cost-efficient manner, providing a significant improvement in performance, quality of service, and security. It has key features such as worldwide roaming capability, compatibility of services within IMT and with fixed networks, capability of interworking with other radio access systems, and high-quality mobile services.

Both IEEE 802.16m[9] and 3GPP LTE-Advanced projects are developing advanced air interfaces to meet the cellular layer requirements of ITU-R IMT-Advanced. They are based on MIMO and OFDMA radio technologies with enhanced QoS and security. This reflects the technology trend from code division multiple access (CDMA) based hybrid circuit/packet switching 3G wireless systems to OFDMA-MIMO-based packet-switching 4G systems.

OFDMA employs orthogonal frequency-division multiplexing (OFDM) digital modulation scheme as a multiuser channel access strategy. It allows assigning subsets of subcarriers to individual users and simultaneously transmits to or receives signals from multiple users, achieving even better system spectral efficiency by leveraging channel frequency selectivity of multiple users and adaptive subcarrier assignment.

Compared to CDMA, OFDMA can better combat multipath and achieve a higher MIMO spectral efficiency because it can have flatter frequency channels than a CDMA RAKE receiver. In addition, OFDM is more flexible in the use of spectrum than CDMA. CDMA requires a wide bandwidth to maintain high chip rates and high spectral efficiency, and it is complex to implement radios with capability of different chip rates and spectrum bandwidths. 3G radio interface such as wideband CDMA (W-CDMA) thus defines the fixed 5 MHz channel spectrum bandwidth. However, this limits the flexibility in system deployment and the maximum bandwidth per handset. OFDMA can easily control the data rate and error probability of each individual user by dynamically allocating resources in the time and frequency domains. It offers a cost-efficient solution for wide bandwidth communications with high peak rates. Therefore, it is considered as more suitable for next-generation broadband wireless networks.

Evolved Universal Mobile Telecommunications System (UMTS) Terrestrial Radio Access (E-UTRA) was introduced in 3GPP Release 8 in 2009. E-UTRA aims at significantly increasing data rates for mobile stations, lowering end-to-end latency for real-time communications, and reducing setup times for new sessions. It uses OFDMA for the downlink and Single Carrier Frequency Division Multiple Access (SC-FDMA) for the uplink and employs MIMO with up to four antennas per station. It supports both single-user MIMO and multiuser MIMO for downlink, and SDMA for uplink. Both frequency-division duplexing (FDD) mode and time-division duplexing (TDD) mode with a number of defined channel bandwidths between 1.25 and 20 MHz are supported to provide system deployment flexibility. The E-UTRA provides a peak downlink rate of 300 Mbps with 4 × 4 MIMO antennas and a peak uplink rate of 75 Mbps for a mobile user over 20 MHz channel, which greatly improves network capacity over 3G systems. MIMO enables ten times as many users per cell as 3GPP's original W-CDMA radio access technology. E-UTRA also increased spectral efficiency by two-to-four times compared to 3GPP CDMA-based UTRA interface. Improvements in architecture and signaling further reduce round-trip latency. It also enhances multicast service capability with single-frequency network support. In addition, E-UTRA improves coverage and battery life. However it is an entirely new air interface and incompatible with W-CDMA. E-UTRA is designed only to connect to 3GPP's new IP-based evolved packet core network.

3GPP is developing further advancements for E-UTRA, also called LTE-advanced, to meet all the IMT-advanced requirements for 4G, which is compatible with E-UTRA and expected to be included in 3GPP Release 10. 3GPP's proposal to ITU-R IMT-Advanced will be based on the LTE-Advanced. Multiple techniques including air interface optimization, scalable system bandwidth up to 100 MHz, enhanced precoding and forward error correction, hybrid OFDMA and SC-FDMA in uplink, relay nodes, advanced inter-eNodeB coordinated MIMO, and so forth are under investigation.

IEEE 802.16m is amending the IEEE 802.16 OFDMA specification to meet the cellular layer requirements of IMT-Advanced, while providing continuing support and upgrade path for IEEE 802.16–2005 based WiMAX OFDMA system. It supports scalable bandwidths from 5 to 40 MHz, with a normalized peak data rate of 15.0 bps/Hz for downlink (4×4 MIMO) and 6.75 bps/Hz for uplink (2×4 MIMO). Both TDD and FDD modes are supported. IEEE 802.16m aims to be the IEEE candidate radio interface for IMT-Advanced 4G mobile networks and compete with 3GPP LTE-Advanced.

Although 802.16m and E-UTRA adopts similar technologies such as OFDMA and MIMO, the differences in detail MAC and PHY layer design make them incompatible. 802.16m will be in conformance with the IEEE 802 architecture defined in 802.1 and provide seamless interworking with other IEEE 802 wired and wireless systems.

2.3 Spectrum Management and Cognitive Radio Networks

Cognitive radio technology allows either a network or a wireless node to dynamically change its transmission or reception parameters to communicate efficiently and to avoid interference with licensed or unlicensed users based on the active monitoring of its operation environment. In general, a cognitive radio system is reconfigurable and can take various external and internal radio environments such as radio frequency spectrum, user behavior, and network state into account to make decision, and adapts various parameters such as frequency spectrum, transmit power, transmit mode, media access method, and so on. More specifically, cognitive radios intelligently access and share radio spectrum by obtaining and sensing spectrum operating environment for efficient usage of licensed/unlicensed spectrum.

The radio frequency spectrum is a limited and valuable resource, but its usage is unbalanced. Some frequency bands are heavily used, for example, cellular network bands. However, a lot of frequency bands are inefficiently utilized, for example, amateur radio and paging frequencies. Furthermore, spectrum utilization depends strongly on time and place. Fixed spectrum allocation prevents the frequency spectrum unused by primary users from being used by unlicensed secondary users. Spectrum utilization can be improved significantly by allowing secondary users to access spectrum holes in the licensed band whenever it would not cause any interference to primary users. Cognitive radio has been proposed as the means for secondary users to utilize the spectrum holes, share the spectrum among them, and avoid the spectrum whenever primary users present.

In November 2008, the Federal Communications Commission (FCC) issued its report and order for unlicensed use of the TV white spaces. The TV white spaces are the frequencies that allocated to TV broadcasting, wireless microphones, and the like, but not used locally. Especially after full-power analog television broadcasts ceased operating in June 2009, many channels had freed up. The new FCC rules allow unlicensed devices to operate in the broadcast television spectrum at locations where that spectrum is not being used, given the secondary white space devices have cognitive radio and dynamic spectrum access capabilities, and shall not interfere the operation of primary users. The FCC currently requires that secondary devices must consult a frequently updated geo-location database to determine which channels are available for use at a given location. Other regulatory bodies such as ITU, European Radio Spectrum Policy Group (ERSPG), U.K. Ofcom, and Japan's Ministry of Internal Affairs and Communication (MIC) are also considering similar regulations.

Various proposals have advocated using TV white spaces to provide different services. The IEEE 802.22[16] working group is developing a standard for wireless regional area network (WRAN) that will operate in unused television channels. 802.22 WRAN mainly aims at providing wireless broadband access in rural areas using vacant TV channels in the VHF and UHF bands while avoiding interference to the broadcast incumbents in these bands. It typically operates with a coverage radius of 17 km to 30 km. 802.22 WRAN system uses TDMA/OFDMA similar to WiMAX networks, but it does not support MIMO because of the large antenna separation requirement at its low operating frequency.

Especially 802.22 specifies cognitive radio capability at the MAC/PHY air interface for dynamic frequency access. It can adjust to the location-dependent and time-variable spectrum availability to avoid interference to incumbents on a real-time basis. Specifically, 802.22 includes two new modules, namely Spectrum Sensing Function (SSF) and Geo-location module. The spectrum-sensing function monitors the RF spectrum of the television channels for a set of signal types and reports the results. The location information is important to protect

TV incumbent transmissions. The TV contours to be protected from interference are stored in a database. The base station (BS) controls the maximum allowed transmit power at individual CPEs using the collective knowledge of channel sensing, the CPE location, the TV operation database information, and so on. The standard also specifies the protocols for coexistence of multiple 802.22 cells.

Several other working groups in IEEE 802 are also studying TV white space. 802.11 has formed a task group 802.11af[7] for WLAN operation in TV white space 802.19, which has started studying coexistence of two or more unlicensed wireless networks such as WLANs, WMANs, WRANs, and ad hoc networks when they operate in the TV white space. Possible coexistence mechanisms under consideration include dynamic frequency selection and transmit power control, listen-before-talk media access or time division multiplexing of different wireless technologies, message-based on-demand spectrum contention based on coexistence beaconing or backhaul, as well as control through a centralized coexistence manager, coexistence database, or spectrum broker.

IEEE Standards Coordinating Committee (SCC) 41 is also developing standards related to dynamic spectrum access networks. The focus is on spectrum management, coexistence, reconfiguration, and dynamic spectrum access for cognitive radio. ITU and ETSI have also started the standard activities related to cognitive radio. In particular, ETSI's Reconfigurable Radio System (RRS) technical committee is defining the system functionalities related to spectrum management and joint radio resource management across heterogeneous access technologies, developing a functional architecture, and studying the concept of a Cognitive Pilot Channel (CPC) as an enabler to support the management of the reconfigurable radio systems.

2.4 All IP Mobile Networks

As part of LTE/System Architecture Evolution (SAE) effort, 3GPP defined the Evolved Packet System (EPS), an IP-based flat mobile network, to meet the increasing user and service demands, and to conform to Internet protocols for converging mobile and fixed network services. It aims at providing improved experience for users and increased performance and reduced cost for network operators. 3GPP All IP Network (AIPN) architecture represents its vision that next-generation mobile networks are based on core Internet protocols.

The existing 2G/3G networks consist of two subdomains: circuit switching for voice and packet switching for data, as shown in Figure 2.2.[18] The EPS unifies these two subdomains into a single end-to-end AIPN, in which voice calls are handled by VoIP using IP Multimedia Subsystem (IMS). EPS is able to integrate and support different radio access systems such as 3GPP radio access (LTE, 3G, and 2G) and non-3GPP radio access (CDMA 2000, WLAN, WiMAX),

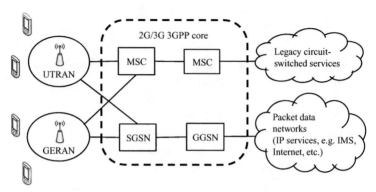

Figure 2.2. Simplified architecture of the 2G/3G 3GPP network.

as well as fixed access (Ethernet, DSL, cable, and fiber) with one common packet core network. It provides diversified mobile services with convergence to IP and enables the introduction of new business models and services, for example, partnering and revenue sharing with third-party content and application providers. It also supports incremental deployment because at the beginning, LTE may be only deployed at most needed areas and coexistence with legacy networks.

The IMS was originally standardized by the 3GPP to deliver IP multimedia services over cellular access networks (UMTS/GPRS networks). It was later enhanced to support other network accesses including Wireless LAN, CDMA 2000, and fixed networks. The IMS includes various control function components such as call session control functions (CSCF) and application servers, for example, the session initiation protocol (SIP) application server, service centralization and continuity (SCC) application server, with standard interfaces based on SIP and many related protocols. It controls the services with user registration, origination, termination, transfer, and release of multimedia sessions. The IMS provides a horizontal control layer that isolates the access networks from the service layer, and is able to maintain the services even when the user is moving across different access networks and terminal types. The user can connect to an IMS system from any access network through IP connectivity as long as it runs a SIP agent. The 2G or 3G circuit-switched network can also be supported as an access network to the IMS through gateways.

As shown in Figure 2.3,[18,19] the flat EPS architecture consists of two parts: the access network and the core network. 3GPP LTE specifies a new access network, E-UTRAN, which offers higher bandwidth, better spectrum efficiency, and better coverage. The core network is called evolved packet core (EPC), which consists of several major elements, including Serving Gateway (S-GW), Packet Data Network (PDN) Gateway (P-GW) and Enhanced Packet Data Gateway (ePDG), Mobility Management Entity (MME), Policy and Charging Rules

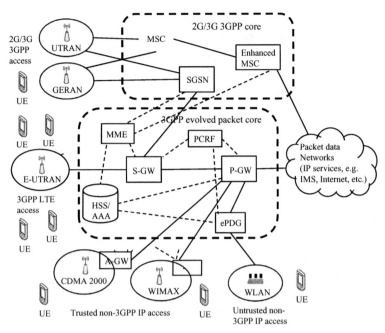

Figure 2.3. Simplified architecture of the 3GPP evolved packet system.

Function (PCRF), Home Subscriber Server (HSS), and Authentication, Authorization, and Accounting (AAA) Server.

The S-GW and P-GW are user data plane elements. S-GW manages user-
planes mobility and also serves as a layer 2 mobility anchor as the User Equipment (UE) moves within 3GPP access. It maintains the IP data paths between
eNodeBs and P-GW and separates the radio access network (RAN) and the core
network. The eNodeBs are the LTE base stations in E-UTRN and combine the
functionality of Node-B and RNC in 3G RAN. The P-GW provides access to
different external packet data networks (PDNs) such as Internet, a corporate
network, or an operator private network. It assigns an IP address to the UE from
the address space of the PDN that can be an IPv4 address, an IPv6 prefix, or both.
The P-GW performs policy and charging enforcement (PCEF) function such as
packet filtering, service flow detection, dynamic policy and QoS enforcement,
and flow-based charging control. It also serves as the mobility anchor point for
the UE as the UE moves between access technologies. A UE may connect to
multiple PDNs through EPC.

MME, PCRF, and HSS/AAA are control plan entities. MME is responsible
for the signaling and control functions for UE access to network, session and
mobility state management, authentication, and security by interacting with HSS.
PCRF makes policy decisions for a user's IP data service flow and provides the

QoS policy and charging rules to the enforcement entities at P-GW and S-GW. HSS maintains user's subscription information, and AAA server supports authentication, authorization, and accounting.

Various access networks can connect to the EPC via different interfaces. Mobility management is an integrated part of the EPS. It provides seamless mobility at IP layer for intra- and interaccess system handover, and ensures the service continuity and QoS for a user's session as the user moves from one access technology to another as described in next section.

In EPS, data plane traffic is carried over virtual connections called service data flows (SDFs). SDFs are carried over bearers, that is, virtual IP transmission containers with unique QoS characteristics such as capacity, delay, packet loss rate, and so forth. A data path between a UE and a PDN, an end-to-end bearer, consists of three segments: a radio bearer between UE and eNodeB, a data bearer between eNodeB and S-GW, and a data bearer between S-GW and P-GW. A bearer exists per combination of QoS class and IP address of the terminal and identified by a unique identifier. The terminal may have multiple IP addresses, for example, when it is connected to multiple IP networks, each assigning it an IP address. It is possible that a terminal has multiple separate bearers associated with the same QoS class to multiple different PDNs. A packet flow is typically specified by an IP quintuple packet filter, that is, the source and destination IP addresses, source and destination port number, and protocol ID. Other filters can also be set up. The terminal (for uplink traffic) and the P-GW (for downlink traffic) classify the packets and map them into the corresponding bearers based on the packet filters. All the packets mapped into the same bearer receive the same packet-forwarding treatment such as scheduling, queuing management, rate shaping, and the like, in the network. The GPRS tunneling protocol (GTP) or proxy mobile IP can be used to implement the bearers in the EPC. Each IP packet entering the network is provided with a tunnel or proxy mobile IP header to route the packet to the destination and provide proper QoS.

The bearer-level QoS control enables network operators to manage the QoS for the different services, for example, mobile TV, telephony, Internet access, and the like, with different QoS requirements, and for each of its subscriber groups, for example, post- versus prepaid subscribers, home versus roaming subscribers. There are two types of bearers: guaranteed bitrate (GBR) and nonguaranteed bitrate (non-GBR) bearers. A GBR bearer typically is established on demand and may require for admission control and resource reservation. A non-GBR bearer can remain established for a long period of time because it does not reserve the resource. Once a terminal attached to the network, one default non-GBR bearer is set up per terminal IP address and remains as long as the terminal retains this IP address. The default bearer provides the basic connectivity and its QoS level is assigned based on the user subscription.

A set of policy and charging control (PCC) procedures has been specified in 3GPP release 8 to manage bearers, provide QoS to subscribers, and charge for the provided resources. PCC in EPS supports multiple-access technologies, IMS and non-IMS services, roaming, and mobility. In the PCC architecture, the application function (AF) – for example, a call-state control function in the IMS architecture – extracts the service-related information for a session by interacting with the applications that requires dynamic policy and charging control. It provides the PCRF with the service information, including traffic parameters such as IP addresses and port numbers, and QoS parameters such as type of traffic, data rate, and the like. The PCRF also obtains user-specific policies and information from the subscription profile repository, as well as user access information from S-GW and P-GW. The PCRF then makes the session policy decisions and provides the charging and policy rules to the policy and charging enforcement function (PCEF) at P-GW, and the policy rules to the bearer-binding and event-reporting function (BBERF) at S-GW. The PCC rules contain uplink and downlink packet filters to identify the service data flow, the gate control information to block or allow the IP flow, and its QoS behavior to be enforced such as QoS class, guaranteed bitrate, and so on.

The PCEF and BBERF are responsible for enforcing the PCC rules to ensure appropriate QoS for a service data flow. Once the PCEF or BBERF receives new or modified PCC rules for a service data flow, it creates or modifies the bearer and initiates resource reservation in the network. The PCEF also interacts with online charging system (OCS) for service access such as prepaid charging and reports the resource usage to the offline charging system.

The PCC provides seamless roaming support. The operators can apply the same dynamic policy and charging control and provide the same QoS to the user no matter whether the user accesses the home or visited networks. There are two different roaming scenarios in the LTE/SAE, namely home-routed access and visited access. In the home-routed roaming scenario, an IP connection with the outside PDN is established through a P-GW in the home public-land mobile network (H-PLMN) and the S-GW in the visit PLMN (V-PLMN). The home PCEF is responsible for the PCC enforcement. In the visited access, an IP connection with the outside PDN is established through a P-GW in the V-PLMN and a S-GW in the V-PLMN. The user data packets are routed through the visited P-GW between the outside PDN and the visited S-GW. The visited P-GW is connected to H-PCRF through V-PCRF to receive the PCC rules. It is also possible to use AFs in the V-PLMN for the visited-access roaming in which the signaling is proxied through the V-PRCF to the H-PRCF. Online charging is also connected to the home OCS through a proxy OCS.

3GPP LTE/SAE also specifies new security mechanisms to handle more diverse architecture with multiple access technologies and improves security robustness. EPS specifies four levels of security. Network access security (level I) protects the radio link and provides users with secure access to the EPC and the backend networks. Network domain security (level II) protects the wireline networks using the IPSec. User domain security (level III) provides the mutual authentication of the Universal Subscriber Identity Module (USIM) and the UE. Application domain security (level IV) enables the applications in the UE and the network to exchange data in a safe manner. The enhancements over UMTS include, among other things, increased security functions for protecting the confidentiality and integrity of signaling messages in access network, more secure key management and identity protection, and better interworking security between 3GPP and non-3GPP networks.

2.5 Mobility and Vertical Handover

It is expected that multiple access technologies will be seamlessly integrated into the global Internet. Both 3GPP and IEEE are defining standards to support mobility and vertical handover, that is, the handover from one network access technology to another. Vertical handover can greatly enhance the user experience. For mobile users, handovers can occur when wireless link conditions change due to the users' movement. For the stationary user, handovers become imminent when the surrounding network environment or application changes, making one network more attractive than another. In the handover, service continuity should be maintained. As an example, when making a network transition during a phone call, the handover procedures should be executed in such a way that any perceptible interruption to the conversation is minimized. Handover can be classified as hard and soft; hard handover is "break before make" regarding the exchange of data packets between the mobile terminal and the network, whereas the soft handover can achieve "make before break."

Generic Access Network (GAN), also called Unlicensed Mobile Access (UMA) defines a secure, managed connection from the 3GPP mobile core network to different devices/access points over IP, which was initially introduced in 3GPP Release 6. It allows extending the services and applications in an operator's mobile core (voice, data, IMS) over IP and Internet to other access technologies. One of applications of GAN is that with a dual-mode mobile phone, users can seamlessly roam and hand over between wireless LANs and cellular networks. When the mobile phone detects a wireless LAN, it establishes a secure IP connection to a GAN Controller (GANC) on the carrier's network. The GANC presents itself to the mobile core network as a standard cellular base station. The handset communicates with the GANC over the secure connection using existing GSM/UMTS protocols. Thus, when a mobile device moves from a GSM to an 802.11 network, it appears to the core network as if it simply attaches to a different base station. Femtocells, analog terminal adaptor for fixed line phone services, and UMA-enabled mobile VoIP clients for PCs are other GAN applications.

3GPP LTE/SAE further advances mobile networking technology by integrating various radio access networks under a single mobile core network. It specifies various interworking and mobility mechanisms based on all IP architecture to enable seamless handover between different access technologies and maintain IP services and voice calls continuity, which facilitate different deployment scenarios and support a flexible evolution path toward 4G. Multiple 3GPP or non-3GPP access networks can connect to the EPC through various access gateways. The EPS specifies different IP mobility mechanisms depending on the access technologies. For 3GPP-defined access technologies such as UTRAN, GERAN, E-UTRAN, either the GPRS tunneling protocol (GTP) or proxy mobile IPv6 (PMIPv6) can be used. For other accesses to connect to the EPC, any of PMIPv6, dual stack mobile IPv6 (DSMIPv6), or Mobile IPv4 (MIPv4) can be used.

PMIPv6 provides a network-based mobility mechanism. The mobile access gateway (MAG), that is, the 3GPP S-GW or the non-3GPP mobile access gateway (A-GW) acts as the proxy/foreign agent for the UE and handles the mobility management signaling. Once the UE has changed its point of attachment to a new mobile access gateway, the new MAG provides the UE with the same IP address as the UE had at its previous point of attachment. The new MAG also handles updating the mobility anchor in the network so that the packets arrive at the new point of attachment. The UE is not aware of the mobility management signaling. On the other hand, DSMIPv6 and mobile IPv4 are client-based. The UE obtains a new care-of address when it moves to a new point of attachment. The UE is responsible for updating its home agent about the binding between the care-of address and the home address of the UE. Compared to the client-based mobility management, the network-based mobility management reduces the UE involvement in mobility management, leading to better UE battery life, less wireless resource usage, and reduced latency in handover.

When terminals move across areas served by eNodeB elements within E-UTRAN, the S-GW serves as a local mobility anchor. The S-GW also serves as an anchor for the mobility within other 3GPP-specific access technologies. S-GW relays packets among eNodeB, P-GW, and legacy SGSN for intra E-UTRAN mobility and mobility with other 3GPP technologies, such as 2G GSM and 3G UMTS.

All data paths from the access networks are combined at the P-GW and routed to the external packet networks. Mobility management between 3GPP and non-3GPP access systems are involved by multiple data plane and control plane entities, including P-GW, S-GW, non-3GPP access gateway, PCRF, and MME based on mobile IP technology. For interaccess handover, 3GPP defines nonoptimized handover and optimized handover procedures, depending on whether the source network is involved in preparing resource in the target network during the handover. Optimized handover is more suitable when the UE cannot simultaneously access the source network and the target network.

Figure 2.4 illustrates the high-level message flow for nonoptimized handover when an UE hands over a VoIP call from a trusted non-3GPP access to a 3GPP LTE E-UTRAN access.[19] PMIPv6 is used in the EPC for this example. The UE initially decides to attach to the trusted non-3GPP access. It initiates an attachment request toward the access gateway via the base station. The UE and the network perform the mutual authentication. After the authentication, an IP address is assigned to the UE, a PMIPv6 tunnel is setup between the P-GW and A-GW, the default access bearer is established, and the UE attaches to the trusted non-3GPP access network. When the subscriber places an IMS VoIP call, the SIP protocol is used to set up the call. The end-to-end signaling is intercepted by the IMS CSCF function in the network. The CSCF extract and pass the session information to the PCRF. The PCRF makes the decisions on charging and QoS rules and sends them to the PCEF at P-GW and BBERF at A-GW. A voice bearer is then set up to carry the call. When the UE decides to hand over to the 3GPP access from the trusted non-3GPP access, it initiates the handover attach procedure to the 3GPP S-GW through eNodeB using its E-UTRAN interface. The 3GPP S-GW obtains the QoS rules for both the default traffic and the VoIP traffic from the PCRF and prepares the resource with the appropriate QoS in the 3GPP access network. Through the proxy binding update between the 3GPP S-GW and P-GW, the P-GW provides the same IP address used by the UE in the non-3GPP access to the S-GW. Meanwhile, the P-GW also updates the PCRF with the UE's handover request and obtains the corresponding charging rules. The default bearer and the dedicated bearer are established in the 3GPP access network, and a PMIPv6 tunnel is set up between the P-GW and 3GPP S-GW. The UE then completes the attachment to the 3GPP. The tunnel is then switched and the traffic is routed through the 3GPP access between the UE and the P-GW. The resource in the non-3GPP access is released.

Dual-radio-capable UEs can simultaneously access both the source and the target networks, and seamless handover can be achieved using the previously mentioned nonoptimized handover. However, if the UE cannot access the two networks simultaneously, a "make-before-break" handoff cannot be achieved with nonoptimized handover, leading to service interruption during interaccess handover. Therefore, optimized handover is specified in 3GPP LTE/SAE to enable seamless handover even for single-radio UEs. In the optimized handover, a tunnel is established between the source system and the target system so the UE can communicate with the target system through the source system and prepare for the handover before the real handover occurs.

In E-UTRAN, voice calls are carried with VoIP technologies and offered as IMS-based services. However, in legacy 2G/3G networks, voice calls are

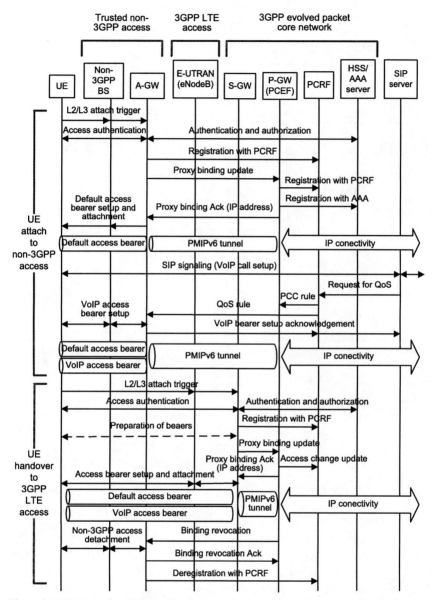

Figure 2.4. Message flow diagram for the scenario when a UE attaches to non-3GPP access and then hands over to 3GPP LTE E-UTRAN access.

carried with traditional circuit switching (CS) technologies. Mobile IP itself cannot meet the voice call continuity requirement. 3GPP LTE/SAE also specifies seamless voice call handover mechanisms between E-UTRAN and various 2G/3G radio accesses, which transfer the call between the CS and IMS domains. It supports the call continuity for single-radio terminals and the handover of

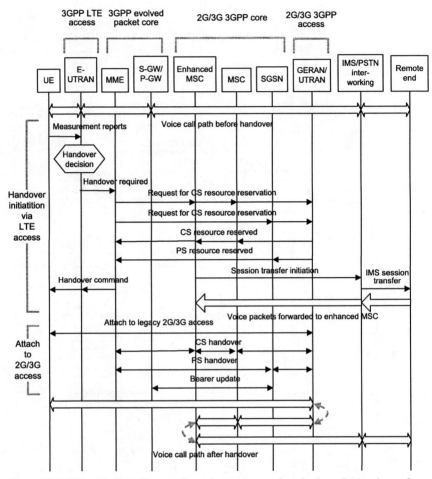

Figure 2.5. Message flow diagram for voice and nonvoice single-radio handover from 3GPP LTE E-UTRAN access to 2G/3G 3GPP access.

associated nonvoice sessions. As an example, Figure 2.5 shows the high-level message flow of the voice and nonvoice single-radio handover procedure from E-UTRAN to UTRAN/GERAN.[18] At least one MSC in the traditional CS domain is enhanced with interworking functionality and a new interface Sv. The MME in the EPC also needs additional functionality to support the Sv interface and the associated single-radio voice call continuity procedure. Due to the make-before-break approach, the voice interruption in the handover procedure is normally less than several hundreds of milliseconds, which should be imperceptible to the user. However, it cannot be guaranteed that the QoS of the nonvoice session is sustained after the handover because of the bandwidth limitation in the UTRAN/GETRAN.

It is possible that voice services are not initially supported over the E-UTRAN access due to the cost of VoIP service deployment. 3GPP also defines the fallback

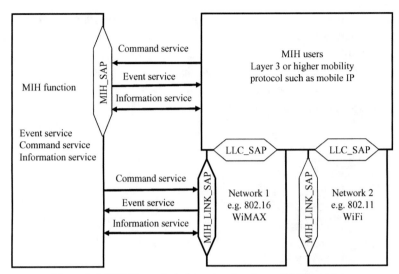

Figure 2.6. 802.21 media independent handover (MIH) services.

mechanisms to handle this case. It hands over the user to the legacy 2G/3G access when a voice service is requested. After the user falls back on the 2G/3G access, the standard CS voice call setup procedure is used to establish the call. Furthermore, 3GPP also specifies the IMS service continuity procedures to hand over a multimedia session above the transport layer based on SIP protocol. One of the benefits with the IMS is that the same service continuity procedures can be used no matter what the source and target accesses are. The IMS-based application layer handover mechanisms provide additional tools for mobility, especially when vertical handover is not supported by the network layer. Note that the pure application layer handover such as that supported by the IMS normally takes a longer time, especially for single-radio terminals, and may lead to perceptible service interruption.

IEEE 802.21[17] is also developing media independent handover (MIH) standards to enable handover and interoperability between heterogeneous networks including both wired and wireless, 802 (e.g., 802.11, 802.16, Ethernet), and non-802 networks (such as cellular). Compared to similar technologies defined by 3GPP (UMA and SAE vertical handover), 802.21 does not assume a 3GPP core network. It intends to provide generic link-layer intelligence independent of the specifics of mobile nodes or radio networks.

As shown in Figure 2.6,[17] 802.21 defined a framework and a set of control functions to facilitate the media independent handover. Specifically, it defined a new logical control entity, called the MIH function (MIHF), in the framework that locates on the mobile nodes or in the network, and provides the event, command, and information services to facilitate seamless handovers between heterogeneous networks. It also standardizes a generic MIH service access point, called the MIH_SAP, as well as associated primitives that provide MIH users with access to the MIHF services. For support of remote MIHF services and communications between the peer MIHF entities, 802.21 specifies a MIH protocol. However, the MIHF in 802.21 depends on the presence of a mobility management protocol stack, for example, mobile IP, within the network elements that support the handover. Enhancements at the higher layer and link layers are required to support the function abstractions of this standard and carry the messages defined in this standard. In addition, handover policies and other algorithms involved in handover decision making do not fall within the scope of the standard, which are left to the network operators and applications. Handover decision making involves cooperation of mobile nodes and network infrastructure. The 802.21 standard supports both hard and soft handover procedures. 802.21 WG is working on the extensions to the basic 802.21 specification to add security mechanisms and support of handover for downlink-only broadcast networks such as DVB network.

2.6 Multihop Wireless Networks

Multihop wireless mesh networks (WMNs) are a promising technology to extend wireless coverage in a flexible and cost-effective way. WMNs can be infrastructure-based or infrastructureless. In infrastructureless WMNs, client stations such as laptops, smart phones, and so on are equipped with mesh-routing functions and form a network on an ad hoc basis to forward the traffic to each other without dedicated infrastructure, in which each node is a mesh router and an end device. In infrastructure WMNs, mesh routers or mesh access points (MAPs) constitute a multihop wireless infrastructure. One or more mesh router/MAP can be connected to the other wired or wireless networks or the Internet, acting as the mesh gateway. Client stations without mesh functions do not participate in the packet relay, but associate with a MAP to obtain the network access. The MAPs forward traffic for the client stations in the mesh.

Industry standards are being developed in the IETF for mobile ad hoc networks (MANET) routing protocols, in the IEEE 802.11s[5] for WiFi-based mesh networks, and in 802.15.5[15] for wireless PAN mesh. Next-generation WiMAX networks based on 802.16j[8] and 802.16m[9] will support multihop relay. 3GPP LTE-advanced is also considering multihop relay architecture for next-generation cellular networks.

2.6.1 IETF MANET Routing Protocols

Radio nodes in a multihop WMN self-organize themselves in a mesh topology and self-heal from failures using discovery and routing protocols, which enhances the network reliability. The nodes cooperatively make forwarding decisions based on a routing protocol. Many mesh-routing schemes have been proposed in research literatures. IETF MANET working group (WG) has standardized a few of IP routing protocols that can be applied for general wireless mesh networks consisting of mobile routers or fixed routers, or a mixture of both. IPV4 and IPv6 are both supported. The WG has developed two tracks of routing protocol specifications: reactive/on-demand MANET protocol (RMP) and proactive MANET protocol (PMP). In a proactive routing protocol, each node establishes and maintains routes to all reachable destinations at all times, whether or not there is currently any need to deliver packets to those destinations. In contrast, an on-demand routing protocol discovers and maintains routes only when they are needed.

Ad hoc On-Demand Distance Vector (AODV) routing[29] is a typical on-demand routing protocol specified by IETF MANET WG. In AODV, when a route to a new destination node is needed, the originating node floods a Route Request (RREQ) message to discover the route to the destination. The intermediate nodes propagate the RREQ hop by hop and also create a reverse route to the originator in its routing table based on the distance vector. When the target receives the RREQ, it responds with a Route Reply (RREP) sent hop by hop in unicast toward the originator over the reverse route. Each intermediate node that receives the RREP creates a route in its routing table to the destination. When the originator receives the RREP, the route has been established between the originator and the destination in both directions. It is also possible that an intermediate node with a valid route to the destination responds to the RREQ with a RREP to reduce route setup time. To maintain the active route and react to changes in the network topology, nodes monitor the link status through optional Hello messages and traffic flow over the link. If a node detects a link break for the next hop of an active route or receives a data packet for forwarding to a destination for which it does not have an active route, it sends the Route Error (RERR) toward the originator of the packet to notify the loss of the link to the other nodes that use this route. The originating node will delete the route when it receives the RERR and initiate a route discovery again if it needs to send the packet to the same destination. The WG later specified a modified version of AODV, called Dynamic MANET On-demand (DYMO) Routing. DYMO uses a more generic and flexible message format, and enables DYMO routers to perform routing on behalf of other attached nodes. The Dynamic Source Routing (DSR)[32] is another on-demand routing protocol defined by the IETF MONET WG. Unlike the AODV, DSR uses source routing to forward the packets.

The IETF MONET WG standardized two proactive routing protocols: Optimized Link State Routing Protocol (OLSR)[30] and Topology Dissemination Based on Reverse-Path Forwarding (TBRPF).[31] OLSR is a table-driven, proactive protocol. It optimizes the classic link state protocol by considering the

MONET requirements and broadcast wireless media characteristics to reduce the number of transmissions in the process of control traffic flooding. In OLSR, each router selects a subset of its neighbor routers as "MultiPoint Relays" (MPRs) to retransmit the broadcast messages from it so that the broadcast messages, retransmitted by these selected MRPs, will reach all nodes two hops away. A node only forwards the broadcast messages directly received from its MPR selectors, that is, the nodes that have selected it as an MPR. The neighbors of a node N that are not in its MPR set receive and process broadcast messages but do not retransmit the broadcast messages received from node N. Hello messages are used between neighbor nodes for link sensing, 1-hop and 2-hop neighbor detection, and MPR selection. This technique facilitates efficient flooding of control messages in the network as compared to a classical flooding mechanism, where every node retransmits each message when it receives the first copy of the message. To reduce the number of control message transmissions further, the link state information may only be generated by nodes elected as MPRs, that is, MPRs declare the link state information for their MPR selectors. In addition, an MPR node may chose to report only links between itself and its MPR selectors. Then in route calculation, the MPRs are used to form the route from a given node to any destination in the network. The WG has also specified an updated version of the OLSR, OLSR version 2, which retains the same basic mechanisms and algorithms while providing a more flexible signaling framework and some simplification of the messages being exchanged.

TBRPF is another proactive, link-state routing protocol standardized by the MONET WG. In TBRPF, each node computes and updates a source tree that provides the shortest paths to all reachable destinations, based on partial topology information stored in its topology table. Instead of disseminating the link states for all the links, each node reports only part of its source tree to neighbors, that is, the status of the links consisting of this reported subtree, to minimize the overhead.

Compared to proactive routing with on-demand routing, proactive routing protocols generally have the advantage of routes immediately available when needed because a node continuously maintains routes to all destinations in the network. The proactive protocols are beneficial for traffic patterns where a large subset of nodes are communicating with another large subset of nodes, and the source and destination pairs are changing over time. However, the proactive protocols incur more routing overhead to keep the routing information current, especially when the nodes are moving or the network topology changes frequently. On the other hand, the on-demand protocols require less routing overhead because they do not maintain the unused routes. However, they require more time to discover and establish a route when the route is needed, leading to extra route discovery delay and data buffering at the source.

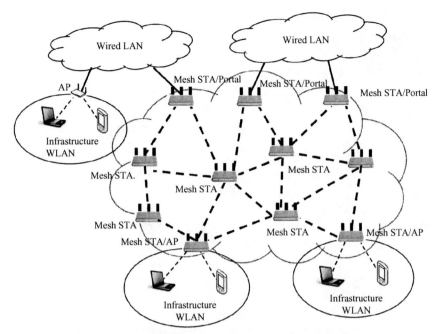

Figure 2.7. An example of a multihop wireless mesh network.

2.6.2 IEEE 802.11s WLAN Mesh Networking

Wireless mesh networks can be implemented with various radio technology including 802.11, 802.16, 802.15, cellular radios, or combinations of more than one type. IEEE 802.11s is developing a mesh networking standard using the IEEE 802.11 MAC/PHY layers. Within the scope of the IEEE 802 standards, IEEE 802.11s addresses layer 2 operations. Compared to IP routing, layer 2 routing uses the MAC address. There is no IP address assignment issue in the use of layer 2 mesh routing protocols. In terms of implementation, the layer 2 mesh software can be incorporated in the drivers and offered by the IC vendors.

The IEEE 802.11s extends the 802.11 MAC layer by defining the architecture and a set of mechanisms to support multihop mesh networking. An example of a 802.11 wireless mesh network is illustrated in Figure 2.7.[5] A mesh station (STA) may be collocated with one or more other entities (e.g., AP, portal, etc.). A mesh AP that incorporates a mesh STA with one or more access points can provide both mesh functionalities and AP functionalities simultaneously. Client STAs associate with APs to gain access to the network, and do not participate in mesh functionalities such as path selection and forwarding. Mesh portals interface the mesh network to other networks.

The main functionalities of IEEE 802.11s include mesh discovery, authentication and link security, peering management, channel selection, routing and forwarding, interworking, congestion control, synchronization and beaconing, and power save. Traditional 802.11 EDCA is the basic medium access protocol. 802.11s also defines an optional channel access mode, called Mesh Coordinated Channel Access.

IEEE 802.11s standardizes the procedures for mesh STAs to discover one another and to organize themselves into a mesh network. A mesh STA discovers candidate peer mesh STAs and their configuration by listening to the beacons sent by its neighbors or using proactive probe request/probe response message exchanges. After discovering a candidate peer mesh STA, a mesh STA can establish peering with the candidate peer mesh STA, secured or not, depending on the local policy. If the peering does not require security, the mesh STA will initiate the peering management protocol to the candidate mesh STA. The peering management protocol allows exchanging and confirming the configuration parameters such as mesh ID, active path selection protocol and metric IDs, and so on, and establishes a peer session through handshake. If the peering requires security, the mesh STA shall initiate a secure authentication protocol. IEEE 802.11s defines an authentication protocol, called Simultaneous Authentication of Equals (SAE), to provide mutual authentication between two mesh STAs using a shared key or password. If the secure authentication protocol succeeds, the two mesh STAs obtain a common pairwise master key (PMK). The IEEE 802.11s further defines two mesh link security protocols, called the authenticated peering exchange and the mesh group key handshake. The mesh link security protocols rely on the existence of the common PMK at the two mesh STAs established by executing the authentication protocol. The authenticated peering exchange protocol is used to authenticate a peering using the PMK, to establish session keys for protecting unicast traffic between two peers, and exchange the group keys. A group key is assigned by the broadcast/multicast source and is used to protect broadcast/multicast traffic from that source. The mesh group key handshake allows a mesh STA to update its group key.

IEEE 802.11s also specifies the channel switching procedures that can be used to satisfy regulatory requirements such as radar signal detection, or to reassign the mesh STA channel to ensure the network connectivity, or other reasons. A mesh STA that determines the need to switch the channel transmits a Mesh Channel Switch Announcement frame to each of its peer mesh STAs to announce its intent through unicast or broadcast. It also includes the channel switch announcement information in its beacon frames and probe response frames during the channel switch process.

IEEE 802.11s specifies an extensible routing framework to enable flexible implementation of path selection protocols and metrics. The standard includes a default mandatory path selection protocol, Hybrid Wireless Mesh Protocol

(HWMP), and default mandatory path selection metric (Airtime Link Metric) for all implementations, to ensure interoperability. Even though the extensible framework allows multiple protocol and metric implementations, only one path selection protocol and one path selection metric shall be actively used by a mesh STA at a time, which is announced by the mesh network in the beacons and probe responses.

The default HWMP protocol specified by 802.11s combines the flexibility of on-demand path selection with proactive topology tree extensions, which takes advantages of both proactive and reactive routing approaches, and enables efficient path selection in a wide variety of mesh networks. HWMP can concurrently operate in on-demand mode and proactive tree-building mode. The on-demand mode of the HWMP protocol is based on the AODV with many enhancements and adapted to the MAC address-based path selection and radio link metric awareness. It allows mesh STAs to communicate using P2P paths. The proactive tree building mode is used to establish the paths between a root mesh STA and the rest of the mesh STAs in advance, so that the communications can begin instantly without executing the path selection operation. This can be performed by configuring a particular mesh STA as a root mesh STA and periodically broadcasting its existence to the rest of the mesh so that every mesh station could create a path to the root station. Typically root STAs are the STAs that act as portal to Internet access. One example of concurrent usage of on-demand and proactive modes is that two mesh STAs begin communicating using the proactively built tree via the root, but subsequently perform an on-demand discovery for a direct path between each other. This type of concurrent usage of the proactive and on-demand modes allows communication to begin immediately while an on-demand discovery finds a more optimal path between two mesh STAs.

The default routing metric specified by 802.11s is the airtime link metric, which takes the link data rate and frame error rate of a wireless link into account. Airtime reflects the amount of channel resources consumed by transmitting a frame over a particular link. The total cost for a path is the sum of the cost of the links on the path.

A 802.11s mesh network functions like an IEEE 802 LAN segment. It can have zero or more portals that can be connected to one or more LAN segments. If two portals connect a mesh to an external LAN segment, broadcast loops may occur, and the IEEE 802.1D bridging protocol can be used to turn off the LAN port of one of the portals for preventing from traffic looping. A portal can send the portal announcement (PANN) to advertise its presence. Portal Announcements allow mesh STAs to select the appropriate portal and build a path toward it.

A mesh STA can serve as a proxy for nonmesh STAs, transmitting and receiving the frames on behalf of the proxied STAs through a tunnel. For example, a mesh AP can serve as a proxy for the client stations associated with it, and a mesh portal can serve as a proxy for an entity behind of it. 802.11s defines the signaling protocol for a proxy mesh station to send the proxy information, including the MAC address of the proxied entities, to a destination mesh STA.

802.11s does not define a multicast routing protocol. Multicast frames are forwarded as broadcast frames. To improve multicast reliability, the standard allows an implementation of multiple unicast transmissions to transmit a multicast frame, which are individually acknowledged. In such a case, the multicast frame can be converted to individually addressed frames and transmitted as individually addressed frame to each of the peer mesh STAs.

IEEE 802.11s also specifies a Congestion Control Signaling protocol. A mesh STA that detects congestion and the incoming traffic sources causing this congestion may transmit a Congestion Control Notification frame to the source mesh STAs or other neighboring mesh STAs. Of course, specific algorithms for local congestion monitoring and congestion detection, as well as local rate control, are beyond the scope of the standardization and are left to the implementers for innovation.

To detect and mitigate the collisions of beacons transmitted by different stations on the same channel within 2 hop range, a Mesh Beacon Collision Avoidance (MBCA) mechanism is specified by IEEE 802.11s. A mesh station reports the target beacon transmission time and beacon interval of its neighboring STAs in its beacon or probe response frames. Using this information, a mesh STA can select and adjust its target beacon transmission time and beacon interval so that its beacon frames do not collide with the beacon frames transmitted by other STAs in a 2-hop range. In addition, a mesh STA can send a message to request its neighbor to adjust the target beacon transmission time.

IEEE 802.11s specifies power save operation. A mesh STA has the capability to buffer frames and track the power mode of a peer mesh STA. A mesh STA uses the peer service periods for unicast frame transmissions to a neighboring peer mesh STA in power save mode. A peer service period is directional. To trigger a peer service period, a mesh STA in power save mode can send a peer trigger frame to its peer, and a mesh STA can also send a peer trigger frame to the mesh STA in power save mode during its Mesh Awake Window. The Mesh Awake Window of a mesh STA is announced in its beacon and probe response frames.

In addition to the traditional EDCA, 802.11s standardizes an optional medium access method, called Mesh Coordinated Channel Access (MCCA), which allows MCCA-capable mesh STAs to access the wireless medium at selected time periods with lower contention. MCCA can be used by a subset of mesh STAs in a mesh network. However, MCCA connections can only be set up among MCCA-enabled mesh stations and their performance may be impacted by the devices that do not respect MCCA reservations. MCCA uses management frames to determine a series of target transmission starting times and durations, called MCCA Opportunities (MCCAOPs), between an MCCAOP owner and one (for individually addressed frames) or more (for group addressed frames) MCCAOP responders for frame transmissions. These MCCAOPs are advertised in the neighborhood around the MCCAOP owner and responders. The MCCA mesh STAs in this neighborhood that could cause interference to transmissions during these MCCAOPs, or that would experience interference from them, shall refrain from accessing the wireless medium during these MCCAOPs. The MCCAOP owner and the MCCAOP responders access the wireless medium during these MCCAOPs using contention-based channel access (EDCA) because some other stations may not respect the MCCA reservations.

Synchronization is needed between mesh STAs that use MCCA, MBCA, or operate in power save mode. IEEE 802.11s introduces an extensible framework to enable implementation of multiple synchronization protocols for mesh STAs. It also includes a default mandatory protocol, called the neighbor offset synchronization protocol, to enable minimal synchronization capabilities and interoperability. With the neighbor offset synchronization protocol, a mesh STA should maintain a timing offset value between its own time synchronization function (TSF) timer and the TSF timer of each neighbor mesh STA with which it synchronizes. A mesh STA can start its TSF timer independently of other mesh STAs, and can update the value of its TSF timer offset based on the time stamps received in the beacon or probe response frames from other mesh STAs.

2.6.3 IEEE 802.16j WMAN Multihop Relay

IEEE 802.16j has specified enhancements to the IEEE 802.16 OFDMA-based PHY and MAC layers to enable the operation of multihop relay stations. However, 802.16j only supports tree topology consisting of one or more relay stations (RS) rooted at a multihop relay base station (MR-BS). Traffic between the subscriber station (SS) and MR-BS is relayed by one or more RS. Each RS is under the supervision of the MR-BS. The RS can be fixed (e.g., attached to a building) or mobile (e.g., traveling with a transportation vehicle). The SS can also communicate directly with the MR-BS. However, it does not allow the P2P communications between relays. The standard specifies new functionality on the relay link to support the multihop relay features. But the protocols on the access link between the SS and RS/MR-BS are not changed from 802.16. 802.16m will also support multihop relay based on the techniques developed in 802.16j.

Two different modes, namely centralized and distributed scheduling, are specified for controlling the allocation of bandwidth for an SS or an RS. In centralized scheduling mode, the bandwidth allocation for an RS's subordinate stations is determined by the MR-BS; in contrast, for distributed scheduling mode, the bandwidth allocation of an RS's subordinate stations is determined by the RS, in cooperation with the MR-BS. Note that the standard only provides the signaling support for centralized and distributed scheduling. The scheduling algorithms themselves are out of the scope of this work and are left for the innovation by the implementers. Two different security modes, namely centralized security mode and distributed security mode, are also defined in 802.16j. The centralized security mode is based on the key management between an MR-BS and an SS. The distributed security mode incorporates the authentication and key management between an MR-BS and an access RS and between the access RS and an SS.

The MAC layer enhancements defined in 802.16j include signaling extensions to support functions such as network entry of an RS, and of an SS through an RS, bandwidth request, packet forwarding, connection management, and handover. The PHY enhancements include extensions to the OFDMA-PHY for transmission of data across the relay links between the MR-BS and the RS.

In 802.16, connections are identified by a 16-bit connection ID (CID), and the CID is carried in the MAC header. At a SS or RS initialization, the management connections are established between the SS/RS and the MR-BS, which are used to carry management traffics. An RS may be configured to operate either in normal CID allocation mode, where management CIDs are allocated by the MR-BS, or in local CID allocation mode where the MR-BS allocates the management CID range to a subordinate RS that assigns CIDs from this range to its subordinate stations. Data traffic connections are established dynamically. Connections may span multiple hops and may pass through one or more intermediate RSs. The CIDs will be unique within an MR cell. In addition, a tunnel connection can be established between the MR-BS and an access RS. Tunnel connections can be used for transporting relay traffic from one or more connections between the MR-BS and the access RS, and can pass through one or more intermediate RSs.

In the network entry, an RS scans the preambles transmitted by the existing MR-BS(s) or RS(s), synchronizes with the MR-BS, and selects a temporary RS/MR-BS to access the network. It then obtains transmission parameters, performs ranging, negotiates basic capabilities, and performs authorization, security key exchange, and registration with the MR-BS. Then the MR-BS obtains the neighbor station measurement reports and selects the final access station for this new RS. After that, the path is created, and the tunnel and IP connectivity are also established. The MR-BS transmits the operation parameters to the RS to configure it. The SS uses similar procedures for the network entry. The differences are the SS will select the RS/MR-BS as the access station once after scanning. It does not perform the neighbor station measurement and the second stage of access station selection, and the path and tunnel establishment.

The MR-BS can instruct the RSs to perform complete neighborhood discovery and measurement. Based on the topology information obtained from topology discovery or update process, MR-BS makes centralized calculation for the path between the MR-BS and an access RS for both the uplink and downlink direction. The path creation is subject to the constraints of tree topology, that is, an RS shall have only one superordinate station and other constraints such as the availability of radio resource, quality of the link, load condition of an RS, and so forth. The specific path calculation algorithms are out of scope of the standardization and left to the implementers for innovation.

802.16j defines two path management modes: the embedded path management and explicit path management. In the embedded path management mode, the MR-BS systematically assigns CIDs to its subordinate stations such that the CIDs allocated to all subordinate RSs are a subset of the allocated CIDs for that station. The network topology is embedded into a systematic CID structure to help RSs find routing paths without storing all CIDs of subordinate RSs in the routing table, which means the packets for a connection are routed based on the CID assignment structure. This is similar to a telephone call being routed based on the telephone number. The CIDs are assigned systematically, using either contiguous integer block or bit partitioning methods. In the explicit path management, after a MR-BS discovers the topology between a newly attached MS or RS and itself, or detects a topology update due to events such as mobility, MR-BS may remove an old path, establish a new path, and notify the new path information to all the RSs on the path. When connections are established or removed, MR-BS may distribute the mapping information between the connection and the path to all the RSs on the path. This is similar to packet routing in data networks and requires for routing table in each node. With this method, it is possible to have multiple paths between a SS/RS and the MR-BS.

In general, while a relay station is transmitting a signal, other neighboring stations do not transmit using the same time-frequency resource. However, a receiver may experience improved decoding performance through diversity gain if it receives the same information from multiple sources. 802.16j standardizes the cooperative relaying technique for downlink transmission. Either an MR-BS and one or more RSs or multiple RSs can transmit the same signal and/or space-time-code encoded signals for the same data to a subordinate subscriber station using the same time-frequency resource in cooperative manner to achieve diversity gain. Cooperative relaying can be seen as a distributed MIMO system in multihop environments. It requires for appropriate MAC scheduling of the transmissions from the MR-BS and multiple RSs, and the data needs to be sent to the cooperative RSs before the cooperative transmission can occur.

2.6.4 IEEE 802.15.5 WPAN Mesh

IEEE 802.15 TG5 has defined a recommended practice to provide the framework that enables WPAN devices for interoperable and scalable wireless mesh networking. The standard consists of two parts: low-rate (LR) WPAN mesh and high-rate (HR) WPAN mesh networks. The low-rate mesh is built on IEEE 802.15.4 MAC, whereas the high-rate mesh utilizes IEEE 802.15.3 MAC.

In LR WPAN mesh, a mesh coordinator (MC) can start a mesh network by scanning all the channels to gather the information from existing networks, deciding the channel and PAN ID, and sending beacons. To join a mesh, a device simply discovers existing channels and networks, and selects a channel, network, and parent device to associate. A logic tree rooted at the MC is first formed for both addressing and routing purposes. The logical address of a device is assigned based on its level on the tree and the number of its children. By binding logical addresses to the network topology, routing can be carried out easily without going through route discovery. After a device is assigned an address block for it and its children, it should broadcast several hello messages to its neighbors, and the number of hops that the hello message will propagate is carried in the time-to-live (TTL) field. By exchanging hello messages, mesh links are established.

Similar to an LR mesh network, an HR mesh network usually gets started by a device that is capable of operating as a mesh coordinator. A device first scans to gather information about the existing networks in its neighborhood. If there is at least one mesh network found from the scan process, the device may join one of them. Otherwise, the device determines to operate as an MC. The MC then determines the mesh operation parameters such as mesh ID, tree IDs, operating channel, and so on. It starts a piconet and sends beacons containing mesh information. A mesh network is constructed on the basis of tree topology with the MC as its root. To join a mesh, a device searches for existing channels/piconets and then selects one of the discovered mesh piconet controllers (MPNCs) to associate, request a block of TREEIDs, and create a new child piconet. When constructing a tree topology, the unique TREEID block for a MPNC is assigned from its parent MPNC in a top-down manner starting from the MC. The TREEIDs are conveyed in the beacons and enable the tree-based routing.

The HR WPAN mesh also supports an alternative routing method based on the optional topology servers to provide the optimal route between two MPNCs. With the existence of topology servers, MPNCs may consult these servers for routing information instead of forwarding the packets on the tree-based route. Every MPNC in the tree network can be allowed to play the role of a topology server. A MPNC initiates the link state registration process by broadcasting a link state request command to its descendants. When an MPNC in a tree receives a link state request command from its parent, the MPNC sends a link state registration command to its parent. Then, the MPNC forwards the link state request command to its children. Based on the received link state information, an MPNC calculates the optimal route between any source-destination pair on the subtree consisting of its descendent MPNCs. When a source MPNC wants to deliver a frame by using server routing, it can seek the help from one of its ancestors to locate the optimal route toward the destination using the route discovery command. To provide the optimal route between any source-destination pair, the common ancestor that is closest to both of them calculates the shortest path and sends the calculated explicit path to the destination MPNC. The destination MPNC sends a route formation command toward the source MPNC to update/establish the routing table entries of the relay MPNCs along the derived optimal route.

2.7 Concluding Remarks

Advanced physical and MAC layer techniques such as MU-MIMO, OFDMA, and SDMA have emerged for increasing network capacity and improving bandwidth efficiency and coverage range. These wireless technologies are viewed as the key components in improving the performance of next-generation wireless networks. However, to obtain the full benefits of these technologies, the higher-layer networking protocols should exploit their capabilities in a systematic way due to the interdependence. The standardization effort to achieve overall system optimization is important.

Wireless access and mobility will be an integrated part of the future Internet. New architecture and protocols for the future Internet will enhance network performance in terms of QoS and security, improve cost efficiency, meet increasing user demand, and facilitate various services/applications and the fixed-mobile convergence with seamless mobility and global roaming capability. The research advance in this area certainly has an impact to the standardization effort.

References

[1] IEEE 802.11 Wireless LAN Medium Access Control (MAC) and Physical Layer (PHY) Specifications, June 2007.
[2] IEEE 802.11n Wireless LAN MAC and PHY Specifications – Amendment 5: Enhancements for Higher Throughput, October 2009.d
[3] IEEE 802.11ac PAR for Wireless LAN MAC and PHY Specifications – Amendment: Enhancements for Very High Throughput for Operation in Bands below 6GHz. September 2008.
[4] IEEE 802.11ad PAR for Wireless LAN MAC and PHY Specifications – Amendment: Enhancements for Very High Throughput in the 60 GHz Band, December 2008.
[5] IEEE 802.11s Wireless LAN MAC and PHY Specifications – Amendment 10: Amendment: Mesh Networking, Draft 3.0, May 2009.
[6] IEEE 802.11p Wireless LAN MAC and PHY Specifications – Amendment 7: Wireless Access in Vehicular Environments, Draft 8.0, July 2009.
[7] IEEE 802.11af PAR for Wireless LAN MAC and PHY Specifications – Amendment: TV White Spaces Operation, December 2009.

[8]   IEEE 802.16j Air Interface for Broadband Wireless Access Systems – Amendment 1: Multihop Relay Specification, June 2009.
[9]   IEEE 802.16m Air Interface for Fixed and Mobile Broadband Wireless Access Systems – Amendment: Advanced Air Interface, Draft 3.0, December 2009.
[10] IEEE 802.15.4f PAR for Wireless MAC and PHY Specifications for Low Rate WPANs – Amendment: Active RFID System PHY, December 2008.
[11] IEEE 802.15.4g PAR for Wireless MAC and PHY Specifications for Low Rate WPANs – Amendment: Physical Layer Specifications for Low Data Rate Wireless Smart Metering Utility Networks, December 2008.
[12] IEEE 802.15.6 Wireless MAC and PHY Specifications for WPANs Used in or around a Body, 2009.
[13] IEEE 802.15.7 PAR for PHY and MAC Layer Standard for Short-Range Wireless Optical Communication Using Visible Light, December 2008.
[14] IEEE 802.15.3c Wireless MAC and PHY Specifications for High Rate WPANs: Amendment 2: Millimeter-wave based Alternative Physical Layer Extension, Draft 13.0, July 2009.
[15] IEEE 802.15.5 Mesh Topology Capability in Wireless Personal Area Networks (WPANs), May 2009.
[16] IEEE 802.22 Draft Standard for Wireless Regional Area Networks Part 22: Cognitive Wireless RAN Medium Access Control (MAC) and Physical Layer (PHY) Specifications: Policies and Procedures for Operation in the TV Bands, Draft 1.1, May 2008.
[17] IEEE 802.21 Media Independent Handover Services, January 2009.
[18] Salkintzis, A. K., M. Hammer, M., Tanaka, I., and Wong, C. 2009. Voice Call Handover Mechanisms in Next-Generation 3GPP Systems. *IEEE Communications Magazine*, **47**(2), Pages 46–56.
[19] Ali, I., Casati, A., Chowdhury, K., Nishida, K., Parsons, E., Schmid, S., and Vaidya, R. 2009. Network-Based Mobility Management in the Evolved 3GPP Core Network. *IEEE Communications Magazine*, **47**(2), Pages 58–66.
[20] Pastor Balbas, J., Rommer, S., and Stenfelt, J. 2009. Policy and Charging Control in Evolved Packet System. *IEEE Communications Magazine*, **47**(2), Pages 68–74.
[21] Ekstrom, H. 2009. QoS Control in the 3GPP Evolved Packet System. *IEEE Communications Magazine*, **47**(2), Pages 76–83.
[22] Sankaran, C. B. 2009. Network Access Security in Next-Generation 3GPP Systems: A Tutorial. *IEEE Communications Magazine*, **47**(2), Pages 84–91.
[23] Irmer, R., Mayer, H., Weber, A., Braun, V., Schmidt, M., Ohm, M., Ahr, N., Zoch, A., Jandura, G., Marsch, P., and Fettweis, G. 2009. Multisite Field Trial for LTE and Advanced Concepts. *IEEE Communications Magazine*, **47**(2), Pages 92–98.
[24] Astely, D., Dahlman, E., Furuskar, A., Jading, Y., Lindstrom, M., and Parkvall, S. 2009. LTE: The Evolution of Mobile Broadband. *IEEE Communications Magazine*, **47**(4), Pages 44–51.
[25] Larmo, A., Lindstrom, M., Meyer, M., Pelletier, G., Torsner, J., and Wiemann, H. 2009. The LTE Link-Layer Design. *IEEE Communications Magazine*, **47**(4), Pages 52–59.
[26] ITU-R M.2133 Requirements, evaluation criteria and submission templates for the development of IMT-Advanced, December 2008.
[27] ITU-R M.2134 Requirements related to technical performance for IMT-Advanced radio interface(s), December 2008.
[28] ITU-R M.2135 Guidelines for evaluation of radio interface technologies for IMT-Advanced, December 2008.
[29] IETF RFC 3561 Ad hoc On-Demand Distance Vector (AODV) Routing, July 2003.

[30] IETF RFC 3626 Optimized Link State Routing Protocol (OLSR), October 2003.
[31] IETF RFC 3684 Topology Dissemination Based on Reverse-Path Forwarding (TBRPF), February 2004.
[32] IETF RFC 4728 Dynamic Source Routing Protocol (DSR) for Mobile Ad Hoc Networks for IPv4, February 2007.

3

Ad Hoc and Mesh Network Protocols and Their Integration with the Internet

Suli Zhao and Shweta Jain

Abstract

Ad hoc and multihop wireless networks are becoming increasingly important for a variety of applications ranging from tactical military networks, to metro area WiFi networks, to sensor applications. Multihop wireless is motivated by the fact that many embedded wireless devices are power-limited and cannot communicate directly with a distant base station or access point. In addition, ad hoc network formation is motivated by mobile service scenarios, such as tactical or vehicular. Protocol design considerations are given for both mobile ad hoc network (MANET) and static (planned) mesh network scenarios. These include self-organization, resource discovery, medium access control, and routing. Existing routing protocols for MANETs, including Destination Sequenced Distance Vector (DSDV), Dynamic Source Routing (DSR), and Ad hoc On-demand Distance Vector (AODV), are described and performance comparisons are given. More recent work on cross-layer mesh-routing protocols is introduced, including cross-layer metrics such as Airtime or PHY/MAC Awate Routing Metric for Ad hoc networks (PARMA), as well as Integrated Routing and Medium Access (IRMA) control. The chapter concludes with implications for future IP protocols that would allow for seamless integration of multihop wired and wireless networks.

3.1 Introduction and Motivation

Wireless ad hoc and mesh networks have been an important research area for about two decades. Research topics like the network architecture and design, integration with TCP/IP, routing, and medium access control in the shared wireless medium have been discussed at length. However, the mobile and dynamic nature of the network introduces new challenges in self-organization, including

54

neighbor and topology discovery, network management, and disconnected operation. Unlike wired communication, wireless channels suffer from intermittent losses due to various environmental effects like multipath fading and shadowing. Therefore, the lossless assumptions on which protocol layers are designed had to be abandoned and a new research topic in cross-layer protocol design emerged to design better wireless communication protocols. This chapter studies the interesting roadmap of research in wireless ad hoc and mesh networks covering topics in network architecture, protocol design, self-organization, and cross-layer adaptation mechanisms. The integration of ad hoc and mesh networks with the Internet is briefly discussed.

3.2 Network Architecture

Ad hoc network is a completely unplanned deployment of mobile wireless nodes whereas a mesh network is a semiplanned deployment of fixed wireless routers that provide Internet connectivity to mobile wireless devices. We describe the two architectures and point out some subtle distinctions in terms of the challenges faced by each.

3.2.1 Mobile Ad Hoc Networks and Mesh Networks

A mobile ad hoc network (MANET) is a self-configuring wireless network of mobile devices connected by wireless links. Each radio device in a MANET can move independently and therefore change its link to other devices. Each device can work as a router to help forward traffic for other nodes. Ad hoc networks may operate by themselves or may be connected to the Internet. The primary challenge in a MANET is to maintain network topology and provide multihop routing in spite of physical connectivity changes and without a centralized controller.

Mesh network 802 (2006) is an example of wireless ad hoc networks. Many mesh networks consist of static devices and can be connected to the Internet. Mesh networks are being deployed in cities to provide ubiquitous wireless coverage for general population or, in many instances, as a network for common use by different first-responder agencies such as the police, firefighters, or emergency medical services (ACG and Meshdynamics; PacketHop). The challenge in the mesh network is due to its dense deployment and integration with the Internet.

3.2.2 Flat Ad Hoc Networks

A traditional ad hoc network has a flat structure in which all nodes in the network have identical functionalities. The control functions, such as routing, are performed on the flat network without any central controller. This flat structure

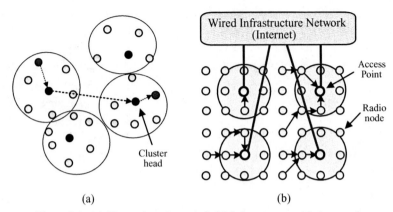

Figure 3.1. (a) Cluster-based network (b) Infrastructure-aided network.

has potential problems, such as poor scalability as network size is getting bigger; Gupta and Kumar's well-known theoretical result (Gupta and Kumar 2000) for multihop wireless networks indicates that achievable end-to-end per-node throughput decreases in proportion to the square root of the number of radio devices.

3.2.3 Cluster-Based Network Structure

A natural approach to facilitate network control functions is to self-organize ad hoc nodes into cluster-based hierarchical structures (shown in Figure 3.1[a]). This is especially motivated by the properties of ad hoc wireless networks: changing topology and shared wireless links, which can be managed effectively by cluster-heads by applying distributed link cluster algorithms (Baker and Ephremides 1981). Previous work (Perkins 2001) also demonstrates that clustering increases network availability and reduces delay in response to network state changes because it reduces sensitivity to small network-state changes and localizes control in response to significant changes. In Lin and Gerla (1997), Gerla proposes a MAC-layer "clustering" that provides a framework for code separation, channel access, and bandwidth allocation, and improves system performance.

3.2.4 Infrastructure-Aided Network Structure

Introducing infrastructure to ad hoc mesh networks further overcomes the problem of scalability bottleneck of flat ad hoc networks. The infrastructure-aided network structure is also motivated by the fact that realistic mesh network scenarios involve predominant traffic flows from mobile or sensor devices to and from the wired Internet, thus requiring effective integration of wired "access points" with the ad hoc wireless network. By introducing some proportion of wired "infrastructure" nodes, the network can be organized into a hierarchy with "shortcut" paths for traffic that would have required larger numbers of hops in a flat ad hoc network (see Figure 3.1[b]). Results of Liu et al. (2003), Kozat and Tassiulas (2003b), and Zemlianov and de Veciana (2005) have shown that adding infrastructure nodes to ad hoc networks can effectively reduce the average number of end-to-end hops and ultimately help achieve better performance relative to flat ad hoc networks.

3.2.5 Hierarchical Hybrid Wireless Network Structure

As described earlier, ad hoc mesh networks benefit from a hierarchical "hybrid" wired or wireless architecture both in terms of scalability and effective integration with the Internet. However, with two-tier architecture, wired infrastructure costs can be high, especially for dense usage scenarios. In Liu et al. (2003), Liu and Towsley proved that linear scaling of throughput can be approached in a two-tier hybrid network as long as the number of access points grows asymptotically faster than the square root of the number of radio nodes.

A network with more than one tier of ad hoc radio nodes is motivated by the previously described considerations. Lower tiers in the network aggregate traffic up to intermediate radio relays, while continuing to use robust ad hoc self-organization and routing protocols as in flat ad hoc networks. A general K-level hierarchy with $(K - 1)$ tiers of radio nodes and a top tier of access points (see Figure 3.2) is presented in Zhao et al. (2003), Ganu et al. (2004), and Zhao and Raychaudhuri (2009). The access points at the top tier provide access to infrastructure and interconnections with the Internet. A key technology enabler for the generalized hierarchical wireless network is the so-called "radio forwarding node" or "radio router," equipped with two or more radio interfaces to

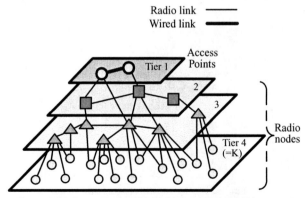

Figure 3.2. Concept of multitier hierarchical hybrid network.

permit it to handle packets going to or from one layer of the hierarchy to another. The end-user devices at the lowest tier of the network, owing to typical energy constraints, have limited routing capability, which helps reduce processing and transmission power.

Multiple tiers of radio-forwarding nodes can provide performance improvements by facilitating shorter routes between distant nodes, improving MAC efficiency via traffic aggregation and less stringent transmit power constraints. Meanwhile, multiple tiers of radio-forwarding nodes potentially reduce the required number of wired access relative to the two-tier network case (Zhao et al. 2004; Zhao and Raychaudhuri 2006b, 2007).

3.3 Protocol Design

Large-scale deployment of ad hoc and wireless networks calls for a protocol design that is both adaptive and robust to changes in the environment as well as to hardware failures. Some challenges that a protocol designer should consider are medium access control, self-organization and automatic resource discovery, routing, and transport control. The protocol stack is shown in Figure 3.3.

3.3.1 Medium Access Control

Medium access control (MAC) layer is responsible for brokering channel access between contending devices. A good design would minimize collisions, provide a reasonable guarantee of reliable transmission across the link, and ensure fair medium sharing between contending devices. In ad hoc and mesh networks, MAC layer needs to perform these functions in the presence of rapidly varying channel conditions owing to multipath fading, interference, hidden and exposed terminals, and multihop topology. In addition, ad hoc networks pose the challenge of dynamic changes in the topology due to mobility. MAC layer on each

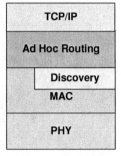

Figure 3.3. Protocol stack of an ad hoc node.

node must also support quality-of-service (QoS) requirements set by the upper layers without the knowledge of QoS requirements of traffic carried by the contending nodes. Due to the absence of infrastructure in ad hoc networks, carrier-sensing-based distributed MAC protocol design has been adopted, and IEEE 802.11, popularly known as WiFi, has been enhanced with an ad hoc mode for this purpose. Mesh networks, on the other hand, may have a planned deployment with some infrastructure support. Therefore, hierarchical or centrally controlled medium access is possible. 802.16, popularly known as WiMAX, is the upcoming standard for mesh networks.

In this section, we will discuss the basic access mechanism in the IEEE 802.11 standard and the fundamental problems with the 802.11 MAC scheme. We will also present an overview of the WiMAX protocol and provide a discussion of some scheduling algorithms proposed for WiMAX.

3.3.1.1 802.11 Standard

The 802.11 standard is the most widely used medium access technique in wireless multihop and access point controlled networks. After the first published IEEE 802.11 standard in 1999, a variety of extensions have been studied to support higher data rate (802.11b, g), mesh and ad hoc architecture (802.11a, b, g, s), mobility (802.11r), vehicular communication (802.11p), cross-layer design (802.11k), multimedia communication (802.11aa), higher throughput (802.11b, g, n, ad, ac), and quality of service (802.11e). The 802.11-2007 revision 802 (2007) combines several amendments including 802.11 a, b, g, d, h, i, j, e providing a single document specifying one medium access and several physical layer specifications for fixed wireless LANs, mesh and mobile ad hoc networks. The most popular versions of the standard is 802.11b/g whereas the 802.11n standard, providing higher throughput, is the next upcoming standard.

Basic Access Mechanism
The basic access mechanism in the 802.11 standard is called the distributed coordination function (DCF), which is based on carrier-sensing multiple access with collision avoidance (CSMA/CA). DCF specifies that every transmitting station must sense the medium for a random duration before initiating a transmission. The station deems it safe to start a transmission if the medium remains idle for the entire random duration. All stations that overhear a transmission must refrain from initiating their own transmissions and continue sensing the medium.

Hidden Terminal Problem
The random backoff mechanism may reduce the probability of collision but does not eliminate it completely. For example, consider Figure 3.4 where station

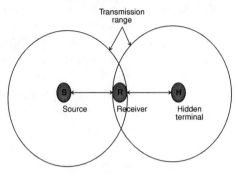

Figure 3.4. Hidden terminal problem in multihop wireless networks.

'H' is beyond the receiving range of the transmitter 'S,' therefore when 'S' transmits, 'H' does not sense the medium as busy. However, the receiver 'R' is within the receiving range of both 'S and 'H'. If 'S' and 'H' simultaneously start transmissions, the data may not be received correctly at 'B'. Retransmission will be required if either data was intended for B. This is known as the hidden terminal problem in multihop literature, and 'H' is known as the hidden node.

Virtual Carrier Sensing and RTS/CTS
802.11 introduces the concept of virtual carrier sensing and short-control message exchange to alleviate the hidden terminal problem. First, all transmitted packets contain a time duration that specifies the duration for which the medium must be occupied by the packet and any subsequent control packet exchange required for successful completion of the transmission. All overhearing stations set a network allocation vector (NAV) to the specified transmission duration and refrain from any transmission even if the medium is sensed as idle. Second, the transmitter must first send a Request to Send (RTS) frame before transmitting the data. The intended receiver must respond to the transmission with a Clear to Send (CTS) frame. The RTS and CTS frames notify all overhearing nodes about the duration of the impending transmission. The hidden terminal 'H' in our example in Figure 3.4 overhears the CTS and holds all transmissions for the entire duration specified in the RTS or CTS frame. This mechanism was designed to solve the hidden terminal problem.

The RTS/CTS mechanism also provides a fast collision detection by inferring lack of reception of CTS within the expected RTS/CTS exchange time as a collision or a busy medium at the receiver. If a larger data packet was sent without the RTS/CTS exchange, a collision would be detected after a longer time duration, wasting more bandwidth and energy. However, the overhead of RTS/CTS is not justified in short message transmissions, and therefore this is an optional feature. We present this control overhead with respect to transmitted data size in Figure 3.5.

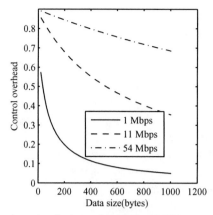

Figure 3.5. Control overhead when using RTS and CTS to protect data packets.

Contention Window

The random backoff duration in 802.11 is selected as a number between 1 to $2^{cw} - 1$, where cw is known as the contention window. The minimum size of the contention window is set to 32 in 802.11b and the maximum size is 1024. Different versions of the standard adjust the backoff window size to support priority channel access for real-time data packets. 802.11 follows a binary exponential backoff (BEB), which essentially means that after each collision, the size of the contention window is doubled, and after a successful data transmission, the window size is reset to the minimum value. The BEB method of setting the window size is sometimes very inefficient as shown in the analytical results from G. Bianchi (2000). In Figure 3.6, we recreate the analytical results from this

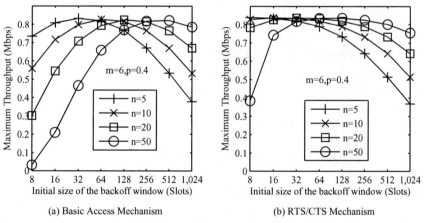

(a) Basic Access Mechanism   (b) RTS/CTS Mechanism

Figure 3.6. Saturation throughput versus initial size of backoff window.
Note: p: conditional collision probability at each time slot, m: number of backoff stages, n: number of contending stations

work to show the maximum achievable throughput with respect to the initial window size. The result shows that for the best throughput performance, the initial window size must be adapted based on the number of contending stations. However, 802.11 does not have any provision to adapt the window size. Therefore, contention is subjective to traffic arrival patterns, which, being a function of user behavior, changes dynamically. An adaptive contention window protocol would thus require a mechanism to continuously sense the environment through passive listening or through the exchange of protocol messages to detect the number of active stations in the network. For example, stations can maintain a moving average of the size of their network allocation vector (Wang and Song 2007), which symbolizes the "busy period" in the neighborhood. With this knowledge, stations may be able to make an approximate "guess" of the contention level and adjust the contention window size accordingly. As an alternative, stations may use the RTS/CTS packets exchange to announce their queue sizes in the neighborhood, thus providing explicit information about future contention levels around their broadcast range. Finally, the beacons used in the ad hoc mode in 802.11 may be enhanced to include fields like "intent to contend" and number of contenders in the neighborhood to exchange the learned expected contention levels (Kim 2005).

Reliable Broadcast and Multicast
The RTS/CTS and ACK mechanism ensures reliable unicast transmission in 802.11, but there is no provision in the standard to ensure reliable multicast and broadcast messages. Several research efforts have suggested extending the RTS/CTS exchange to improve broadcast/multicast reliability. Some suggest protocols in which a leader is selected to transmit CTS on behalf of all receivers (Tourrilhes 1998). Another reliable broadcast mechanism, Broadcast Medium Window (BMW) (Tang and Gerla 2001), maintains neighbor lists and the sequence numbers of missing broadcast data packets. An RTS message is transmitted to each neighbor to inquire of the next missing broadcast sequence number. The CTS returned by the neighbor indicates the requested information. When the requested data is sent, all overhearing neighbors take advantage of the broadcast nature of the medium and update their received sequence numbers. Finally, BMMM (Min-Te, Lifei, Arora, and Ten-Hwang 2002) suggests exchanging RTS/CTS with each next hop receivers, and once all receivers have responded, the sender commences the data transmission. Such simple modifications of 802.11 provide compliance with the standard but may not be always effective. For example, in Tourrilhes (1998), the choice of the leader may not be representative of all receivers. The schemes that request multiple CTS may incur substantial delay depending upon the number of receivers and contention level.

3.3.1.2 802.16 WiMAX Standard

The IEEE 802.16 standard, popularly known as WiMAX, specifies physical and medium access control layers of fixed broadband wireless access systems. Unlike 802.11, which is based purely on contention, 802.16 subscribers initially compete for transmission and then are assigned fixed time slots. The standard specifies the provision for implementing scheduling algorithms that assign collision-free time slots to contending subscriber stations while leaving the design of any particular algorithm to the vendor. Each frame in the IEEE 802.16 standard is divided into two parts: a control subframe to transmit all packets necessary for establishing schedules, and a time division multiplexing based data subframe in which specific time frames are assigned to and reserved for each station enabling collision-free data transmission.

The standard defines two modes of operation: Point to Multipoint (PMP) and Mesh. In the PMP mode, like single-hop wireless LANs, traffic flows only between the base station and the subscriber stations. In the Mesh mode, which is of more interest in this chapter, stations may communicate directly with one another as well as with base station. Moreover, in the Mesh mode, the base station may not be reachable from all subscriber stations, in which case, communication to the base station is through multihop traffic forwarded by other subscriber stations. There are three different scheduling mechanisms defined for the Mesh mode:

- Centralized Scheduling (CS): The base station assigns schedules to all subscriber stations, including those that are not directly connected, through resource request and grant messages transmitted in a collision-free manner during a control subframe.
- Coordinated Distributed Scheduling (CDS): Stations collectively agree upon coordinated transmission schedules in their two hop neighborhood.
- Uncoordinated Distributed Scheduling (UDS): Schedules are established by direct request and grants between two stations.

Both CDS and UDS exchange their schedules through a three-way handshake in which requests are sent indicating potential slots for replies and schedules, grants are sent indicating a subset of suggested available slots that suit the stations, and a grant from the original requestor indicating the final schedule, if any. All overhearing stations must accept and work within the agreed schedule. The only difference is that the three-way handshake occurs in a collision-free manner in CDS whereas in UDS, the handshake messages can collide. Nodes replying to the Request message in UDS must take care to give priority to those that are listed earlier in the Request to avoid collision.

3.3.1.3 WiMAX Scheduling Algorithms

The 802.16-2004 of the WiMAX standard defines four scheduling classes to differentiate traffic based on QoS requirements. First, the Unsolicited Grant Service (UGS) class designed to support real-time traffics that generate fixed-size periodic packets. Second, the real-time Polling Service (rtPS) designed for real-time applications that generate variable-size packets periodically. The rtPS traffic class requires a minimum traffic rate guarantee and latency bounds. Third, Non-Real-time Polling Service (nrtPS) that supports non-real-time, delay-insensitive traffic classes with a minimum bandwidth requirement. Fourth, the besteffort (BE) traffic class that is delay- and bandwidth-insensitive. The 802.16e revision adds an Extended real-time Polling Service (ertPS) that provides more bandwidth to real-time applications.

Scheduling for Real-time Polling Service
The rtPS scheduler at the subscriber station that has real-time traffic to send must send demands for time slots for data transmission. The base station then grants the requested number of slots. There is generally a time lag between the request/grant process and when the data is actually sent. If more data arrives at the subscriber station during this interval, additional slots need to be requested through another request/grant process. The total delay that the application traffic may incur depends on the traffic arrival pattern as well as the number of contending stations. To reduce this delay, an adaptive rtPS scheduler is proposed (Mukul et al. 2006). The subscriber station performs a stochastic prediction of data that may arrive before the requested time slots become available. This calculation is based on the average arrival rate and the time lag between request and the grant process for the transmission slot.

Scheduling for Best Effort Traffic Class
Wireless transmission rates and success probability largely depend on the SNR experienced at the receiver. WiMAX scheduling algorithms for best-effort service class can prioritize transmissions that traverse higher data rate links in order to improve the overall system efficiency. Therefore, many scheduling algorithms suggest the use of SNR to assign transmission scheduling priorities to links with high signal-to-noise ratio. Care should be taken to prevent starvation of links and to avoid violating packet delivery deadlines, fairness policies and bandwidth guarantees. Several scheduling strategies described further in this chapter have been suggested to achieve the efficiency objectives while staying within the constraints. Based on the assumptions made, these scheduling strategies may be suitable for either uplink or downlink transmissions only.

Temporary-removal scheduling (TRS) (Ball, Treml, Gaube, and Klein 2005) policy, as the name suggests, removes packets for low-rate links from the transmission queues. If the link does not improve after a fixed maximum number of removals, the packet must be scheduled anyway. In the worst case, a packet under TRS may be delayed by an additional amount. However, if the link conditions improve, the overall system throughput would benefit from this scheme. Opportunistic Deficit Round Robin (ODRR) scheduling (Rath, Bhorkar, and Vishal Sharma 2006) policy is a polling-based uplink-scheduling policy. Like TRS, ODRR prefers links with higher SNR as it transmits packets if they pass certain eligibility criteria, one of the criteria being SNR, while a packet of a slow SNR link is considered ineligible for transmission. The base station periodically determines an eligible set of subscriber stations. Results show that the polling interval k affects the delay, efficiency, and fairness of the system. When k increases, the system becomes more efficient but less fair. The choice for k is left to the WiMAX service provider.

Scheduling for All Traffic Classes
Due to different requirements of traffic classes, most algorithms concentrate on a single class. However, Wongthavarawat and Ganz (2003) presented a framework for QoS scheduling for all traffic classes in WiMAX using a combination of strict priority services. The UGS connection uses a fixed bandwidth, rtPS uses earliest deadline first (EDF), and weighted fair queuing (WFQ) is used for nrtPS traffic, whereas equal bandwidth distribution is applied to best-effort traffic class. However, in strict priority assignments, it is possible that high-priority traffic may starve the low-priority one. To circumvent this problem, a traffic policing module is introduced, which ensures that a subscriber station does not exceed its total bandwidth allocation. The overall contribution in this work is that it uses well-known scheduling policies applicable to different traffic classes and applies them to each traffic class in WiMAX with appropriate scheduling to prevent starvation.

3.3.2 Self-Organization and Discovery

Because there is no central controller in the ad hoc wireless network, it is required that ad hoc nodes be capable of self-organizing to perform desired tasks. Examples of self-organization include neighbor discovery, construction of connectivity, clustering, and formation of hierarchical structures. A basic discovery protocol is used for the radio nodes to discover each other and organize themselves into a topology in a distributed manner.

In an ad hoc network where there are no explicit discovery mechanisms, the routing protocol is responsible for building the topology by exchanging and disseminating neighboring information. Although this may be sufficient for small networks, it results in excessively high routing overhead as the network size increases. This problem becomes more severe in a multichannel network because the routing messages have to be propagated across all channels. Meanwhile, forming any kind of structure, such as clusters or hierarchy, requires a suitable control function. With the discovery protocol designed particularly for topology control, the control overhead, including routing overhead, can be greatly reduced.

The design of discovery protocol is also motivated by the heterogeneous network scenario. Note that topology formation based on routing protocols assumes the radio nodes to be homogeneous with the same capabilities. A heterogeneous network consists of different types of devices, for example, the end-user devices at the lowest tier of the hierarchical network in Figure 3.2 that do not have the full routing capability.

A very important characteristic of a mobile ad hoc wireless network is its changing topology (Baker and Ephremides 1981; Ephremides 2002). In particular, the network connectivity is affected by the nodes when they join, leave, and move in the network. The topological changes may also be caused by fluctuating wireless link quality and physical bit-rate adaptation (Holland et al. 2001; Kamerman and Monteban 1997). Under these circumstances, whether a node is a "neighbor" or not depends on a set of physical layer parameters instead of fixed connections as in wired networks. The topological changes may be frequent and/or unpredictable, which makes protocol design of ad hoc wireless networks an important challenge. Therefore, the protocol stack must include management functions that discover and maintain the network topology in spite of physical connectivity changes.

3.3.2.1 Neighbor Discovery

Neighbor discovery (ND) is the determination of what nodes are neighbors when a wireless network is initially deployed. If the network topology is changed during network operation, the ND algorithm could be rerun. The neighbor discovery is an important enabler of network connectivity. In neighbor discovery phase, nodes are required to discover their neighbors quickly and efficiently in order to feed topology information to routing protocols and other topology-control algorithms while conserving energy.

A node A detects the presence of its neighbor B upon successful reception of B's packet when the received signal-to-noise ratio is greater than a defined threshold. Due to asynchronous transmitting and sensing behavior of the nodes in the ad hoc network, some packets may be dropped if the receiver does not listen to the channel while the sender is transmitting or the packet collides with others.

To improve the possibility of successful packet reception during the ND phase, several ND algorithms have been proposed (Borbash et al. 2007;

McGlynn and Borbash 2001; Vasudevan et al. 2005). Those ND algorithms run on top of a broadcast-based physical layer and a random-access MAC protocol to handle the medium contention. To further improve the efficiency of neighbor discovery, some contention-free mechanisms have been proposed; for example, the multiuser detection approach avoids collisions at modulation level (Angelosante et al. 2007).

3.3.2.2 Topology Control and Self-Organization

Through neighbor discovery, radio nodes become aware of each other's presence. Furthermore, ad hoc nodes can execute a suitable distributed control function to create a wireless network via selection of appropriate radio frequencies and nodes to associate with. In particular, the discovery protocol can determine the logical topology based on the physical topology detected by the MAC protocol. By making a subset of wireless links available to routing, the discovery protocol creates an efficient topology and reduces burden on routing and routing overhead.

The discovery protocol may also provide a metric that can be used by the routing protocol for path selection based on some performance objective. The objective function aims to optimize a target performance metric such as maximum throughput, minimum delay, energy efficiency, link fairness, or load balance.

The BEacon Assisted Discovery (BEAD) protocol (Raju et al. 2004) is proposed for the hierarchical ad hoc networks in Figure 3.2. The radio nodes use MAC beacons to identify themselves and exchange information such as node type and link quality. Different metrics are used for different types of associations: The received beacon signal-to-noise ratio is used for the energy-constrained mobile nodes (MNs) to select the association with either a forwarding node (FN) or AP; the hop count is used for the FN to choose its association with another FN or an AP. As a result, the logical topology of the network is formed by taking into account connectivity, throughput, delay, and energy requirements or constraints. Simulation results demonstrate that the BEAD protocol provides the flexibility of topology control, reduces the routing overhead, and achieves the desired performance based on the objective function.

There are other discovery issues under investigation, for example, resource or service discovery. In a heterogeneous network, some nodes may choose to provide a service to other nodes. This requires a resource or service discovery mechanism to locate the resource (Dekar and Kheddouci 2009; Kozat and Tassiulas 2003b).

3.3.3 Routing

Throughout the history of wireless protocol research, several routing protocols have been discussed. These protocols may be broadly classified as on-demand

and link-state. We will discuss various routing protocols in this section, including CNF/DTN routing protocols.

3.3.3.1 Distance Vector and Link State Routing

A routing algorithm is used to generate a decision-making procedure for each node to select one or more of its neighbors to forward a packet on its way to the correct destination. Most ad hoc routing protocols are based on ideas from routing methods in conventional wired computer networks. Two of the most popular routing algorithms in computer networks are *distance vector* and *link state* routing (Tanenbaum 1996).

Distance vector routing algorithms maintain a table in each router, which gives the best known distance to each destination and the link to use to get there. Pure distance vector algorithms such as distributed Bellman-Ford (Bertsekas and Gallager 1992) do not perform well in mobile networks because of slow convergence and count-to-infinity problem. Therefore, these algorithms need to be modified and enhanced when used in ad hoc network scenarios. Examples are Destination Sequenced Distance Vector (DSDV) (Perkins and Bhagwat 1994) and Ad hoc On-demand Distance Vector (AODV) (Perkins and Royer, 1999).

In link-state routing algorithms, each router discovers its neighbors and measures the link cost to each of them, then distributes the link-state information to all other routers and finally computes the shortest path to every other router. Optimized Link State Routing (OLSR) (Jacquet et al. 2000) falls into this category. Compared to link-state routing, distance vector routing is easier to implement and requires less storage space because of its computation efficiency. But link-state routing records the entire path and enables nodes to gather more link-state information, which facilitates route selection corresponding to different criteria.

3.3.3.2 Proactive and Reactive Routing

Ad hoc routing protocols may also be categorized as *proactive* (or *table-driven* e.g., DSDV) and *reactive* (or *on-demand*, e.g., AODV and Dynamic Source Routing [DSR] [Johnson and Maltz 1996]) routing protocols, and combinations thereof (e.g., Zone Routing Protocol [ZRP] [Haas and Pearlman 1997]). Proactive routing protocols continuously compute routes to all nodes so that a route is readily available when a packet needs to be sent to a particular node. On the other hand, on-demand routing protocols start a route computation process only when a packet needs to be sent to some other node. Therefore, on-demand routing protocols save bandwidth and reduce power consumption in mobile environments without periodic route advertisements, but data packets may experience larger delay than using proactive routing protocols (Lee et al. 2000).

3.3.3.3 Location-Aided, Directed Diffusion, and Geographic Routing

Location information may improve routing performance. For example, as an extension of DSR, Location-Aided Routing (LAR) (Ko and Vaidya 1998) sends location information in all packets to decrease the overhead of a future route discovery.

In addition to the ad hoc routing protocols extended from wired networks, presented here, there are some other routing approaches specifically designed for dense sensor networks. *Directed diffusion* (Estrin et al. 1999) is an example of such. Directed diffusion incorporates attribute-based naming, data-centric routing, and application-specific processing inside the network. In particular, each node in the sensor network names data that it generates with one or more attributes, and other nodes may disseminate interests based on these attributes. The propagation path of interest then sets up a reverse data path for data that matches the interest.

Another routing approach designed for ad hoc networks is geographic routing. Geographic routing identifies nodes by their locations and uses these coordinates to forward packets (if possible) toward the destination in a greedy manner. This type of routing scales well because it only keeps local information. The challenge of geographic routing is how to get through dead ends when greedy routing fails. Greedy Perimeter Stateless Routing (GPSR) (Karp and Kung 2000) solves this problem by routing around the perimeter of such regions.

3.3.3.4 Adaptive Routing and Adaptive Routing Framework

There are many studies on specific classes of ad hoc routing protocols, but no single routing protocol performs well across the full range of parameters associated with a complex real-world environment. For example, previous work (Broch et al. 1998) shows that DSDV, as a proactive routing protocol, is preferable for latency-sensitive traffic; but DSR, as an on-demand routing protocol, outperforms DSDV in high-mobility environment.

Adaptive routing has been proposed to dynamically adapt routing to changing network topology and external service needs. For example, the Sharp Hybrid Adaptive Routing Protocol (SHARP) (Ramasubramanian et al. 2003) automatically finds the balance point between proactive dissemination and reactive discovery of routing information and dynamically adapts to changing network characteristics and traffic behavior. Another example that dynamically combines

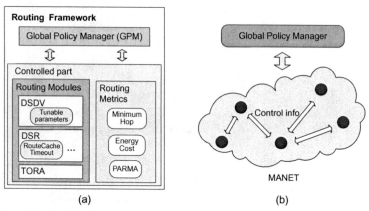

Figure 3.7. (a) Adaptive routing framework (b) Distributed global policy manager.

table-driven and on-demand routing is the strategy presented in McDonald and Znati (2000). It is adaptive to node mobility by balancing the tradeoff between path optimality and routing overhead. Tuning routing algorithm parameters can also help achieve adaptive behavior. Such an example is Adaptive Zone Routing Protocol (AZRP) (Giannoulis et al. 2004) that uses variable zone radius and controllable route update interval.

Adaptive routing framework proposed in Zhao and Raychaudhuri (2006a) is another step further. This unified adaptive framework aims to solve the routing efficiency problem in a systematic approach, and various adaptive mechanisms can be deployed in it. The architecture of the adaptive routing framework is shown in Figure 3.7(a). It implements self-adaptation by a control loop, which collects the information about routing states from the system, makes decisions, and adjusts the system as necessary. The control loop consists of two parts: the controlling part and the controlled part. The Global Policy Manager (GPM) is the controlling part that implements particular adaptation operations such as selecting the routing module, tuning the routing algorithm parameters, and adjusting the routing metric variables. The routing modules and routing metrics are the controlled elements.

Several routing modules are available in the framework. The routing module that would produce the best desired performance is selected by the GPM. The parameters of the selected routing module can also be tuned by the GPM. According to service requirements and traffic behavior, the GPM decides a routing metric and its variables. To achieve global optimization, the GPM, when making decisions, needs not only local information but also information from other nodes of the network. The control information, including state variables and management information, is disseminated through the network. Thus the

GPM entities of all the nodes in the network construct a distributed system and perform the global controlling functionalities, instead of being isolated, as shown in Figure 3.7(b).

According to the implementation of adaptation operations, the adaptive mechanisms can be classified into two types: One is switching between available routing modules or routing metrics, and the other is implementing an integrated adaptive algorithm to control a particular routing element, such as a routing metric or a routing algorithm parameter. The integrated adaptive approach is more interesting because it implements the controlled and controlling parts of the self-adaptation function in a distributed algorithm. The control information including time-varying state variables can be propagated over the network by routing messages, and each node makes decisions based on its local information without a consensus protocol. Therefore, the integrated adaptive algorithms are practically flexible to implement with reduced adaptation overhead and do not involve service interruptions.

Integrated adaptive algorithms include adaptive routing protocols (e.g., SHARP described earlier), adaptive routing parameters (e.g., the cache time-out of DSR [Johnson and Maltz 1996]), or adaptive routing metrics. In addition, cross-layer adaption is achieved when cross-layer information such as physical data rate and wireless medium busy level is incorporated into routing decision. Cross-layer adaptive mechanisms are discussed in details in Section 3.4.

3.3.3.5 DTN Routing

The routing protocols that we discussed so far work on the underlying assumption that there is always an end-to-end path in the ad hoc network. Delay/Disruption Tolerant Networks (DNTs) relax this assumption and propose partially, intermittently connected networks where an end-to-end path between two parts of the network may not always exist. Figure 3.8 shows a typical DTN network where regions A and B are never connected. However, low-earth orbiting relay satellite or a motorbike/bus might be available to make the necessary connections periodically. DTN routing has been classified based on complete, partial, and zero knowledge of periodic connectivity through mobile message carriers (Fall 2003). When mobility models of such entities are perfectly known, a message ferry may be used to carry data to a designated server, as illustrated in Figure 3.8.

Vahdat and Becker (2000) propose epidemic routing for DTN networks when no knowledge regarding mobility behavior is known. Each node maintains a list of all the messages in its buffer, called summary vector. This contains both the messages it has initiated and the messages in transit. When two nodes come in communication range of each other, an antientropy session is initialized in

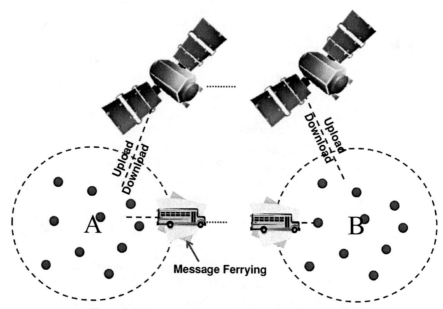

Figure 3.8. Example of a delay-tolerant network scenario.

which the nodes exchange their summary vectors. Based on the vectors, each node in the communication pair may decide to "download" messages that it has not encountered yet. By spreading messages in such an "epidemic" manner, the probability that a message reaches its destination increases with the number of replications. However, this scheme is resource-intensive, and buffer management techniques must be employed to manage the large number of messages in the network.

The MaxProp (Burgess, Gallagher, Jensen, and Levine 2006) routing protocol provides an improvement over the epidemic routing protocol by reducing the amount of storage space used. Like in epidemic routing, messages are replicated to improve the likelihood of delivery; however, a cost metric called "estimated delivery likelihood" is used to "contain" the epidemic instead of disseminating messages to all nodes encountered. Each node in the network keeps an account of its probability of meeting its peers. This probability is estimated by averaging the number of contacts between nodes over time. Nodes that were in contact further in the past have a lower likelihood of delivery compared to those encountered recently. The cost for a path is the sum of the probabilities that each connection on the path does not occur, estimated as one minus the likelihood of delivery. The cost for a destination is the lowest path cost among all possible paths.

When peers meet, the following information is exchanged. First, all messages from the peer are transmitted. Second, routing information, that is, a vector listing estimations of the probability of meeting every other node, is exchanged. Third, acknowledgments for all delivered packets are exchanged so that peers can delete delivered packets from their buffers. Fourth, new packets, that is, those that have not traversed too far in the network, are given priority when selecting packets from the set of packet to forward. These strategies have been shown to reduce delivery latency.

3.3.4 Transport Control Protocol

End-to-end protocols such as TCP developed on wired networks perform poorly over ad hoc and mesh networks. This section discusses the transport control protocol design considerations for ad hoc and mesh networks.

TCP is the most widely used transport protocol in the Internet today. In the early 1980s, when connection to the Internet was over slow dial-up links, a set of thinwire protocols (Farber, Delp, and Conte 1984) were suggested for computers connecting to the ARPA-Internet over a data path of 9,600 baud or less. Wired broadband at home and office has replaced the slow wired dial-up access to the Internet; however, the demand for mobile wireless broadband brings the same challenge with an order-of-magnitude data rate difference between wired, wireless, and cellular networks. Therefore, cross-layer transport layer design is still required to efficiently navigate bottleneck wireless links. Header compression and transport proxy are two different optimizations often suggested to improve the network performance when the rate information is available. Mowgli data channel protocol (MDCP) (Alanko, Kojo, Liljeberg, and Raatikainen 1997) employs header compression, reduced control overhead, use of transmission rates based on the speed of the transmission link, and transport proxy nodes to improve the transport performance. Proxy nodes act as the mobile user and receive data from the network on behalf of the mobile. The mobile may then retrieve the requested content from the proxy at a later time. Indirect TCP (I-TCP) uses the same proxy concept for the wireless endpoint connection. Several optimizations over TCP have been suggested to improve transport layer performance in wireless multihop networks. We will discuss some of these protocols in this section.

3.3.4.1 Cross Layer Aware Protocol (CLAP)

The Cross Layer Aware Protocol (CLAP) was designed for wireless networks with link rates that fluctuate with time in response to changes in signal-to-noise ratio in the link. The main objectives of CLAP is to adapt the transport layer flow rate with the current detected physical layer data rate, reduce self-interference by minimizing control transmissions in the opposite direction, and decouple flow control from error control by removing the dependence of flow control on roundtrip time. The transport layer flow control depends solely on the transmission rate to the destination. Before initiating a transmission, the transport layer observes the data rate to the next hop and the length of the link layer queue. The number of packets to be transmitted during the next fixed interval is computed based on the two parameters. Per-hop per-packet reliability is considered redundant at the transport layer, and the existing MAC layer per-hop reliability in wireless networks is leveraged. The transport layer only sends an aggregate list of packets that were not received successfully. These design choices improve the transport layer performance for wireless networks, and simulation results show significant improvement when compared to TCP Selective Acknowledgment Protocol (TCP-SACK).

3.3.4.2 Freeze TCP

Freeze TCP was proposed for last-hop wireless links and when the destination is mobile. The destination is responsible to measure signal strengths received from the access point to detect an impending disconnection. In case the signal strength weakens and the mobile determines that a disconnection or handoff is about to happen, it sends an advertisement of zero window size to the source, forcing the source into the ZWP mode and preventing the dropping of packets from the congestion window. When the connection is reestablished, the mobile may send back three acknowledgments of the last received packets. This scheme prevents the TCP session from breaking when the mobile node is disconnected for a short time duration. Simulation results show performance gain of about 38 percent when disconnections last for ten seconds.

3.3.4.3 Hop-by-Hop Transport

Cache and Forward architecture suggests a novel hop-by-hop transport protocol where large files traverse the network as single transport layer entity. At each network hop, the file is transferred in its entirety to the next hop router before being forwarded further downstream, as shown for the media file C1 moving from router A to destination M1 in Figure 3.9.

While the file is transferred through the cache and forward (CNF) network, an en-route router may decide to cache it. The caching decision may be based on popularity, availability of the file in caches nearby, and the access frequency in the local network. In Figure 3.9, router B caches the file C1 while forwarding it to C. Similarly requests to retrieve a file also result in hop-by-hop transport from the end-user to the file's location. Every en-route router checks its cache to locate the requested file. If an intermediate router finds the file in its cache, it sends a cache hit response to the requestor followed by the requested file. In

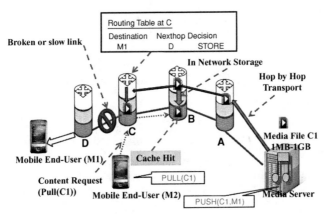

Figure 3.9. Conceptual diagram of CNF.

our example (Figure 3.9), router B responds to the mobile M2 with a cache hit message and serves the request (Pull[C1]) from its local cache.

There is a large improvement in file delay performance shown in comparison to end-to-end streaming concept of TCP under simulation scenarios where wireless endpoints communicate with one another through a wired backbone.

3.4 Cross-Layer Adaptive Mechanisms

Wireless medium access control layers are designed to shield physical layer variations from the upper layers in the protocol stack. The intention is to present to the upper layer an illusion of error-free physical medium so that the wireline protocol stacks may function in wireless access networks without any modification. Over the years, research has shown that this design choice is inhibitive to efficient wireless communication. Today wireless consumers of the Internet outnumber wired clients, and research interest is highly skewed in favor of cross layer designs that expose the properties of layers to one another. Distance or hop-based routing cost computation do not work well in practice in wireless networks (De Couto, Aguayo, Bicket, and Morris 2005). Instead, physical data rate, congestion, collision, and retransmissions are conveyed by the medium access control layer to the network layer to facilitate expected transmission count (De Couto, Aguayo, Bicket, and Morris 2005), transmission time (Draves et al. 2004) and data rate (Park and Kasera 2005) routing costs. Similarly, TCP being designed for wired networks considers any packet loss to be a sign of congestion. This assumption is not valid in the error-prone wireless medium (Shen and Zhao 2006), and therefore variations have been suggested to improve TCP performance in wireless networks (Gerla et al. 1999). In some cases, new transport protocols suitable for both wired and wireless networks have been suggested (Paul, Yates, Raychaudhuri, and Kurose; Jain et al. 2009).

3.4.1 Cross-Layer Routing Metric

Most conventional ad hoc routing protocols, including DSDV, AODV, and DSR, use the minimum hop (MH) as the metric to make routing decisions. This is primarily a carry-over from wired networks where the transmission rate of a link does not dynamically change and the link rate is independent of the physical transmission range. However, in case of wireless networks, the MH metric tends to choose paths with fewer hops, and each hop in paths tends to have a longer physical span and also is associated with a lower bit rate than an alternative path with more hops. Meanwhile, note that rate control has been implemented readily, such as auto-rate feedback (ARF) (Kamerman and Monteban 1997) and receiver-based auto-rate (RBAR) (Holland et al. 2001) schemes proposed for the IEEE 802.11 devices 802 (1999). Therefore, to take advantage of the wireless multirate capability and make better use of available network capacity, transmission rate needs to be incorporated into the routing metric.

Some examples of routing metrics that incorporate the physical layer parameters include the Medium Time Metric (MTM) (Awerbuch et al. 2004), the expected transmission count metric (ETX) (De Couto et al. 2003), the Expected Transmission Time (ETT) (Draves et al. 2004), and airtime link 802 (2006). The MTM aims to find a path with the minimum total transmission time. It is a static solution that only handles the multirate capability. Upon observing that using the shortest path would result in poor throughput, De Couto et al. (2002) propose the ETX to incorporate the effects of link loss ratios. The ETX introduces extra routing overhead of the dedicated link probe packets to measure the delivery ratio. Like the MTM, the ETX is independent of network load and does not attempt to route around congested links. The ETT incorporates both the link loss rate and the link speed, and is used as the weight associated to each link. The individual link weights are combined into a path metric called Weighted Cumulative ETT (WCETT) that explicitly accounts for the interference among links that use the same channel. The WCETT metric tends to choose channel-diverse paths to improve throughput in multiradio multihop wireless networks. The airtime link metric is specified in the IEEE 802.11s draft 802 (2006) and is in fact equivalent to the ETT.

In addition to the physical layer parameters, it is also possible to provide an awareness of congestion at each node in order to avoid bottleneck regions with high link utilizations. The wireless link is usually shared with other links in the same neighborhood, whereas in a wired network, links operate independently of each other, and channel access on one link has no effect on any of the adjacent links. Thus it makes sense to devise metrics that account for both congestion and transmission rate in a combined manner. For example, a link may provide for a high transmission rate but could appear congested because neighboring links have a high link utilization. Thus, if we account for only the link rate, this link would show up as a "good" link, but when combined with a congestion metric, it may turn out to be just the opposite, which is a more accurate reflection of the PHY/MAC layer. The next section introduces a PHY/MAC-aware routing metric and discusses techniques needed to handle different variations of changes of different layers in cross-layer design.

3.4.1.1 PARMA: A PHY/MAC-aware Routing Metric for Ad Hoc Wireless Networks with Multi-Rate Radios

The PHY/MAC-aware routing metric for ad hoc networks (PARMA) (Zhao et al. 2005) incorporates both the PHY bit-rate and MAC congestion information.

Rate-Adaptive PHY and Auto Multirate Mechanism
The widely used IEEE 802.11x standard uses adaptive selection of physical layer bit rate as a function of observed channel quality. 802.11b radios can choose different physical rate (1, 2, 5.5, 11 Mbps) whereas 802.11a/g radios select between 6, 9, 12, 18, 24, 36, 48, or 54 Mbps as the physical channel rate. This automatic PHY bit-rate adaptation feature is considered to be useful in most systems because it permits end-users to take advantage of good-quality short-range links when available. When such multirate radios are used to build ad hoc networks, the network topology and link speed change more dynamically than in radio networks with a single mode radio with fixed bit rate and range.

Figure 3.10 depicts the way in which an 802.11b radio device experiences different bit rates when connecting to its neighbors at various distances. As

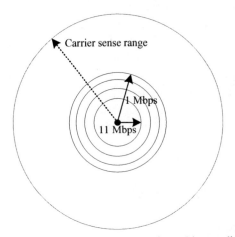

Figure 3.10. Transmission range of a multirate radio.

shown in the figure, if a node wants to use the 11 Mbps rate, only nodes in the innermost circle can decode its frame correctly with sufficient signal-to-noise ratio (SNR). However, if it chooses to use the lower 1 Mbps rate, the transmission range would be much larger. The outermost circle indicates the threshold of carrier sense in 802.11 MAC. If there is a radio outside this circle, then the signal level of this radio's transmission received at the central node is not large enough, so the central node would sense the channel as "idle." Note that in the above description, a two-ray path loss channel model (Goldsmith 2005) is assumed and the received signal strength is simply compared to a series of fixed thresholds.

Two popular auto-rate schemes are the auto-rate feedback (ARF) (Kamerman and Monteban 1997) and receiver-based auto-rate (RBAR) (Holland et al. 2001) proposed for the IEEE 802.11 devices 802 (1999). A multirate device working with ARF scheme adjusts the rate according to the positive or negative feedback. For the RBAR scheme, the receiver decides the rate based on the measured signal strength and piggybacks the rate information to the sender via RTS/CTS exchanges. Another auto-rate scheme eliminates the extra overhead by choosing the rate for each outgoing packet based on the SNR measurements of the packets received on the reversed link (Zhao et al. 2005). This SNR-based auto-rate scheme can be easily incorporated with periodic routing updates.

MAC Channel Congestion

The MAC channel congestion can be measured by the channel access delay. The channel access delay correlates to the traffic at the MAC layer by taking into account both the locally offered traffic and that forwarded by neighboring nodes. Because the wireless medium is shared, whether a packet can access the channel immediately is determined by not only the states of the two end-nodes of the link, but also those of all neighboring nodes.

To measure this effect, a "virtual access delay" estimation based on physical layer information is introduced. To avoid unnecessary overhead introduced by periodic probes, a passive estimation method is employed. Every node records every channel event (i.e., transmission) sensed from the physical channel and makes an estimation of the "expected delay if a packet has to be sent." Suppose an M/M/1 queueing system (Bertsekas and Gallager 1992) with the common channel as the server, where packets arrive from the nodes in the neighborhood of the channel to obtain service (i.e., access the channel and get transmitted). According to the results of queueing theory, the average waiting time in queue, i.e., the *channel access delay*, is given by

$$T_W = T_S \frac{\rho}{1 - \rho}, \tag{3.1}$$

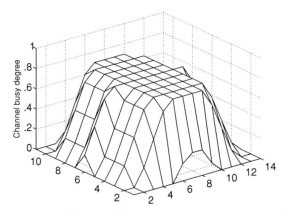

Figure 3.11. Channel busy degree around a flow with saturated load.

where ρ represents the utilization factor of server (i.e., the *channel busy degree*), and T_S is the service time for the channel event, corresponding to the packet transmission time. The estimated T_W is given to the routing protocol for use in the routing metric.

Each node can estimate ρ by sensing the occupancy of the channel. Figure 3.11 illustrates the channel busy degree ρ around a flow with saturated load. Note that the region with high busy degree ($\rho \simeq 0.75$) actually lose the ability to support any flows further.

PHY/MAC-aware Route Selection
PARMA aims to optimize the packet end-to-end delay. The end-to-end delay of a packet of size L_{pkt} transversing a path p_i is calculated as

$$Delay_{p_i} = \sum_{\forall links \in p_i} (T_{transmit} + T_{access} + T_{queuing}), \tag{3.2}$$

where $T_{transmit}$ denotes the packet transmission time in the link, T_{access} the medium access time spent by the packet getting access to the link, and $T_{queuing}$ the queueing time required for the packet waiting before trying to access the channel.

The packet transmission time can be calculated as

$$T_{transmit} = N_{transmit} \times \frac{L_{pkt}}{R_s}, \tag{3.3}$$

where R_s is the link speed, which would be one of the rates the multirate devices provide, and $N_{transmit}$ is the number of transmissions, including retransmissions, needed for the packet to be received correctly. When the link quality is poor, packet retransmissions will be carried out by the MAC protocol. With adaptive multirate PHY, around 90 percent of packets get transmitted successfully in

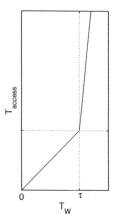

Figure 3.12. A nonlinear mapping from T_W to T_{access}.

the first attempt (Gopalakrishnan 2004). Hence $N_{transmit}$ can be set to 1 as an approximation.

The medium access time, T_{access}, is used to indicate the medium busy level around the sending node of the link. When the medium is busy, it takes a relatively long time for a packet to get the chance to transmit. Incorporating the medium access time into the routing metric, the routing algorithm can choose a route with light traffic load in addition to high-speed links, and thus spread the traffic over the network to achieve load balance, avoid congestion, and increase effective bandwidth. Notice that the highly busy region loses the ability to support any flows (see Figure 3.11). A nonlinear mapping from T_W, the access delay estimated by the MAC layer, to T_{access}, as shown in Figure 3.12, is applied. It is one of the techniques introduced to smooth the link layer change effect on routing and also to improve route convergence.

A large access delay reflects a growing interface queue length when the network is congested. When a system below saturation is considered, $T_{queueing}$ can be omitted.

With the above assumptions and simplifications, the routing metric computation can be summarized as

$$Delay_{p_i} = \sum_{\forall links \in p_i} \left(\frac{L_{pkt}}{R_s} + T_{access} \right). \tag{3.4}$$

The system performance with the PARMA metric is compared with the MH and MTM (Awerbuch et al. 2004) metrics using the *ns*-2 network simulator (Fall and Varadhan 2002). Those routing metrics are plugged into DSDV. To make DSDV work well with the PHY/MAC-aware routing metric, specific enhancements to the routing protocol are required, such as maintaining two routing tables to avoid missing the best route when it arrives later than the first route of the next new sequence number. In addition, smoothing techniques for the link portion of the proposed metric are introduced to adjust the different variations of changes between the PHY/MAC layer and the network layer, and also to improve route convergence. The simulation results indicate that, with both the PHY rate and the MAC occupancy level taken into account in the routing metric, packets can choose high-rate links while avoiding congested areas in the network, thus improving system throughput and reducing average end-to-end delay.

3.4.2 Integrated Routing and MAC Scheduling

The PHY/MAC-aware routing metric optimizes the routing function by incorporating MAC contention and interference effects. It belongs to the layered implementation of multihop 802.11 MAC and routing protocol. However, the performance achieved by optimizing an individual layer such as routing or MAC in multihop wireless environment is limited by a certain point (Barrett et al. 2002). Meanwhile, the fundamental inefficiency caused by CSMA/CA MAC (Bertsekas and Gallager 1992; 802 [1999]) in multihop scenarios need to be solved. Therefore, it makes good sense to treat MAC and routing jointly to improve MAC efficiency and further improve system performance.

Several studies on joint optimization of routing and link scheduling provide performance bounds (Jain et al. 2003; Kodialam and Nandagopal 2003; Tassiulas and Ephremides 1992). A recent work proposes IRMA (Integrated Routing and MAC Scheduling Algorithm) to integrate routing and MAC as a single algorithmic framework (Wu et al. 2006; Wu and Raychaudhuri 2008). The IRMA uses joint optimization techniques to establish end-to-end path and TDMA schedules for flows across the network.

3.4.2.1 IRMA: Integrated Routing and MAC Scheduling for Single-Channel Wireless Mesh Networks

The IRMA approach is proposed to overcome the problems presented in multihop wireless networks, for example, hidden nodes contending for channel (Xu and Saadawi 2001), poor spatial reuse due to channel sensing-based backoffs in the extended neighborhood of an ongoing transmission (Li et al. 2001), and self-interference among packets of the same flow transmitted at each hop along the path (Gerla et al. 1999).

The nodes in a traditional 802.11-based mesh network randomly access the shared medium based on the locally observed information. The IRMA algorithm attempts to solve the fundamental inefficiency caused by the CSMA/CA mechanism by creating conflict-free TDMA link schedules based on traffic demand across all end-to-end routed paths.

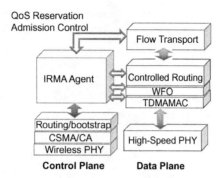

Figure 3.13. IRMA protocol architecture.

Control Plane

Because of lack of coordinations, the neighboring nodes using CSMA/CA contend to access the medium in a distributed manner. To improve the efficiency of medium access, the IRMA uses a centralized algorithm to allocate schedules and paths simultaneously for each source-destination pair of traffic in the network. In particular, the network entities execute online control procedures to collect necessary information, run optimization, and distribute the MAC and routing parameters. The IRMA framework places the control processes in a *control plane*, separated from the *data plane* over which packet data are transferred. The protocol architecture consisting of control plane and data plane is shown in Figure 3.13.

The control plane may be implemented using either a dedicated portion of TDMA slot or a separate channel or frequency to exchange the control information including topology, bandwidth, and traffic flow specifications. Based on the information, the centralized control algorithm determines the route and TDMA slot assigned for each source-determination pair.

The medium contention is eliminated by arranging conflicting transmissions in different TDMA slots. Spatial reuse is maximized by scheduling a maximum number of interference-free transmissions simultaneously in the same time slot.

Traffic Aware Scheduling

Link scheduling in a single-channel packet radio network has been formulated as a vertex-coloring or edge-coloring problem (Cidon and Sidi 1989; Ramaswami and Parhi 1989; Ephremides and Truong 1990; Gandham et al. 2005). Meanwhile, several distributed MAC schemes (Zhu and Corson 1998; Bao and Garcia-Luna-Aceves 2001) have been proposed to create interference-free TDMA schedules. However, these approaches are based on oversimplified interference models and are per-packet scheduling approaches. Relying on the control plane to collect and disseminate the topology information and traffic specifications, the IRMA assigns traffic flows to alternative paths based on actual end-to-end traffic demand.

Joint MAC and Routing Algorithms: IRMA-MH and IRMA-BR

The IRMA establishes nonconflicting radio resources by considering all relevant traffic in the interference neighborhood collectively. The routing table and link access schedules for each involving node are solved jointly in one algorithm.

Two alternative joint MAC and routing algorithms are designed. Link scheduling with minimum hop count (IRMA-MH) algorithm solves min-hop routing and optimizes link scheduling based on routing results and real-time flow demands. Link scheduling with bandwidth-aware routing (IRMA-BR) algorithm optimizes routing and scheduling decisions simultaneously by using available MAC bandwidth information to route around congested areas.

It is worth mentioning that the IRMA-BA algorithm conducts bandwidth-aware path selection to route around medium busy areas. As illustrated in Figure 3.14, node A will choose node B as the next hop and its packets will enter a highly interfered region (shown as the shaded region in the figure), when the min-hop routing is applied. The IRMA-BA takes into account the available bandwidth and chooses a link (from node A to C) with light interference while still leading to the destination.

To provide upper and lower bounds, the integrated routing and MAC optimization is formulated as a linear programming (LP) problem. The objective function maximizes the aggregate throughput of all end-to-end traffic. After conducting a certain large number of iterations, if the upper and lower bound converge, the converged value is used as the analytical throughput of the LP solution as a reference to compare with the simulation results.

Both IRMA-MH and IRMA-BR with centralized and distributed algorithm variations are evaluated using *ns*-2 simulations (Fall and Varadhan 2002) and

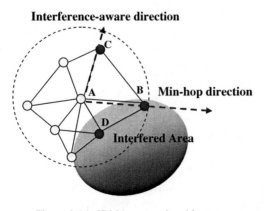

Figure 3.14. IRMA protocol architecture.

compared with two baseline schemes (DSDV and AODV). The results show significant two-to-three times improvements in network throughput over baseline 802.11-based mesh networks with independent routing protocols such as DSDV and AODV. The control overhead is also much lower than baseline schemes.

3.5 Integration with the Internet

It is noted that most applications involve traffic flows to and from the Internet in addition to peer-to-peer communication between ad hoc nodes. As the Internet users turn toward mobile Internet access using mobile broadband, 3G, and WiFi, there are worldwide initiatives to rethink the Internet architecture, layering and services (FIND; Lemke 2007; FP7 2007). The focus is on designing solutions that provide seamless integration of wired Internet with wireless, ad hoc, sensor, mesh, and cellular networks. Having studied the nature of different types of wireless deployments earlier in this chapter, we know that they are very different from the traditional fixed infrastructure. Therefore, we need to answer several questions while looking for an interface that integrates them into one network. Can the IP-based addressing be used to identify each device in the Internet? How far should information like changes in the network organization and mobility-based disconnections be propagated into the core network? Should BGP be extended for wireless networks, disseminating summary information across all connected networks? Should the architecture be designed for end-to-end connectivity, or is the hop-by-hop protocol a better option?

An easy approach is to make minimal changes and preserve the existing protocol stack, then simply provide a Proxy server to glue together these different types of networks to the wired core. A second approach would be to allow more awareness across the networks and let BGP like gateway routing disseminate long-term events, like a mobile end-user disappearing from the network, while hiding the short-lived changes like transient link variations. A third approach is for seamless integration, where a control plane in which any router can query the routing information for any end-point across the network, regardless of the wired or wireless nature of the links that constitute the path. These ideas are elaborated in the following paragraphs.

3.5.1 Glue Network

A glue network requires minimal changes to the current state of the art. It is based on the idea that simple interfaces may be designed to translate data at the boundaries so that it can be represented in the format suitable for the destination network type. Thus, simple "Proxy Gateway" nodes can be designed with multiple communication interfaces. To the wired network, all data coming from or going to the wireless end nodes appears to originate/terminate at the Proxy Gateway and vice versa, just as transport models like I-TCP and CLAP described earlier. Concerns for cross-layer design, self-organization, addressing, routing, and so on become local to the individual networks and the wired Internet is oblivious to all the idiosyncracies of the wireless networks. However, this creates a narrow waist of congestion and excessive delays at the entry and exit point to the wireless "cloud." Perhaps local in-network caching of popular content may be used to reduce the amount of traffic that must cross the network boundaries.

3.5.2 Extended Glue Network

The glue network described earlier has several performance issues in terms of localized congestions both in the wired and wireless end of the glue/proxy gateway. Another design may solve this problem by allowing some summarized routing information to percolate into the wired core. For example, a Border Gateway Protocol-like method may be used to exchange summarized longer-term variations and local congestion information in wireless routes so that TCP-like flow control may be used at further upstream points rather than closer to the wireless gateway.

However, by design, ad hoc networks are expected to appear and tear down dynamically at any place in the world. Therefore, the nodes sending these summary information may also appear at random places in the network. Moreover, these ad hoc gateways may appear via cellular, satellite, or multihop mesh networks and even through a combination of multiple heterogeneous network hops. It is unclear if the summarized information regarding such a network should reach all the way to the wired core, or should the information be further summarized each time it traverses a different network type. Therefore, this approach calls for at least a redesign of the BGP protocol to handle these points.

3.5.3 Seamless Integration

Perhaps the answer for a seamlessly and fully integrated Internet lies in the cross-layer design philosophy that has been so popular in wireless network designs. The protocol stack may be redesigned so that the upper layer can implement optimizations depending on the link-layer technology used for communication. Perhaps a control plane may be developed to make information like routing, location, sessions, and so on available on an on-demand basis so that protocols can opportunistically select the best mode of operation. For example, if an application server notices a slow-end link to the destination, it may change the format of the data being delivered. Thus, high-definition video transmission may be scaled down so that the content is delivered but at a lower rate. The transport protocol may dynamically switch to I-TCP/CLAP/Hop-by-Hop transport when a weak wireless link is encountered and revert back to normal TCP when a series of good wired links are being traversed. Network layer may temporarily store instead of forwarding over slow links, especially if it is possible to judge through past observations that the link may improve soon.

Similarly, self-organization tactics can be used for dynamic connection with the control plane for information dissemination. When a new ad hoc network appears and needs to connect to the Internet, it may perform a resource discovery to select the best set of gateway nodes for this connection. The gateway nodes may then announce their presence to the control plane and start feeding relevant local state information to the Internet. Mobile devices that can have several heterogeneous interfaces may be able to dynamically select the best network for communication. Similarly naming, addressing, authentication, and dynamic entry and exit to and from the broader Internet should be resolved in terms of physical location coordinates of devices rather than network-level IP addresses.

This design requires a complete overhaul of the network architecture and therefore seems like a daunting task. But a seamless redesign will perhaps be the most sustainable way into the future. It is also the first opportunity for researchers to design the Internet based on experiences from the past, rather than from making good guesses and imaginations regarding how the network to be used in the future.

3.6 Conclusion

The research in wireless ad hoc and mesh networks has come a long way. It is a maturing field where several important topics are now understood quite well. However, several areas are still lightly explored and therefore important for current and future works. Some important current research topics include integration of various wireless technologies and architectures, as well as improvement in communication capacity using MIMO techniques and multichannel operation. Opportunistic channel access using cognitive radios and integration with the Internet will form the major research effort in the coming years.

References

1999. *Wireless LAN Medium Access Control (MAC) and Physical Layer (PHY) Specification.*

2005. *Lecture Notes in Computer Science: Information Networking.* Vol. 3391/2005. Springer Berlin / Heidelberg. Chap. Adaptive Window Mechanism for the IEEE 802.11 MAC in Wireless Ad Hoc Networks, 31–40.

2006. *IEEE 802.11s Tutorial: Overview of the Amendment for Wireless Local Area Mesh Networking.*

2007. *FP7 Information and Communication Technologies: Pervasive and Trusted Network and Service Infrastructures.* FP7-ICT-2007-2.

2007 (June). *IEEE Std 802.11-2007 (Revision of IEEE Std 802.11-1999) Part 11: Wireless LAN Medium Access Control (MAC) and Physical Layer (PHY) Specifications.* IEEE Computer Society Sponsored by the LAN/MAN Standards Committee.

ACG and Meshdynamics. *MeshDynamics Structured Mesh.*

Alanko, T., Kojo, M., Liljeberg, M., and Raatikainen, K. 1997. Mowgli: Improvements for Internet Applications Using Slow Wireless Links. *The 8th IEEE International Symposium on Personal, Indoor and Mobile Radio Communications (PIMRC)*, vol. 3, pages 1038–1042.

Angelosante, D., Biglieri, E., and Lops, M. 2007. Neighbor Discovery in Wireless Networks: A Multiuser-Detection Approach. *Proc. Information Theory and Applications Workshop*, pages 46–53.

Awerbuch, B., Holmer, D., and Rubens, H. 2004. High Throughput Route Selection in Multi-Rate Ad Hoc Wireless Networks. *Proc. 1st IFIP TC6 Working Conference on Wireless On-demand Network Systems (WONS 2004)*, pages 251–268.

Baker, D. J., and Ephremides, A. 1981. The Architectural Organization of a Mobile Radio Network via a Distributed Algorithm. *IEEE Trans. Commun.*, **29**(11), 1694–1701.

Ball, C. F., Treml, F., Gaube, X., and Klein, A. 2005. Performance Analysis of Temporary Removal Scheduling Applied to Mobile WiMax Scenarios in Tight Frequency Reuse. *IEEE 16th International Symposium on Personal, Indoor and Mobile Radio Communications (PIMRC)*, vol. 2, pages 888–894.

Bao, L., and Garcia-Luna-Aceves, J. J. 2001. A New Approach to Channel Access Scheduling for Ad Hoc Networks. *Proc. 7th Annu. Int. Conf. Mobile Computing and Networking (ACM MobiCom 2001)*, pages 210–221.

Barrett, C., Marathe, A., Marathe, M. V., and Drozda, M. 2002. Characterizing the Interaction between Routing and MAC Protocols in Ad-Hoc Networks. *Proc. The 3rd ACM International Symposium on Mobile Ad Hoc Networking and Computing MobiHoc 2002*, pages 92–103.

Bertsekas, D., and Gallager, R. 1992. *Data Networks*. Second ed. Prentice Hall.

Bianchi, G. 2000. Performance Analysis of the IEEE 802.11 Distributed Coordinationfunction. *IEEE Journal on Selected Areas in Communications*, **18**(3), 535–547.

Borbash, S. A., Ephremides, A., and McGlynn, M. J. 2007. An Asynchronous Neighbor Discovery Algorithm for Wireless Sensor Networks. *Ad Hoc Networks*, **5**(sep), 998–1016.

Broch, J., Maltz, D. A., Johnson, D. B., Hu, Y.-C., and Jetcheva, J. 1998. A Performance Comparison of Multi-Hop Wireless Ad Hoc Network Routing Protocols. *Proc. ACM/IEEE MobiCom'98*, pages 85–97.

Burgess, J., Gallagher, B., Jensen, D., and Levine, B. N. 2006. MaxProp: Routing for Vehicle-Based Disruption-Tolerant Networks. *Proc. IEEE INFOCOM*.

Cidon, I., and Sidi, M. 1989. Distributed Assignment Algorithms for Multihop Packet Radio Networks. *IEEE Trans. Comput.*, **38**(10), 1353–1361.

De Couto, D. S. J., Aguayo, D., Bicket J., and Morris, R. 2005. A High-Throughput Path Metric for Multi-hop Wireless Routing. *Wireless Networks*, **11**(4), 419–434.

De Couto, D. S. J., Aguayo, D., Chambers, B. A., and Morris, R. 2002. Performance of Multihop Wireless Networks: Shortest Path Is Not Enough. *Proc. ACM 1st Workshop on Hot Topics in Network (HotNets-I)*.

De Couto, D. S. J., Aguayo, D., Chambers, B. A., and Morris, R. 2003. A High-Throughput Path Metric for Multi-Hop Wireless Routing. *Proc. ACM/IEEE MobiCom 2003*.

Dekar, L., and Kheddouci, H. 2009. A Resource Discovery Scheme for Large Scale Ad Hoc Networks Using a Hypercube-Based Backbone. *Proc. International Conference on Advanced Information Networking and Applications AINA 2009*, pages 293–300.

Draves, R., Padhye, J., and Zill, B. 2004. Routing in Multi-Radio, Multi-Hop Wireless Mesh Networks. *Proc. ACM/IEEE MobiCom 2004*, pages 114–128.

Ephremides, A. 2002. Ad Hoc Networks: Not an Ad Hoc Field Anymore. *Wireless Communications and Mobile Computing*, **2**, 441–448.

Ephremides, A., and Truong, T. 1990. Scheduling Broadcasts in Multihop Radio Networks. *IEEE Transactions on Communications*, **38**, 456–460.

Estrin, D., Govindan, R., Heidemann, J., and Kumar, S. 1999. Next Century Challenges: Scalable Coordination in Sensor Networks. *Proc. ACM/IEEE MobiCom'99*, pages 263–270.

Fall, K. 2003. A Delay Tolerant Network Architecture for Challenged Internets. *Pro. SIGCOMM*.

Fall, K., and Varadhan, K. 2002. *The ns Manual*.

Farber, D. J., Delp, G. S., and Conte, T. M. 1984. *RFC 914 – Thinwire protocol for connecting personal computers to the Internet*. http://www.faqs.org/rfcs/rfc914.html

FIND. *NSF NeTS FIND Initiative*. http://www.nets-find.net/

Gandham, S., Dawande, M., and Prakash, R. 2005. Link Scheduling in Sensor Networks: Distributed Edge Coloring Revisited. *Proc. IEEE Conference on Computer Communications INFOCOM*, pages 2492–2501.

Ganu, S., Raju, L., Anepu, B., Zhao, S., Seskar, I., and Raychaudhuri, D. 2004. Architecture and Prototyping of an 802.11-Based Self-Organizing Hierarchical Ad-Hoc Wireless Network (SOHAN). *Proc. IEEE Int. Symp. Personal, Indoor and Mobile Radio Commun. (PIMRC 2004)*, pages 880–884.

Gerla, M., Tang, K., and Bagrodia, R. 1999. TCP Performance in Wireless Multi-hop Networks. *Proc. The 2nd IEEE Workshop on Mobile Computer Systems and Applications WMCSA 1999*, page 4.

Giannoulis, S., Katsanos, C., Koubias, S., and Papadopoulos, G. 2004. A Hybrid Adaptive Routing Protocol for Ad Hoc Wireless Networks. *Proc. 2004 IEEE International Workshop on Factory Communication Systems*, pages 287–290.

Goldsmith, Andrea. 2005. *Wireless Communications*. Cambridge University Press.

Gopalakrishnan, P. 2004. *Methods for Predicting the Throughput Characteristics of Rate-Adaptive Wireless LANs*. M. Eng. thesis, Rutgers, The State University of New Jersey.

Gupta, P., and Kumar, P. R. 2000. The Capacity of Wireless Networks. *IEEE Trans. Inf. Theory*, **46**(Mar.), 388–404.

Haas, Z. J., and Pearlman, M. R. 1997. A New Routing Protocol for the Reconfigurable Wireless Networks. *Proc. IEEE International Conference on Universal Personal Communications*.

Holland, G., Vaidya, N., and Bahl, P. 2001. A Rate-Adaptive MAC Protocol for Multi-Hop Wireless Networks. *Proc. ACM/IEEE MobiCom 2001*.

Jacquet, P., Muhlethaler, P., and Qayyum, A. 2000. *Optimized Link State Routing (OLSR) Protocol*. Internet Draft, draft-ietf-manet-olsr-01.txt.

Jain, K., Padhye, J., Padmanabhan, V. N., and Qiu, L. 2003. Impact of Interference on Multi-hop Wireless Network Performance. *Proc. ACM/IEEE MobiCom 2003*, pages 66–80.

Jain, S., Saleem, A., Liu, H., Zhang, Y., and Raychaudhuri, D. 2009. Design of Link and Routing Protocols for Cache-and-Forward Networks. *IEEE Sarnoff Symposium (SARNOFF '09)*, pages 1–5.

Johnson, D. B., and Maltz, D. A. 1996. Dynamic Source Routing in Ad Hoc Wireless Networks. In Imielinski, T., and Korth, H. (eds.), *Mobile Computing*. Chap. 5, pages 153–181. Kluwer Academic Publishers.

Kamerman, A., and Monteban, L. 1997. WaveLAN-II: A High-Performance Wireless LAN for the Unlicensed Band. *Bell Labs Technical Journal*, 118–133.

Karp, B., and Kung, H. T. 2000. GPSR: Greedy Perimeter Stateless Routing for Wireless Networks. *Proc. ACM/IEEE MobiCom 2000*, pages 243–254.

Ko, Y., and Vaidya, N. H. 1998. Location-Aided Routing (LAR) in Mobile Ad Hoc Networks. *Proc. ACM/IEEE MobiCom'98*, pages 66–75.

Kodialam, M., and Nandagopal, T. 2003. Characterizing Achievable Rates in Multi-Hop Wireless Networks: the Joint Routing and Scheduling Problem. *Proc. ACM/IEEE MobiCom 2003*, pages 42–54.

Kozat, U. C., and Tassiulas, L. 2003a. Network Layer Support for Service Discovery in Mobile Ad Hoc Networks. *Proc. IEEE INFOCOM 2003*.

Kozat, U. C., and Tassiulas, L. 2003b. Throughput Capacity of Random Ad Hoc Networks with Infrastructure Support. *Proc. 9th Annu. Int. Conf. Mobile Computing and Networking (ACM MobiCom 2003)*, pages 55–65.

Lee, S.-J., Gerla, M., and Joh, C.-K. 2000. A Simulation Study of Table-Driven and On-Demand Routing Protocols for Mobile Ad Hoc Networks. *Proc. IEEE International Conference on Communications (ICC 2000)*, vol. 3, pages 1702–1706.

Lemke, M. 2007. *Position Statement: The European FIRE Initiative Washington DC*.

Li, J., Blake, C., De Couto, D. S. J., Lee, H. I., and Morris, R. 2001. Capacity of Ad Hoc Wireless Networks. *Proc. 7th Annu. Int. Conf. Mobile Computing and Networking (ACM MobiCom 2001)*, pages 61–69.

Lin, C. R., and Gerla, M. 1997. Adaptive Clustering for Mobile Wireless Networks. *IEEE J. Sel. Areas Commun.*, 15(7), 1265–1275.

Liu, B., Liu, Z., and Towsley, D. 2003. On the Capacity of Hybrid Wireless Networks. *Proc. IEEE INFOCOM 2003*, vol. 2, pages 1543–1552.

McDonald, A. B., and Znati, T. 2000. A Dual-Hybrid Adaptive Routing Strategy for Wireless Ad-Hoc Networks. *Proc. IEEE Wireless Communications and Networking Conference 2000 (WCNC 2000)*.

McGlynn, M. J., and Borbash, S. A. 2001. Birthday Protocols for Low Energy Deployment and Flexible Neighbor Discovery in Ad Hoc Wireless Networks. *Proc. The 2nd ACM International Symposium on Mobile Ad Hoc Networking and Computing MobiHoc 2002*, pages 137–145.

Mukul, R., Singh, P., Jayaram, D., Das, D., Sreenivasulu, N., Vinay, K., and Ramamoorthly, A. 2006. An Adaptive Bandwidth Request Mechanism for QoS Enhancement in WiMax Real Time Communication, *Wireless and Optical Communications Networks*.

PacketHop. *Infrasructure Free Broadband Communications*.

Park, J. C., and Kasera, S. K. 2005. Expected Data Rate: An Accurate High-Throughput Path Metric for Multi-hop Wireless Routing. *Second Annual IEEE Communications Society Conference on Sensor and Ad Hoc Communications and Networks (SECON 05)*, pages 218–228.

Paul, S., Yates, R., Raychaudhuri, D., and Kurose, J. 2008. The Cache-and-Forward Network Architecture for Efficient Mobile Content Delivery Services in the Future Internet. *First ITU-T Kaleidoscope Academic Conference on Innovations in NGN: Future Network and Services*, pages 367–374.

Perkins, C. E. 2001. *Ad Hoc Networking*. Addison-Wesley.

Perkins, C. E., and Bhagwat, P. 1994. Highly Dynamic Destination-Sequenced Distance-Vector Routing (DSDV) for Mobile Computers. *Proc. ACM SIGCOMM'94 Conf. Commun. Architectures, Protocols and Applicat*, pages 234–244.

Perkins, C. E., and Royer, E. M. 1999. Ad Hoc On-Demand Distance Vector Routing. *Proc. 2nd IEEE Workshop Mobile Computing Syst. and Applicat. (WMCSA'99)*, pages 90–100.

Raju, L., Ganu, S., Anepu, B., Seskar, I., and Raychaudhuri, D. 2004. Beacon Assisted Discovery Protocol (BEAD) for Self-Organizing Hierarchical Wireless Ad-Hoc Networks. *Proc. IEEE Global Commun. Conf. (GLOBECOM 2004)*, pages 1676–1680.

Ramasubramanian, V., Haas, Z. J., and Sirer, E. G. 2003. SHARP: A Hybrid Adaptive Routing Protocol for Mobile Ad Hoc Networks. *Proc. The Fourth ACM International Symposium on Mobile Ad Hoc Networking and Computing MobiHoc 2003*, pages 303–314.

Ramaswami, R., and Parhi, K. 1989. Distributed Scheduling of Broadcasts in a Radio Network. *Proc. IEEE Conference on Computer Communications INFOCOM*, vol. 2, pages 497–504.

Rath, H. K., Bhorkar, A., and Sharma. 2006. NXG02-4: An Opportunistic Uplink Scheduling Scheme to Achieve Bandwidth Fairness and Delay for Multiclass Traffic in Wi-Max (IEEE 802.16) Broadband Wireless Networks. *Global Telecommunications Conference GLOBECOM 06*, pages 1–5.

Shen, M., and Zhao, D. 2006. TCP Throughput Performance in IEEE 802.11-based Multihop Wireless Networks. *QShine '06: Proceedings of the 3rd International Conference on Quality of Service in Heterogeneous Wired/Wireless Networks*. New York, NY, USA: ACM, page 23.

Sun, M.-T., Huang, L., Arora, A., and Lai, T.-H. 2002. Reliable MAC Layer Multicast in IEEE 802.11 Wireless Networks. *International Conference on Parallel Processing*, pages 527–536.

Tanenbaum, A. S. 1996. *Computer Networks*. Third ed. Prentice Hall.

Tang, K. and Gerla., M. 2001. MAC Reliable Broadcast in Ad Hoc Networks. *Military Communications Conference MILCOM*, pages 1008–1013.

Tassiulas, L., and Ephremides, A. 1992. Jointly Optimal Routing and Scheduling in Packet Radio Networks. *IEEE Transactions on Information Theory*, 165–168.

Tourrilhes, J. 1998. Robust Broadcast: Improving the Reliability of Broadcast Transmissions on CSMA/CA. *The Ninth IEEE International Symposium on Personal, Indoor and Mobile Radio Communications*, vol. 3, pages 1111–1115.

Vahdat, A., and Becker, D. 2000. *Epidemic Routing for Partially-Connected Ad Hoc Networks*. Tech. rept. Duke University.

Vasudevan, S., Kurose, J., and Towsley, D. 2005. On Neighbor Discovery in Wireless Networks with Directional Antennas. *Proc. IEEE INFOCOM 2005*, vol. 4, pages 2502–2512.

Wang, J., and Song, M. 2007. An Efficient Traffic Adaptive Backoff Protocol for Wireless MAC Layer. *International Conference on Wireless Algorithms, Systems and Applications*, pages 169–173.

Wongthavarawat, K., and Ganz, A. 2003. Packet Scheduling for QoS Support in IEEE 802.16 Broadband Wireless Access Systems. *International Journal of Communication Systems Special Issue: Wireless Access to the Global Internet: Mobile Radio Networks and Satellite Systems*, 16(1), 81–96.

Wu, Z., Ganu, S., and Raychaudhuri, D. 2006. IRMA: Integrated Routing and MAC Scheduling in Multihop Wireless Mush Networks. *Proc. The 2nd IEEE Workshop on Wireless Mesh Networks WiMesh 2006*.

Wu, Z., and Raychaudhuri, D. 2008. Integrated Routing and MAC Scheduling for Single-Channel Wireless Mush Networks. *Proc. 9th IEEE Int. Symp. World of Wireless, Mobile and Multimedia Networks (WoWMoM 2008)*.

Xu, S., and Saadawi, T. 2001. Does the IEEE 802.11 MAC Protocol Work Well in Multihop Wireless Ad Hoc Networks? *IEEE Communications Magazine*, 39(6), 130–137.

Zemlianov, A. and de Veciana, 2005. Capacity of Ad Hoc Wireless Networks with Infrastructure Support. *IEEE J. Sel. Areas Commun.*, 23(3), 657–667.

Zhao, S., and Raychaudhuri, D. 2006a. Policy-Based Adaptive Routing in Mobile Ad Hoc Wireless Networks. *Proc. 2006 IEEE Sarnoff Symp.*, pages 1–4.

Zhao, S., and Raychaudhuri, D. 2006b. On the Scalability of Hierarchical Hybrid Wireless Networks. *Proc. IEEE Conf. Inform. Sci. and Syst. (CISS 2006)*, pages 711–716.

Zhao, S., and Raychaudhuri, D. 2007. Multi-Tier Ad Hoc Mesh Networks with Radio Forwarding Nodes. *Proc. IEEE Global Commun. Conf. (GLOBECOM 2007)*, pages 1360–1364.

Zhao, S., and Raychaudhuri, D. 2009. Scalability and Performance Evaluation of Hierarchical Hybrid Wireless Networks. *IEEE/ACM Trans. Networking*, 17(5), 1536–1549.

Zhao, S., Seskar, I., and Raychaudhuri, D. 2004. Performance and Scalability of Self-Organizing Hierarchical Ad Hoc Wireless Networks. *Proc. IEEE Wireless Commun. and Networking Conf. (WCNC 2004)*, pages 132–137.

Zhao, S., Tepe, K., Seskar, I., and Raychaudhuri, D. 2003. Routing Protocols for Self-Organizing Hierarchical Ad-Hoc Wireless Networks. *Proc. 2003 IEEE Sarnoff Symp.*

Zhao, S., Wu, Z., Acharya, A., and Raychaudhuri, D. 2005. PARMA: A PHY/MAC Aware Routing Metric for Ad-Hoc Wireless Networks with Multi-Rate Radios. *Proc. 6th IEEE Int. Symp. World of Wireless, Mobile and Multimedia Networks (WoWMoM 2005)*, pages 286–292.

Zhu, C. X., and Corson, M. S. 1998. Five-Phase Reservation Protocol for Mobile Ad-Hoc Networks. *Proc. IEEE Conference on Computer Communications INFOCOM*, pages 322–331.

4

Opportunistic Delivery Services and Delay-Tolerant Networks

Sanjoy Paul

Abstract

The number of endpoints connected wirelessly to the Internet has long overtaken the number of wired endpoints, and the difference between the two is widening. Wireless mesh networks, sensor networks, and vehicular networks represent some of the new growth segments in wireless networking in addition to mobile data networks, which is currently the fastest-growing segment in the wireless industry. Wireless networks with time-varying bandwidth, error rate, and connectivity beg for opportunistic transport, especially when the link bandwidth is high, error rate is low, and the endpoint is connected to the network in contrast to when the link bandwidth is low, error rate is high, and the endpoint is not connected to the network. "Connected" is a binary attribute in TCP/IP, meaning one is either part of the Internet and can talk to everything or is isolated. In addition, connecting requires a globally unique IP address that is topologically stable on routing timescale (minutes to hours). This makes it difficult and inefficient to handle mobility and opportunistic transport in the Internet. Clearly we need a new networking paradigm that avoids a heavyweight operation like end-to-end connection and enables opportunistic transport. In addition to the these scenarios, given that the predominant use of the Internet today is for content distribution and content retrieval, there is a need for handling dissemination of content in an efficient manner. This chapter describes a network architecture that addresses the previously mentioned unique requirements.

4.1 Introduction

Caching[1,2] and Content Distribution Networks (CDNs)[3,4] have proven to be extremely useful on the Internet today. However, the mechanisms used to leverage the usage of caches on the Internet today are not very clean. For example, to use an institutional proxy cache, typically, the browsers have to be *configured* to point to the proxy cache, or a special device like a Layer-4 switch has to be used to transparently redirect Web requests to the institutional cache, or some automated scripts are run to identify the proxy cache for the corresponding browser. Multiple mechanisms exist because each has its own pros and cons, and none of these techniques is a clear winner. Similarly, to redirect a user request to the nearest mirror server of a CDN, different CDN vendors use different mechanisms. Moreover, the details of the mechanism and signaling used by a CDN vendor like Akamai[3] and/or Limelight Networks[4] are *proprietary* even though we know it is most likely based on Domain Name System (DNS). Whereas the DNS-based redirection is best for CDN vendors like Akamai and Limelight networks who do not *own* the network, it may not be the best way out for Network Service Providers like AT&T, who own their network, for building a CDN. Once again, just as in the case of caching, multiple techniques are used in CDNs to redirect an end-user request to the "nearest" mirror server. In summary, *several* complex parallel *control and signaling* infrastructures have been built on top of the Internet to make use of the caches (or storage nodes). The question is, *if* we had the luxury of building a *clean-slate* next-generation Internet, would it make sense to maintain status quo or to design a simpler unified mechanism to leverage the well-proven benefits of caching (storage) in the network.

A parallel development has been happening in the Internet community in the context of Delay/Disruption-Tolerant Networking (DTN)[5] whose objective is to deal with *disruption* or *intermittent* connections on the Internet that the traditional TCP/IP paradigm cannot handle efficiently. Interestingly enough, DTN community recognized the need for hop-by-hop transport combined with *caching* as a way of mitigating the effect of disruption in communication. DTN community has proposed a different control and signaling mechanism on top of the Internet.

Yet another community, mostly driven by the researchers in the field of mobile communications and networking,[6,7,23] had realized the benefit of hop-by-hop transport in multihop wireless communications to improve performance of content delivery, and once again caching plays a central role. To take advantage of caching, this community is designing yet another control and signaling mechanism.

Given that caching is so central to multiple communities and that it is being used to serve a variety of needs, and given that due to the limitations of the current Internet design, each community has to come up with its own control and signaling mechanism, and also given the luxury of designing the next-generation Internet from scratch, there is tremendous benefit in designing a unified protocol for leveraging the caches to meet the needs of these diverse communities.

The architecture proposed in this chapter is not an alternative to what the CDN community has deployed, or DTN community has proposed, or mobility community has proposed; rather it is an attempt to leverage the best ideas from these communities and to put them together into a unified framework. In the context of the current Internet, this framework can be thought of as an overlay network on top of the Internet. In a clean-state design of the next-generation Internet, the unified framework may very well be integrated into the network itself.

4.2 Design Principles

There are several reasons why a new architecture is needed for opportunistic transport and delay-tolerant networking. First, the Internet architecture assumes that there exists an end-to-end path between the endpoints that need to communicate and exchange information. This is certainly not true for mobile endpoints that may not be within the range to communicate or for sensor nodes that wake up intermittently to communicate. Second, the Internet architecture computes a single path from the source to the destination for routing packets between the two endpoints. However, there are several scenarios where computing a path from the source to the destination is not possible ahead of time, especially when the source or the destination is not connected to the network. In addition, in the event of congestion along the precomputed path, packets get delayed. It may be a better approach to decide on the route dynamically as opposed to statically before the communication begins. Third, packet switching is assumed to be the most appropriate abstraction for interconnecting heterogeneous systems. However, when the end-users are mostly interested in content, the appropriate switching entities need not be packets, but rather messages or contents themselves. Fourth, the Internet architecture assumes that packet loss rate is small and the lost packets can be recovered through end-to-end retransmissions. However, when such assumptions fail, as in time-varying wireless links where packet loss rate could be significantly high from time to time, or in systems where an end-to-end path does not exist, the end-to-end performance suffers badly. In general, these shortcomings of the Internet Architecture need to be addressed for the following types of networks:

1. Hybrid Fixed and Mobile Networks
2. Military Ad hoc Networks
3. Vehicular Networks
4. Mobile Wireless Networks
5. Media Distribution Networks
6. Sensor Networks

All the previously mentioned factors lead to the design of a new architecture for opportunistic communication and delay-tolerant networking with the following characteristics:

1. Network elements should have *persistent memory* or *storage (cache)* integrated in them. This is important because the intended destination may be out of reach and the message may need to be stored at an intermediate network element until the intended destination gets connected. The intermediary carrying the message to the final destination could also be mobile and hence may need to hold on to the message until it gets back into the connected network or gets a chance to hand over the message to the destination. The side-effect of storing content in the network is the efficient delivery that can be achieved by virtue of delivering the content from the network itself as opposed to from a server outside the network.

2. The network should not be built on packet-switching technology but rather on message-switching technology where a message could be as big as the entire content file itself.

3. Messages should be transmitted between two successive intermediate network elements using a reliable *virtual link layer* protocol. The link between two successive network elements is called virtual because it consists of multiple hops in the underlying physical network but behaves as a single link between two nodes in the overlay network. The link layer protocol should be configurable so that it can be tuned to the characteristics of the virtual link.

4. *Routing* decisions should not be made at the source at the time of transmission but rather should be made at each intermediate network element as the message is transmitted hop by hop.

5. In addition to address-based routing, there is a need for *content-based* routing.

6. *Network layer* should support multiple classes of service so that some messages are treated with higher priority compared to others based on the urgency of message delivery.

7. *Naming and late binding* should be two of the most important support services in the network. Late binding is useful because resolving names upfront makes sense only when the routing needs to be decided at the source. However, when the destination may not even be connected to the network or the exact location of the destination is not known ahead of time, it makes sense to resolve names to addresses toward the end of the delivery process.

8. Semantics of *multicasting* needs to be defined differently because the members of a multicast group may not be online when the multicast session starts

or ends. Moreover, the source and/or destinations may be mobile leading to dynamic formation of the multicast tree.

9. *Transport layer* becomes minimal in this case because the network itself provides reliable transmission between network elements. Moreover, since the final destination may not be connected, it may be difficult, if not impossible, to have a timely end-to-end acknowledgment as in the case of TCP in the Internet.

10. Acknowledgment continues to make sense for the *Application layer* protocol. However, the semantics may vary depending on the circumstances.

4.3 Alternative Architectures

Several network architectures and associated protocols have been proposed to handle disruptive communication. However, the driving factors behind these architectures have been different and hence, despite significant functional commonality, these architectures evolved slightly differently as described next.

4.3.1 Delay and Disruption Tolerant Networking (DTN) (RFC 4838)

Delay and Disruption Tolerant Networking (DTN)[5,30] was the result of combining research in the fields of mobile and ad hoc networking (MANET), vehicular ad hoc networking, and the DARPA-funded research on Interplanetary Internet (IPN). The IPN architecture that was developed to cope with significant delays and packet corruption of deep-space communications laid the foundation of DTN architecture. However, it evolved significantly from the initial IPN architecture as the focus shifted from just Interplanetary Internet to more general concept of Delay and Disruption Tolerant Networking.

4.3.1.1 Architecture

DTN (RFC 4838) architecture consists of endpoints (source and destination) and intermediate nodes, some of which merely forward bundles (bundles are equivalent of packets in DTN architecture) and some, in addition to forwarding bundles, also store them for forwarding at an opportunistic moment some time in the future (such nodes are referred to as custodians). All nodes in the architecture have a common protocol layer, namely the bundle protocol layer that binds together all components of DTN architecture. Bundle protocol layer, as described later, is the equivalent of TCP/IP in the Internet architecture. Architectural highlights of DTN are presented next.

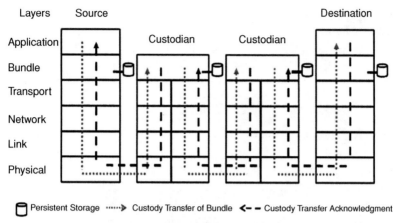

Figure 4.1. DTN architecture (RFC 4838).

Hop-By-Hop Delivery
Fundamental paradigm used in DTN networks is "store and forward" where storing is persistent and not transient as in IP networks. Furthermore, the unit of storage and forwarding in DTN networks is a "bundle" as opposed to a "packet." A bundle is formed by adding relevant header information to an Application Data Unit (ADU) so that the ADU can be routed to the right destination by the bundle layer. A bundle header consists of the original source and final destination endpoint identifier (EID), so that each intermediate node in the DTN network knows where the bundle originated from and where it is headed. Each intermediate node forwards the bundle based on the EID. However, all intermediate nodes are not the same. Some of them simply forward the bundle toward the final destination, whereas some others take on custody of the bundle. Taking custody of a bundle means taking on the responsibility of "reliably" transferring the bundle to the next custodian or to the final destination, whichever may be closer. Reliable transmission requires the custodian to figure out if the bundle has been successfully delivered to the next custodian or not, and if not, retransmit it until the bundle reaches the desired custodian and/or the final destination.

Naming and Late Binding
Endpoints in DTN architecture are identified using EID that follows the syntax of Uniform Resource Identifier (URI) (RFC 3956). Each EID may refer to either a single destination endpoint or a set of destination endpoints. The latter is applicable to anycast and multicast.

Binding refers to mapping an EID to the next-hop EID or the lower-layer address for transmission. For example, in the context of the Internet, the binding happens at the source where the name is mapped into an IP address using DNS. However, in case of DTN architecture, EIDs may be reinterpreted at each intermediate node because the final destination may not be connected to the network or its location in the network may not be known. Thus, DTN nodes perform "name-based" routing with late binding as opposed to "address-based" routing.

4.3.1.2 Protocols

Virtual Link (Bundle Delivery) Layer
In DTN networks, "virtual" link (bundle delivery) layer protocol is responsible for transferring a "bundle" from one DTN node to the next DTN node just as the link layer protocol is responsible for transferring a packet from one router (host) to the next router (host) in the TCP/IP protocol stack. The "virtual" link (bundle delivery) layer in DTN networks rides on top of traditional transport layer protocols (TCP and UDP).

In contrast to the TCP/IP protocol stack where the link layer is usually best effort (no guarantee of delivery, for example in Ethernet), the bundle layer in DTN supports both best-effort as well as reliable delivery mechanisms. Best-effort delivery happens between two nodes when the next-hop DTN node is not a "custodian." However, between two "custodian" nodes, the delivery is expected to be "reliable."

Virtual Network (Bundle Forwarding and Routing) Layer
In DTN networks, "virtual" network (bundle forwarding and routing) layer protocol is responsible for computing the route of a "bundle" from the original source to the final destination. DTN node does the forwarding of a "bundle" to the next-hop node. In fact, the "virtual" network (bundle forwarding and routing) layer resides on top of traditional transport layer protocols (TCP and UDP).

Bundle header contains the original source EID, final destination EID, current custodian EID, and report-to EID in addition to some other fields. Forwarding decisions are made based on the final destination EID, and reports, such as return receipt, among others, are sent to the report-to EID.

Routing is tricky in DTN because the capacity and delay in DTN links vary with time. If link characteristics are known ahead of time, forwarding decisions can be made in an intelligent manner. However, many a time, such information is not available, and then routing becomes challenging. In general, the links could be persistent (DSL line), on-demand (dial-up modem), scheduled intermittent (low-orbiting satellite), opportunistic intermittent (unscheduled low-flying aircraft), or predictive intermittent (based on a previously observed pattern). Different routing protocols are appropriate for different types of links.

DTN architecture supports routing and forwarding of anycast and multicast traffic in addition to that of unicast traffic. However, the semantics of multicast routing in DTN is tricky, because a member of the multicast group might express interest in a content that might have already been delivered to other members of the multicast group. This requires support for storage and forwarding at intermediate nodes for delivery at a later point of time.

Virtual Transport (Bundle Flow Control and Congestion Control) Layer
In DTN networks, "virtual" transport (bundle flow control and congestion control) layer protocol is responsible for ensuring that the average rate at which a sending node transmits data to a receiving node does not exceed the average rate at which the receiving node is prepared to receive data (flow control), and the aggregate rate at which the senders inject traffic into the network does not exceed the maximum aggregate rate at which the network can deliver data to the destinations over time (congestion control). In addition, there are various acknowledgment schemes to guarantee end-to-end delivery. Because the "virtual" transport protocol for DTN network rides on top of transport layer protocols (TCP and UDP), it can leverage both the flow/congestion control of TCP and the acknowledgment scheme of TCP for its own "equivalent" functions at a higher level.

Application Layer
Applications interface with the DTN architecture asynchronously and that is the most appropriate mechanism in long/variable delay environments. Usually the applications register callback actions when certain triggering events occur (such as arrival of an ADU). The application layer protocol generates ADUs and uses the bundle layer for forwarding and delivery.

4.3.2 BBN's SPINDLE

BBN's SPINDLE[8] program was driven by DARPA with the objective of transforming U.S. military into an agile, distributed network-centric force. To achieve that goal, it was critically important to have access to mission-related information even under temporary disruptions to connectivity in the Global Information Grid (GIG). DARPA's Disruption Tolerant Networking (DTN) program, with the above goal in mind, has been developing technologies that enable access to information when stable end-to-end paths do not exist and infrastructure access cannot be assured. DTN technology makes use of persistence within network nodes, along with the opportunistic use of mobility, to overcome disruptions to connectivity. That is the genesis of BBN's SPINDLE architecture.

BBN's SPINDLE architecture is designed on the principle of extensibility with the goal of leveraging the same architecture for serving a variety of next-generation networking needs. A DTN application that focuses on delivering a bundle to the destination in an intermittently connected network would have

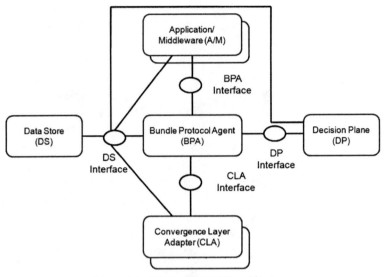

Figure 4.2. BBN's SPINDLE architecture.

different needs compared to a content discovery-and-retrieval solution. However, BBN's SPINDLE network is designed to meet the needs of both of these seemingly disparate types of applications through its extensible architecture. The details of the architecture are described further in this chapter.

4.3.2.1 Architecture

The core of SPINDLE architecture consists of Bundle Protocol Agent (BPA) that implements the main functionality of bundle protocol (RFC 4838). For example, BPA implements forwarding a bundle to the next-hop DTN node, performs delivery of a bundle to the applications, implements custody-transfer mechanism, and so on. However, the routing and forwarding functions, the implementation of reliable delivery of bundle, and such are decoupled from the basic forwarding functionality of the bundle protocol and are designed as separate components. In fact, the other components of the SPINDLE architecture are Data Store (DS), Decision Plane (DP), Convergence Layer Adapter (CLA), and Application/Middleware (A/M). These components are coupled with the core BPA component through Inter Component Communication Protocol (ICCP).

Bundle Protocol Agent (BPA)
Bundle Protocol Agent offers the services of bundle protocol (BP). It executes the procedures of BP and that of bundle security protocol (BSP) in cooperation with other components of the architecture. For example, even though BPA is responsible for implementing the mechanisms of the BP, such as reading,

creating, and updating the fields in the bundle header, it has the flexibility
of leveraging the DP component of the SPINDLE architecture for any key
decisions, such as those related to policy or optimization.

Main functions of BPA are:

1. Forwarding a bundle to the next-hop DTN node, whether it is for unicast,
   anycast, or multicast. However, the next-hop computation is done by the DP
   and passed on to BPA.
2. Doing fragmentation and reassembly of bundle payload as needed to adapt
   the delivery of payload over a link with time-varying capacity.
3. Implementing custody-transfer mechanisms in the bundle header, such as
   sending custody acknowledgment. However, whether to accept or reject custody is determined by the DP again.
4. Delivering a bundle to a "registered" application.
5. Discarding and deleting a bundle. Once again, it is the DP that decides
   whether a bundle should be discarded or not.
6. Implementing all security functions, such as authentication, confidentiality,
   and data integrity.

In addition to the functionalities on the list, BPA implements agent interface
that can be accessed by applications and it uses the interfaces exposed by other
components of the architecture.

Data Store (DS)

DS implements persistent storage used to store not only the bundles, but also the
bundle metadata, network state information, and application state information.
Network-state information includes, among others, routing tables, content meta-
data, and policies, whereas application-state information includes registration
information, application metadata, and so on.

Data Store implements a full Data Base Management System (DBMS) to
enable basic database functions such as query processing.

Knowledge Based (KB) systems can also be integrated with DS to enable
advanced inferencing based on execution of rules.

Decision Plane (DP)

If BPA is the heart of the system, DP is the brain. Specifically, DP is responsible for routing information dissemination, route computation, routing table
updates, late binding and name resolution, policy handling, content caching and
replication decision, content search, and other decisions. DP consists of several
modules:

1. *Routing information dissemination module*: This module is responsible
   for exchanging routing-related information among the network elements.

Specifically, this module decides what information will be shared with whom and when. In addition to disseminating the information, this module also collects the routing-related information in incoming bundles and updates the relevant entries in the knowledge base.

2. *Routing module*: This module is responsible for computing routes for unicast and multicast, for updating the routing table entries, for generating next hops for bundles, for scheduling the bundles, for making decisions about whether to take custody of a bundle or not, and so forth.

3. *Policy module*: This module is responsible for interpreting policies, enforcing policies, dispatching events based on policies, enabling users to add/delete policies, and for subjecting bundles to policies as they pass through the DTN node.

4. *Naming and late binding module*: This module is responsible for resolving names of DTN nodes and feeding the information to the router module so that the right decision about forwarding a bundle can be taken. Usually, this module is invoked and used when the bundle is close to the final destination or close to the care-of address of the final destination where it will be stored for opportunistic delivery to the final destination.

5. *Content module*: This module is used for content-based access, specifically for content search, content caching and replication, content routing, and other content-related functionality.

Convergence Layer Adapter (CLA)
Convergence Layer Adapter is responsible for actual transport of the bundles. CLA leverages whatever transport functionality is available from the underlying network. Status of links (available or not), schedule (for opportunistic delivery), and quality-of-service (QoS) parameters are all monitored by CLA, and the relevant information is passed on to the relevant modules of the architecture for their efficient functioning.

Application/Middleware (A/M)
Application/Middleware module is responsible for sending and receiving bundles based on application needs. This module leverages the services exposed by BPA.

4.3.2.2 Protocols

Virtual Link Layer
In BBN's SPINDLE network, "virtual" link layer functionality is implemented by CLA. The beauty of CLA is that it is not limited to using TCP and UDP, rather it can potentially use any custom protocol (such as, CLAP [23]) that might be available at the corresponding DTN node for use in a specific type of network (for example, CLAP may be available at a DTN node and it may be

the best-suited protocol for wireless links with highly time-varying bandwidth, delay, and error characteristics).

Virtual Network Layer

In the SPINDLE architecture, the "virtual" "network layer" functionality is implemented by the Decision Plane (DP). The beauty of DP is that it is not limited to using any specific protocol; rather it allows usage of any protocol to disseminate and assimilate routing information. Moreover, the routing information is also customizable, meaning the information that will be distributed will depend on the type of routing. For example, routing information to be disseminated for content-based routing could be very different from the routing information needed for traditional address-based routing. The SPINDLE architecture supports this flexibility. Specifically, policy-based routing, content-based routing, late binding, and rich naming, among others, are supported in an extensible manner by DP in The SPINDLE architecture.

Virtual Transport Layer

The "virtual" transport layer protocol is responsible for ensuring flow control and congestion control and is implemented by DP. This is because the network state information, including congestion, is available to and disseminated by the DP module.

Application Layer

Applications interface with BBN's SPINDLE network architecture asynchronously, and that is the most appropriate mechanism in long/variable delay environments. In the SPINDLE architecture, the application-layer functionality is implemented by A/M module. A/M module also provides multiplexed DTN communication service to non-DTN applications running on the node.

4.3.3 KioskNet

KioskNet[9,10] started at the University of Waterloo with the goal of providing very-low-cost Internet access to rural villages in developing countries using the principles of Delay-Tolerant Networking. KioskNet system uses vehicles, such as buses, to ferry data between village kiosks and Internet gateways in nearby urban centers. The data carried by the buses from the rural areas are reassembled at an Intermediary (or Proxy Server) for interaction with legacy servers.

4.3.3.1 Architecture

KioskNet consists of a set of kiosks from which *ferries* (buses) carry data to a set of *gateways* that communicate with a *proxy* on the Internet. The ferries not only carry data from the kiosks, but they also carry data to the kiosks. The main architectural components of KioskNet are described in more detail further in the chapter.

Kiosks

Each kiosk is equipped with a server referred to as kiosk controller, from which one or more PCs can boot off. Kiosk controllers have WiFi connectivity to allow users to connect to them wirelessly. In addition, although Kiosk controllers could have different types of backhaul, such as dial-up, GPRS, or VSAT, the most interesting one from the perspective of DTN is the mechanical backhaul, such as ferries (buses, cars, motorbikes, etc.). The kiosk is expected to be used by two types of users. First type of users use a PC that boots over the network from the kiosk controller and can then access and execute application binaries provided by the kiosk controller. The second type of users uses their own devices such as smart phones, PDAs, and laptops to connect to one or more kiosk controllers or a bus directly and use them as wireless hot spots that provide store-and-forward access to the Internet.

A KioskNet *region* consists of a set of kiosks in the same geographic area administered by the same entity. This means all entities within the region are certified by the same certificate authority. In addition, from the networking perspective, all data bundles are flooded within a region. Figure 4.3 shows a system with two regions, which can be managed either by different administrative entities or by a single administrative entity.

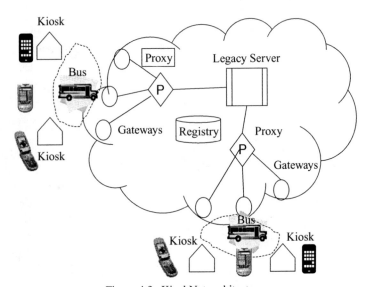

Figure 4.3. KioskNet architecture.

Ferries

Ferries provide internet connectivity to the kiosks via a mechanical backhaul. Examples of ferries include cars, buses, motorbikes, or trains that pass by a kiosk and an Internet gateway. A ferry has a PC with 20–40GB of storage and a WiFi network interface and is powered by the vehicle's own battery. The PC communicates opportunistically with the kiosk controllers and Internet gateways when it comes within their coverage area. During an opportunistic communication session, which may last up to several minutes, hundreds of MB of data can be transferred in each direction. This data is stored and forwarded in the form of self-identifying *bundles*. Ferries upload and download bundles opportunistically to and from an Internet gateway.

Gateways

A gateway is just a PC with WiFi network interface and a broadband (DSL or Cable) Internet access. A gateway collects data opportunistically from a ferry and holds it in local storage before uploading it to the Internet through the proxy. A region may have one or more gateways.

Proxy

Most likely communication between a kiosk user and the Internet would be for existing services such as e-mail, or for accessing back-end systems that provide government-to-citizen services. Legacy servers that provide such services typically are not designed to handle either long delays or disconnections, and most importantly, they cannot be easily modified. Therefore, the architecture requires a disconnection-aware proxy that hides end-user disconnection from legacy servers. A proxy is assumed to exist in every region.

The proxy is resident in the Internet and has two halves. One half establishes disconnection-tolerant connection sessions with applications running on the kiosk controller or on mobile users' devices. The other half communicates with legacy servers on behalf of disconnected users. When a proxy receives application data from a legacy server, it transfers the data to an appropriate gateway that eventually forwards it to a passing ferry. The ferry delivers the data to a kiosk, which in turn passes it to kiosk users.

In the opposite direction, when a kiosk user wants to send data to the Internet, it uses a ferry to transport the data to a gateway that in turn transfers it to a proxy. The proxy passes received data to the legacy Internet servers.

4.3.3.2 Protocols

Virtual Link Layer

In KioskNet, TCP is used as the "virtual" link-layer protocol and is responsible for transferring a "bundle" from one DTN node to the next DTN node. Note that the mobile device (cell phone), kiosk controller, ferry, gateway, as well as the proxy are considered DTN nodes in KioskNet architecture.

Virtual Network Layer

In KioskNet architecture, "virtual" network layer protocol is responsible for routing a "bundle" from the original source to the final destination. Routing within a disconnected region of KioskNet is different from routing from Internet to Kiosk.

Routing within a Disconnected Region

A routing algorithm allows a kiosk to decide which ferry to use to send data to the Internet, and for a gateway to decide which ferry to use to communicate with a particular kiosk. However, ferries may fail and ferry trajectories are not always known beforehand. Therefore, routing in KioskNet is a hard problem. Fortunately, a ferry can transfer several tens of megabytes of data to and from a kiosk as it passes by, and it can store tens of gigabytes of data on its hard drive. Based on these observations, routing is done using flooding, thereby trading off over-the-air bandwidth and storage for reliability and ease of routing. This means, in KioskNet, a kiosk or a gateway transfers all its data to every ferry that passes by and accepts data from every ferry. Clearly, this redundancy maximizes the probability of bundle delivery while eliminating routing decisions altogether. An added benefit is that with flooding, communication between kiosk users in the same region does not require a bundle to go to the proxy. Finally, flooding requires fewer configurations at deployment time, making KioskNet easier to deploy.

KioskNet eliminates the inefficiencies commonly associated with naïve flooding using two optimization techniques. First, before any data is transferred from a kiosk controller to a ferry and vice versa, bundle metadata is exchanged so that each side knows what bundles the other side has, and as a result can avoid accepting bundles it already has.

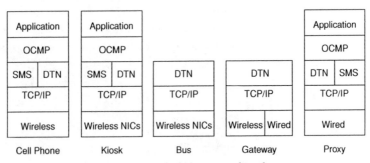

Figure 4.4. KioskNet protocol stack.

Bundle metadata exchange happens as follows:

- The kiosk controller tells the ferry the user GUIDs registered at the kiosk.
- The ferry informs the kiosk controller of the bundle IDs on the ferry belonging to these users.
- The kiosk controller determines the missing bundles and requests them from the ferry.
- The ferry transfers these bundles to the kiosk controller. No metadata exchange is required in the other direction: A kiosk transfers all its bundles to every passing ferry.

In addition, although bundles sent from a kiosk destined to legacy servers on the Internet are flooded to all reachable gateways in the same region, and these gateways accept all bundles from all kiosks, these gateways coordinate with each other to make sure that each bundle will be sent to the proxy by one and only one gateway. This avoids wasting bandwidth on the link between the gateways and the proxy.

With these two optimizations, despite flooding, KioskNet resources, namely kiosk-to-bus communication link and the gateway-to-proxy link, are not unnecessarily wasted.

Routing of Internet-to-Kiosk Bundles
Data from legacy servers destined to kiosk users is first buffered at the responsible proxy, then sent to gateways that transfer bundles to ferries. After a bundle is sent to a gateway, it is flooded to reach its destination kiosk (i.e., handed to all ferries passing by that gateway).

Proxies are located in bandwidth-rich data centers, but gateways are connected to the Internet typically using slow dial-up or DSL links. Given that the link between the gateways and the proxy is the bottleneck, ideally the proxy should choose only *one* gateway in the region to send each bundle to, rather than flooding it to all the gateways in the region.

If the schedules of ferries are known to the proxy, a routing and scheduling algorithm can be used at the proxy that can choose the best gateway for each bundle and decide the order in which they are sent so as to minimize the overall delay. Moreover, this algorithm can also enforce arbitrary bandwidth allocation among kiosks. If bus schedules are not known, then the proxy has no choice but to flood it to all the gateways.

Virtual Transport Layer
In KioskNet, these capabilities are provided by the opportunistic connection management protocol (OCMP) that runs on top of DTN and other available network connections. OCMP can be viewed as a disconnection-tolerant and

policy-driven session layer that runs over both DTN and standard links. Each type of available communication path is modeled as a connection object (CO) within OCMP. For instance, the DTN mechanical backhaul path is a CO, just as a TCP connection over WiMAX or dial-up is.

OCMP allows a policy manager to arbitrarily assign bundles to transmission opportunities on COs. This scheduling problem is complex because it has to manage many competing interests: reducing end-to-end delay while not incurring excessive cost, and maximizing transmission reliability.

Application Layer

Applications, residing on mobile device (cell phone), kiosk controller, and the Proxy, interface with the KioskNet architecture asynchronously, which is the most appropriate mechanism in long/variable delay environments. Usually the applications layer protocol generates ADUs and uses the bundle layer for forwarding and delivery.

4.4 Converged Architecture

The previous section described several alternative architectures for dealing with disruption-tolerant networking. However, each architecture was designed to solve a slightly different problem and hence evolved differently. For example, the DTN architecture (RFC 4838) has been designed primarily to deal with significant delays, including long interruptions, in communications. BBN's SPINDLE architecture evolved from the need to provide access to information to military field force where stable end-to-end paths do not exist and infrastructure access cannot be assured. KioskNet was designed with the goal of providing very-low-cost Internet access to rural areas. DieselNet's[11,12,13] goal has been primarily to deal with the challenges of vehicular DTN. PocketNet[14,15] was designed to enable communications via storage and networking purely at the end-hosts.

Looking back at Section 4.2 in this chapter, shortcomings of the Internet architecture need to be addressed for a variety of networks including Hybrid Fixed and Mobile Networks, Military Ad hoc Networks, Vehicular Networks, Mobile Wireless Networks, Media Distribution Networks, and Sensor Networks.

Whereas DTN architecture primarily addresses the requirements of hybrid fixed and mobile networks, BBN's SPINDLE mostly focuses on military ad hoc networks, KioskNet deals with hybrid fixed and mobile networks, and DieselNet primarily explores DTN in vehicular networks. There are isolated efforts to deal with the time-varying characteristics of mobile wireless networks[23,31,32] and the existence of content delivery networks (CDNs) to address the needs of media distribution.[3,4]

A deeper look into the entire problem space exposes an underlying commonality in the basic building blocks of a converged network architecture that addresses all the previously mentioned problems in a uniform manner. The converged network architecture is referred to as Cache and Forward (CNF) Network Architecture.

4.4.1 Cache and Forward (CNF) Network Design Goals

Cache and Forward (CNF) architecture[16,17] evolved at WINLAB, Rutgers University, in order to solve four main problems: (1) efficient delivery and retrieval of video, (2) improving throughput in multihop wireless network, (3) improving content delivery in a mobile network where the mobile nodes may be intermittently connected to the wired infrastructure, and (4) improving communication in sensor networks. These issues are briefly discussed further in this chapter.

(1) Efficient delivery and retrieval of video (*challenges of media distribution networks*): Video will be driving the need for improved communications infrastructure in the foreseeable future, as is evident from the phenomenal rise of YouTube,[18] Revver,[19] and other video sharing sites in addition to the rise of Internet television, where specialized sites provide niche television content over the Internet. The uniqueness of video as content is the huge size of files that are several orders of magnitude larger than music (audio) files. P2P networking is helping scale the distribution of video, but the P2P delivery mechanism, by itself, cannot optimize the bandwidth usage in the underlying network. Moreover, the existing TCP/IP networking paradigm is not exactly suitable for video retrieval because the TCP/IP paradigm expects the application to figure out through an out-of-band mechanism (such as a search engine) the name/IP address of the server where a given video is hosted and then connect to the server to fetch the desired content, as opposed to allowing the application to query the network for a given video and retrieve it from the network, all operations being done in-band.

(2) Improving throughput in multihop wireless networks (*challenges of mobile wireless networks*): When TCP/IP is used over wireless links, performance is often degraded due to transport layer timeouts, and in-network solutions such as indirect TCP have been proposed in earlier work.[20] In addition, when TCP is used over multiple wireless hops (an increasingly common scenario), the so-called self-interference effect in which packets from the same flow (specifically the data and acknowledgment packets belonging to the same flow but traveling in opposite directions), contending for the same radio resources, can further degrade end-to-end performance.[21,22] For multihop wireless networks, the probability of impairment or disconnection in at least

one radio link can be quite high as the number of hops, n, increases. It can be shown that the probability of failure before the file transfer is completed is increased by a factor of n^2 over the probability of a single hop failure. This is almost an order-of-magnitude increase for $n = 3$ hops and is two orders of magnitude increase for $n = 10$ hops.

(3) Improving content delivery in mobile networks where mobile nodes may be intermittently connected to wired network (*challenges of hybrid fixed and mobile networks, military ad hoc, and vehicular networks*): The existing TCP/IP architecture embraced the concept of mobile IP to reach mobile hosts when the point of attachment of the mobile host (with the wired network) changes due to its mobility. However, the scope of mobile IP is limited to the case when mobile node is not disconnected from the wired network for a significant amount of time (longer than the lifetime of a typical Internet session). At the same time, research has shown that if content is temporarily stored in the network when the destination node is not connected to the wired network, and is ferried via "mobile nodes" to the destination node, the capacity of the wireless network increases substantially.[24,25]

(4) Improving communication in sensor networks (*challenges of sensor networks*): Internet applications involving sensors are expected to grow rapidly in the next ten years. Sensor scenarios have unique networking requirements,[33] including the ability to deal with disconnections due to wireless channel impairments as well as sensor hardware sleep modes. In addition, sensor applications tend to be data-centric and are thus more interested in content-aware services (e.g., querying data) than in connecting to a specific IP address.

CNF architecture was designed to address these issues in an efficient manner.

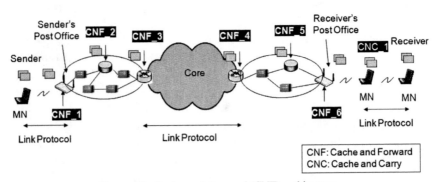

Figure 4.5. Cache-and-Forward (CNF) architecture.

4.4.2 Architecture

The main concepts of CNF architecture are listed in this section:

Post Office (PO): The CNF architecture is based on the model of a postal network designed to transport large objects and provide a range of delivery services. Keeping in mind that the sender and/or receiver of an object may be mobile and may not be connected to the network, we introduce the concept of "Post Office" (PO) that serves as an indirection (rendezvous) point for senders and receivers. A sender deposits the object to be delivered in its PO and the network routes it to the receiver's PO, which holds the object until it is delivered to the final destination. Each sender and receiver may have multiple POs, where each PO is associated with a point of attachment in the wired network for a mobile endpoint (sender/receiver). In the context of DTN network and BBN's SPINDLE, a PO is nothing but a special type of custodian node, whereas in the context of KioskNet, PO is equivalent of a Gateway.

Cache and Forward (CNF) Router: The CNF Router is a network element with persistent storage and is responsible for routing packages within the CNF network. Packages are forwarded hop-by-hop (where a hop refers to a CNF hop and not an IP hop) from the sender's PO to the receiver's PO using forwarding tables updated by a routing protocol running either in the background (proactive) or on demand (reactive). In the context of DTN network and BBN's SPINDLE, a CNF Router is nothing but a DTN node that may or may not be a custodian node, whereas in the context of KioskNet, a Kiosk as well as a Gateway is a CNF Router.

Cache and Carry (CNC) Router: The CNC Router is a mobile network element that has persistent storage exactly as in a CNF Router, but is additionally mobile. Thus a CNC router can pick up a package from a CNF router, another CNC router, or a PO and carry it along. The CNC router may deliver the package to the intended receiver or to another CNC router that might have a better chance of delivering the package to the desired receiver. In the context of DTN network and BBN's SPINDLE, a CNC Router is nothing but a mobile DTN node that may or may not be a custodian node, whereas in the context of KioskNet, a Ferry is a CNC Router.

Content Identifier (CID): To make content a first-class entity in the network, we introduce the notion of persistent and globally unique content identifiers. Thus if a content is stored in multiple locations within the CNF network, it will be referred to by the *same* content identifier. The notion of a CID is in contrast to identifiers in the Internet, where content is identified by a URL whose prefix consists of a string identifying the *location* of the content. CNF endpoints will request content from the network using content identifiers. Since none of the described alternative architectures in Section 3 considered content as a first-class citizen of the network, there was no need to have a specific content ID which is a "network" level ID rather they continued to use the "application" level ID, such as URLs as in the case of traditional Internet. However, CID is a fundamentally important concept in the converged network architecture.

Content Discovery: Since copies of the same content can be cached in multiple CNF routers in the network, discovering the CNF router with the desired content that is "closest" to the requesting endpoint must be designed into the architecture. We discuss this in more detail in the next section. Once again, since none of the described alternative architectures in Section 3 considered content as a first-class citizen of the network, there was no need to discover content within the "network"; rather they continued with the traditional Internet model whereby a search engine is expected to be used to locate the node holding the content and once the node is located, traditional network-based routing is used to access the content. One exception is BBN's SPINDLE architecture where the concept of content module has been conceived as a part of the Decision Plane module for enabling content-based access, specifically for content search, content caching and replication, content routing, and other content-related functionality. However, in the converged architecture, since content is a first-class citizen of the network, content discovery is part of the network layer functionality.

Type of Service: To differentiate between packages with different service delivery requirements (high priority, medium priority, low priority), a Type of Service (ToS) byte will be used in the package header. The ToS byte can be used in selecting the cache replacement policy and in determining the delivery schedule of packages at the CNF routers. This concept exists with both DTN architecture as well as BBN's SPINDLE architecture.

Multiple Delivery Mechanisms: A package destined for a receiver would be first delivered to, and stored in, the receiver's PO. There are several ways in which the package can be delivered from the PO to the receiver:

- A PO can inform the receiver that there is a package waiting for it at the PO and it (the receiver) should arrange to pick it up. The receiver can pick up the package when in range of that PO. Otherwise, it may ask its new PO and/or a CNC router to pick up the package on its behalf.
- A receiver can poll the PO to find out if there is a package waiting for pick up. If it is, and the receiver is within range of the PO, it can pick

up the package itself. Otherwise, it may ask its new PO and/or a CNC router to pick up the package on its behalf.

- A PO can proactively *push* the package to the receiver either directly or via CNC routers.

Routing mechanisms are not prescribed in either DTN architecture or in BBN's SPINDLE architecture as the architecture is expected to provide flexibility for choosing variety of routing techniques, especially at the edge of the wired network. KioskNet, however, does talk about intelligent flooding from the Gateway (equivalent of a PO) to the final destinations (PCs and/or mobile phones) as a way of routing. Nonetheless, the previously mentioned techniques in the converged architecture cover the entire gamut of routing from the wired edge node to the mobile or wirelessly connected end nodes (or final destinations).

Details of protocols used in CNF network are described next.

4.4.3 Protocols

The cache-and-forward architecture represents a set of new protocols that can be implemented either as a clean-slate implementation or on top of IP.

Virtual Link Layer: Virtual Link Layer in CNF architecture uses a reliable link layer protocol referred to as CNF LL in Figure 4.6. CNF LL protocol is used for reliable delivery of packages (bundles) between two adjacent CNF nodes that could be either CNF/CNC routers or CNF hosts. In the traditional TCP/IP network paradigm, two adjacent CNF nodes could be separated by multiple IP router hops or could be connected by a wireless link with highly time-varying bandwidth, delay, and loss characteristics. Because of this diversity, the link-layer protocol used in CNF is *configurable* to suit the characteristics on the underlying link. For example, if the virtual link in CNF consists of a few wired IP router hops, TCP may be the best virtual link-layer protocol for reliable delivery between two adjacent CNF nodes. On the other hand, if the virtual link

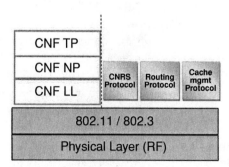

Figure 4.6. Cache-and-Forward protocol stack.

in CNF consists of a wireless link with highly time-varying bandwidth, delay, and loss characteristics, then some proprietary protocol (such as CLAP) may be the best virtual link-layer protocol for reliable delivery between two adjacent CNF nodes.

Virtual Network Layer: Virtual Network Layer in CNF architecture uses a network layer protocol referred to as CNF NP in Figure 4.6. Each node in the CNF network, as described earlier, is assumed to have a large storage cache (~TB) that can be used to store packages (files/file segments) in transit, as well as to offer in-network caching of popular content. CNF routers may either be wired or wireless, and some wireless routers may also be mobile. The basic service provided by the network is that of file delivery either in "push" or "pull" mode, that is, a mobile end-user may request a specific piece of content, or the content provider may push the content to one (unicast) or more (multicast) end-users. Each query and content file transported on the CNF network is carried as a CNF packet data unit or *package* in a strictly hop-by-hop fashion. The package is transported reliably between data stores at each CNF router before being prepared for the next hop toward its destination. The CNF network assumes the existence of a reliable link-layer protocol between any pair of CNF routers, and this protocol can be customized to the requirements of each wireless or wired link in the network. Packages are forwarded from node to node using opportunistic, short-horizon routing and scheduling policies at CNF nodes that take into consideration factors such as package size, link quality, and buffer occupancy. Alternative routing techniques may also be used opportunistically to deal with congestion or link failure. Caches in the network can create more complex scenarios. To retrieve any content, a host would send a query to the network with the location-independent CID, and the query would then be routed to the nearest CNF router using a *content routing* procedure, and the content would then be routed back to the host using the conventional routing capability mentioned earlier.

One unique aspect of CNF network layer protocol is the concept of "query routing" whereby an application may trigger query for a specific content object that may be stored "within" the CNF network. Note that this is possible as content is cached within the network itself. The query is routed from the originating node through the network. Just as a traditional router in the Internet maintains a routing table with IP address of destination networks/hosts, a CNF router maintains a routing table with CID and indicates next hops to follow in order to reach the desired content object stored within the network.

When the query is routed to the CNF node that has the desired content cached, the network layer triggers what is called response routing. Response is routed to the node that originated the query; the response forwarding is similar to traditional TCP/IP routing where packets are routed to a given destination IP

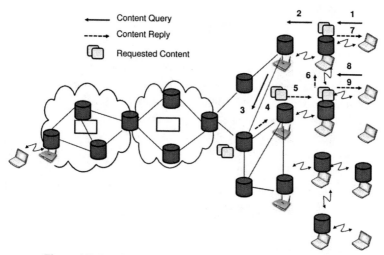

Figure 4.7. Routing of queries and content in the CNF network.

address. Perhaps the only difference is that the response (desired content object) is cached at each CNF router en route to the destination.

Figure 4.7 shows how content queries and responses are routed. Specifically, the content query originated at the top right laptop is routed through the CNF network (steps 1–3) until the query reached a CNF node with the desired content. Content is routed back (steps 4–7) where steps 5–7 are over wireless links that may or may not be mobile (if the link is a mobile wireless link, the corresponding node would be considered a cache and carry [CNC] router). As the content is routed through the CNF network, it is cached at intermediate nodes, and the benefit of caching shows up when another CNF host queries for the same object (step 8), which is now cached at the first hop (thanks to the previous query and content response routing and caching), and the content is immediately sent back to the originating CNF host (step 9).

Virtual Transport Layer: Virtual Transport Layer in CNF architecture uses a transport layer protocol referred to as CNF TP in Figure 4.6. The purpose of virtual transport layer protocol is to provide an end-to-end acknowledgment or notification for the delivery of content where the ends are defined as the original source and final destination. Because of reliable link-layer delivery between CNF nodes, transport layer also includes "intermediate" level acknowledgment and notification, which helps diagnose delivery problems in the same way as the tracking system does in FedEx or UPS delivery networks. In addition, it is the virtual transport layer that needs to deal with congestion and flow control as contents are transported across the CNF network from multiple sources to multiple destinations.

4.4.3.1 Support Services

Content Name Resolution Service (CNRS): Because CNF network is designed to support efficient distribution and retrieval of content, and it allows applications to "query" for content cached in the network (see the virtual network layer), it is useful to have the IDs of CNF routers corresponding to a given file (or CID) that have the corresponding content cached. Since there would be potentially millions of objects, constantly updating the CNRS server for each content would not scale. Hence the idea in CNF is to update the CNRS server with the IDs of caches where a "popular" is cached. This information may be used by the CNF hosts (when originating a query) and/or by the CNF routers when forwarding the query in order to optimize routing.

4.4.4 Performance of Protocols in CNF Architecture

4.4.4.1 CNF and TCP/IP Based Internet in Mobile Content Delivery

The goal of this section is to compare the performance of the proposed converged architecture (referred to as CNF architecture) with that of the traditional TCP/IP-based Internet architecture when it comes to delivery of content, especially large-size content, such as video, files when the sender or the receiver or both are mobile. To compare the performances of these two networks, a 24-node transit-stub network is considered, and the time taken in transferring a file from a source to a destination is computed under varying offered load where Offered Load = arrival rate × file size × no. of source nodes. Specifically, three scenarios are considered: (1) client and server nodes are wired, (2) client nodes are wireless but server nodes are wired, and (3) both client and server node are wireless. The results are shown in Figure 4.8 and are taken from Liu, Zhang, and Raychaudhuri.[26]

For CNF traffic, the transmission delay depends on the number of hops and the bandwidth of each hop. For TCP traffic, the transmission delay depends on the bandwidth of the bottleneck links. If all the links have the same bandwidth, TCP-based data transfer performs better than CNF-based data transfer because there is no store and forward delay in TCP and it is able to take full advantage of "streaming" data. However, if the bottleneck bandwidth is much smaller than that of the other links, TCP throughput is significantly reduced because it is limited by the bottleneck bandwidth. From the plots in Figure 4.8, it is clear that the file delivery time (referred to as file transfer delay in the figure) for TCP/IP network shoots up at much lower offered load compared to the case of CNF networks. Specifically, if the throughput is defined as the offered load with delay limit of 100s, the throughput of TCP is less than 2 Mbps in the case where both clients and servers are wirelessly connected, while CNF throughput is 2–7 Mbps.

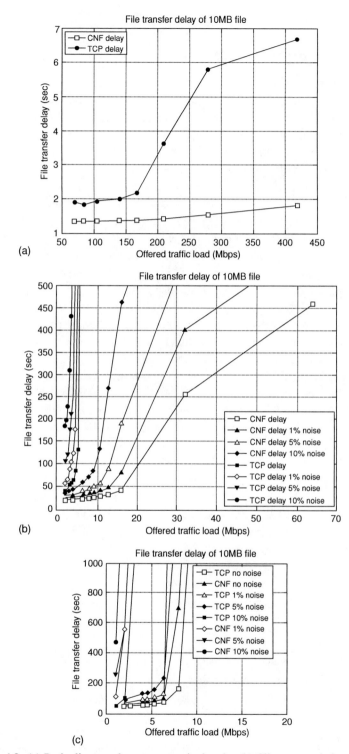

Figure 4.8. (a) Both clients and servers are wired nodes (b) Clients are wireless, servers are wired (c) Both clients and servers are wireless nodes.

117

4.4.4.2 CNF and TCP/IP-Based Internet in Content Retrieval

The goal of this section is to compare the performance of the proposed converged architecture (referred to as CNF architecture) with that of the traditional TCP/IP-based Internet architecture when it comes to content retrieval. To compare the performances of these two networks, a 12-node transit-stub network is considered, the time taken to retrieve a specific content is computed when the network has the intelligence (as in the converged network architecture) versus when the network is dumb (as in the traditional Internet), and the intelligence of locating content resides outside the network at the application level.

Two schemes are compared in Lijun, Yanyong, and Sanjoy[27]: (1) Server Only, or SO (meaning that the content resides only on the servers) and (2) Cache and Capture, or CC (meaning content gets cached in the network as it transits through the network and hence can be retrieved from the network as opposed to from the server only) under three different scenarios: (a) small network, many requests, (b) large network, few requests, and (c) large network, many requests.[27]

It is clear from Figure 4.9 that whereas caching helps in every scenario, request number has a bigger impact on the caching effect than network size. In both scenarios (a) and (c), where a node makes a large number of requests, integrated caching can improve the performance by more than 52 percent, whereas in scenario (b), the improvement is only 24 percent. This is because more requests are served off the cache in cases (a) and (c) compared to that in case (b).

In the histograms shown in Figure 4.10, the first bin corresponds to the number of requests that were satisfied by a one-hop neighbor of the requester, either because that neighbor is the server hosting the content or because the neighbor has a copy of the named content in its cache, and the same explanation holds for the ith bin. Figures 4.10 (a)–(c) show that Caching and Capture (CC) as proposed in the CNF architecture can satisfy many more requests by nodes that are within a small number of hops from the requester than Server Only (SO) where the content is located only at the server outside the network as in the traditional TCP/IP-based Internet architecture. This clearly demonstrates

Scenario	(a)	(b)	(c)
SO (seconds)	0.369	0.737	0.635
CC (seconds)	0.175	0.561	0.304
Improvement	52.6%	23.9%	52.4%

Figure 4.9. Comparison of content retrieval schemes in CNF networks.

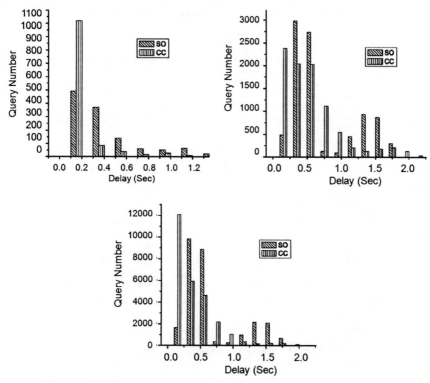

Figure 4.10. Histogram of content retrieval latency (scenarios a, b, and c).

the benefit of integrated caching and routing that is a core part of the proposed converged architecture.

4.4.4.3 CNF and TCP/IP-Based Internet in Routing

The goal of this section is to compare the performance of the proposed converged architecture (referred to as CNF architecture) with that of the traditional TCP/IP-based Internet architecture when the converged architecture uses storage-aware intelligent routing. To compare the performances of these two networks, Storage Aware Routing (STAR) scheme proposed for CNF architecture in Paul et al. [16] is compared with the traditional OLSR routing protocol in a 25-node network in two cases: (1) static network in 500 m-by-500 m grid with intermittently failing links, and (2) network in which the nodes move according to Truncated Levy Walk in 2, 500 × 2, 500 area.[28]

Figure 4.11 shows the number of files transmitted and delivered, average file transfer delays, file streaming throughput, and overall network throughput. File streaming throughput represents the average physical data rate used to transfer

Protocol	OLSR			STAR		
Mean Inter-arrival time (seconds)	50	10	1	50	10	1
Files transmitted/ Delivered	149/ 147	320/ 265	1016/ 691	165/ 164	468/ 438	1147/ 862
Average File Delay (seconds)	0.678	1.047	2.936	1.634	10.883	5.518
File Stream throughput (Mbps)	3.868	3.187	2.305	3.546	3.755	3.281
Network throughput (Mbps)	0.302	0.548	1.417	0.337	0.897	1.765

(a)

Protocol	OLSR		STAR	
Mean Inter-arrival time (seconds)	10	50	10	50
File Delivery Fraction	72.34	66.67	89.66	79.17
Average File Delay (seconds)	100.65	109.65	92.28	38.66
File Stream throughput (Mbps)	1.09	1.39	1.7	1.59

(b)

Figure 4.11. (a) Static network with intermittently failing links (b) Performance with Levy mobility model.

the file at every hop. Unlike the network throughput, streaming throughput does not include delays incurred in queues and storage.

From Figure 4.11(a) showing results for the static case (case 1), it can be observed that the STAR protocol is able to deliver a larger percentage of files that are admitted into the network, and this is because STAR selects faster paths for transmission. The average file delays and streaming throughput in STAR are larger showing the preference for storage over using slow transmission channels.

Although file delays are high, the cache-and-forward concept improves the overall network throughput.

From Figure 4.11(b) showing results for the mobile case (case 2), it can be observed that the file delivery fraction achieved by STAR is 17 percent and 33 percent more when the mean interarrival times are 10 and 50 seconds, respectively. The average file delays are much lower (or equivalently, the file stream throughputs are much higher) in STAR compared to OLSR. These results indeed justify the benefits of STAR in DTN like mobility models.

4.4.4.4 CNF and TCP/IP-Based Internet in Multihop Mobile Wireless Networks

The goal of this section is to compare the performance of the proposed converged architecture (referred to as CNF architecture) with that of the traditional TCP/IP-based Internet architecture when the network consist of multiple wireless hops in an end-to-end connection. To compare the performances of these two networks, a 49-node grid topology was considered under variety of traffic and noise patterns as shown in Figure 4.12 to observe the effect on average file transfer delay.[7]

Since TCP's performance over wireless without MAC layer reliability only becomes worse, the network capacity achieved by TCP/IP without MAC level reliability is not shown. It can be seen that for case 1, the network capacity offered by CNF Link Layer (LL) with MAC level ACKs is about 70 percent higher than that offered by TCP. Disabling the MAC level ACKs increases the CNF capacity gains to 140 percent over TCP. For case 2, the client server model,

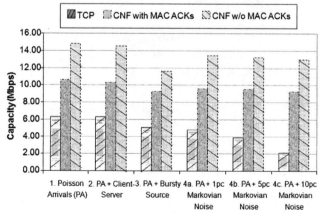

Figure 4.12. Comparison of network capacity (TCP vs. CNF LL).

the capacities achieved remain approximately the same for all three strategies, that is, TCP, CNF LL with MAC ACKs, and CNF LL without MAC ACKs. Hence the capacity gains of CNF over TCP also remain the same. For case 3, the bursty source model, it can be seen that noticeable reductions occur in the capacity achieved by the CNF LL protocol with and without MAC ACKs. This is because burstiness in traffic causes congestion at nodes. Link Layer queue sizes increase and the average file delay becomes longer. Capacity gains of 85 percent and 130 percent can be seen for CNF LL with MAC ACKs and CNF LL without MAC ACKs, respectively, over TCP. For case 4, where Markovian noise was introduced in the links, it should be noted that CNF is much more noise resilient as compared to TCP. For example, in 10 percent Markovian noise, TCP shows a two-third reduction in capacity whereas CNF LL without MAC ACKs suffers from 13 percent reduction in capacity. The capacity gains achieved by CNF LL over TCP in this case become about 650 percent.

4.5 Concluding Remarks

This chapter is motivated by networking scenarios driven by intermittent connectivity, content, and mobility that are not effectively handled by the traditional TCP/IP-based Internet. The proposed converged architecture to deal with these networking scenarios consists of in-network persistent storage and hop-by-hop reliable transport in the data plane, as well as name resolution, late binding, and routing in the control plane. Whereas these architectural components can be built as an overlay on the core networking (TCP/IP in the Internet) infrastructure, in a clean-slate network, these should be built into the core network itself. Given the exponentially dropping cost of processing/MIPS and storage/GB, it is highly conceivable that the network itself will consist of network elements (or future routers) with significant amount of persistent storage, and significant amount of processing power so that the route would be computed in real time at each node (rather than being computed in the background and the forwarding table used in real time for forwarding packets) based on multiple dimensions, such as congestion in the network, available storage in the network elements, availability of cached content, and so forth. Thus it makes a lot of sense for the next-generation clean-slate network architecture to encompass the support for intermittent connectivity, content, and mobility right in the network fabric because these themes together have a very broad scope in the overall area of networking.

References

[1] http://www.squidcache.org
[2] http://www.appliansys.com
[3] http://www.akamai.com

[4] http://www.limelightnetworks.com

[5] Delay-Tolerant Networking Architecture, IETF RFC 4838.

[6] Gopal, S., and Paul, S. 2007. TCP Dynamics in 802.11 Wireless Local Area Networks. *IEEE International Conference on Communications*, June.

[7] Saleem, A. 2008. Performance Evaluation of the Cache and Forward Link Layer Protocol in Multihop Wireless Subnetworks. Master's Thesis, Rutgers University, WINLAB.

[8] Krishnan, R., Basu, P., Mikkelson, J. M., Small, C., Ramanathan, R., et al. 2007. The SPINDLE Disruption-Tolerant Networking System. *IEEE Milcom*, 29–31 Oct., pages 1–7.

[9] Guo, S., Falaki, M. H., Oliver, E. A., UrRahman, S., Seth, A., Zaharia, M. A., and Keshav, S. 2007. Very Low-Cost Internet Access Using KioskNet. *ACM Computer Communication Review*. http://blizzard.cs.uwaterloo.ca/tetherless/images/c/c0/Kiosknet.pdf

[10] Guo, S., Falaki, M. H., Oliver, E. A., UrRahman, S., Seth, A., et al. 2007. Design and Implementation of the KioskNet System. *International Conference on Information Technologies and Development*, December.

[11] Balasubramanian, A., Mahajan, R., Venkataramani, A., Levine, B. N., and Zahorjan, J. 2008. Interactive WiFi Connectivity for Moving Vehicles. *Proc. ACM SIGCOMM*, pages 427–438.

[12] Banejree, N., Corner, M. D., Towsley, D., and Levine, B. N. 2008. Relays, Base Stations, and Meshes: Enhancing Mobile Networks with Infrastructure. *Proc. ACM Mobicom*, pages 81–91.

[13] Soroush, H., Banerjee, N., Balasubramanian, A., Corner, M. D., B. N., and Lynn, B. 2009. DOME: A Diverse Outdoor Mobile Testbed. *Proc. ACM Intl. Workshop on Hot Topics of Planet-Scale Mobility Measurements (HotPlanet)*. Article No. 2.

[14] Hui, P. Chaintreau, A. Gass, R., Scott, J., Crowcroft, J., and Diot, C. 2005. Pocket Switched Networking: Challenges, Feasibility and Implementation Issues. *Proceedings of the Second IFIP Workshop on Autonomic Communications 2005*.

[15] Hui, P., Chaintreau, A., Scott, J., Gass, R., Crowcroft, J., and Diot, C. 2005. Pocket Switched Networks and the Consequences of Human Mobility in Conference Environments. *Proceedings of the SIGCOMM 2005 Workshop on Delay Tolerant Networking*, pages 244–251.

[16] Paul, S., Yates, R., Raychaudhuri, D., and Kurose, J. 2008. The Cache-and-Forward Network Architecture for Efficient Mobile Content Delivery Services in the Future Internet. *Proceedings of the First ITU-T Kaleidoscope Academic Conference on Innovations in NGN: Future Network and Services*, pages 367–374.

[17] Dong, L., Liu, H., Zhang, Y., Paul, S., Raychaudhuri, D. 2009. On the Cache-and-Forward Network Architecture. *IEEE International Conference on Communications*, pages 1–5.

[18] http://www.youtube.com

[19] http://www.revver.com

[20] Bakre, A., and Badrinath, B. R. 1995. I-TCP: Indirect TCP for Mobile Hosts. *Proc. 15th Int'l Conf. on Distributed Computing Systems*, pages 136–143.

[21] Gopal, S., and Raychaudhuri, D. 2005. Experimental Evaluation of the TCP Simultaneous-Send Problem in 802.11 Wireless Local Area Networks. *ACM SIGCOMM Workshop on Experimental Approaches to Wireless Network Design and Analysis (E-WIND)*, pages 17–22.

[22] Gopal, S., Paul, S., and Raychaudhuri, D. 2005. Investigation of the TCP Simultaneous-Send Problem in 802.11 Wireless Local Area Networks. *Proceedings of the IEEE Computer and Communications Conference (ICC)*, **5**, 3594–3598.

[23] Gopal, S., Paul, S., and Raychaudhuri, D. 2007. Leveraging MAC-Layer Information for Single-Hop Wireless Transport in the Cache and Forward Architecture of the

Future Internet. *The Second International Workshop on Wireless Personal and Local Area Networks (WILLOPAN)*, pages 1–6.

[24] Zhao, W., Ammar, M., and Zegura, E. 2004. A Message Ferrying Approach for Data Delivery in Sparse Mobile Ad Hoc Networks. *Proceedings of ACM Mobihoc 2004*, pages 187–198.

[25] Zhao, W., and Ammar, M. 2003. Message Ferrying: Proactive Routing in Highly-Partitioned Wireless Ad Hoc Networks. *Proceedings of the IEEE Workshop on Future Trends in Distributed Computing Systems*, pages 308–314.

[26] Liu, H., Zhang, Y., and Raychaudhuri, D. 2009. Performance Evaluation of the Cache-and-Forward (CNF) Network for Mobile Content Delivery Services. *IEEE International Conference on Communications (ICC) Workshops 2009*, pages 1–5.

[27] Lijun, D., Yanyong, Z., Sanjoy, P. Performance Evaluation of In-network Integrated Caching. Forthcoming.

[28] Shinkuma, R., Jain, S., Yates, R. 2009. Network Caching Strategies for Intermittently Connected Mobile Users. *IEEE PIMRC*, pages 1771–1775.

[29] Jain, S., Saleem, A., Liu, H., Zhang, Y., and Raychaudhuri, D. 2009. Design of Link and Routing Protocols for Cache-and-Forward Networks. *IEEE Sarnoff Symposium*, pages 1–5.

[30] Cerf, V., Burleigh, S., Hooke, A., Torgerson, L., Durst, R., Scott, K., Fall, K., and Weiss, H. 2007. Delay-Tolerant Network Architecture. RFC 4838, MILCOM 2007.

[31] Acharya, A., Ganu, S., and Misra, A. 2006. DCMA: A Label Switching MAC for Efficient Packet Forwarding in Multihop Wireless Networks. *IEEE JSAC Special Issue on Wireless Mesh Networks*, pages 1995–2004.

[32] Wu, Z., Ganu, S., and Raychaudhuri, D. 2006. IRMA: Integrated Routing and MAC Scheduling in Wireless Mesh Networks. *Proceedings of the Second IEEE Workshop on Wireless Mesh Networks, (WiMesh)*, pages 1–8.

[33] Akyildyz, I., Su, W., Sankarasubramaniam, Y., and Cayirci, E. 2002. A Survey on Wireless Sensor Networks. *IEEE Communications Magazine*, pages 102–114.

5

Sensor Networks Architectures and Protocols

Omprakash Gnawali and Matt Welsh

5.1 Introduction

Wireless sensor networks (WSNs) are an important emerging class of embedded distributed systems that consist of low-power devices integrating computation, sensing, and wireless communications. WSNs have been deployed for a wide range of applications, including monitoring microclimates in redwood forests (Tolle et al. 2005), collecting seismic signals from active volcanoes (Werner-Allen et al. 2006), sniper detection in urban settings (Simon et al. 2004), and tracking wildlife (Zhang et al. 2004).

One of the most popular WSN node platforms is the Telos node platform (Polastre et al. 2005a), shown in Figure 5.1. The Telos incorporates a low-power microcontroller (TI MSP430) with 10 KB of SRAM and 48 KB of program ROM; a low-power radio (Chipcon CC2420) that supports the IEEE 802.15.4 standard; and 1 MB of on-board flash memory. Various sensors can be attached to the board; a standard set includes light, temperature, and humidity sensors. An external connector provides digital and analog I/O ports that can be used to mate the node to a wide range of sensors and other devices. The USB connector is used to program the node when plugged into a host, as well as to provide a serial interface. This allows the node to act as a USB wireless transceiver when attached to a *base station* that collects data from and controls the network.

WSN platforms are designed from the ground up for low-power operation. The Telos consumes approximately 41 mW when the CPU and radio are active, but can drop down to a low-power idle state consuming less than 6 μW. Depending on the duty cycle, the device can potentially operate for months on a pair of alkaline AA batteries. However, this low power consumption comes at the cost of extreme limitations on computational horsepower, memory capacity, and radio bandwidth. The CC2420 radio operates at a PHY rate of just 250 Kbps, but

125

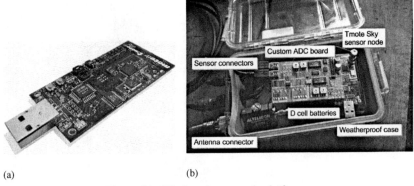

Figure 5.1. Wireless sensor node platforms.

link throughput (including framing and control message overhead) is less than 100 Kbps in practice, and may be much less under lossy conditions. Moreover, the radio consumes far more power than the CPU, mandating careful control over the radio listen, receive, and transmit modes.

Therefore, network protocols for sensor networks must be designed to operate with lossy and low-throughput links and limited memory for storing routing tables and other state. Protocols must also carefully manage the power consumption of the radio in order to ensure long battery lifetimes.

In this chapter, we provide an overview of protocol design for wireless sensor networks, focusing on how these designs differ from those in more conventional networking environments. We begin with link layer protocols (Section 5.2) including low-power MAC and link quality estimation. We move onto tree-based routing (Section 5.3) and efficient dissemination (Section 5.4), which are fundamental primitives for data collection and broadcast in WSNs. In Section 5.5 we describe reliable transport protocols, which differ in design from conventional approaches such as TCP. Cross-layer protocols that optimize for power consumption across the stack are described in Section 5.7. The emergence of IPv6 for low-power wireless networks (6loWPAN) is discussed in Section 5.8. In Section 5.6, we describe auxiliary protocols for time synchronization and localization, two essential services in sensor networks. Finally, in Section 5.9, we present our thoughts on how sensor network protocols might influence the design of the future Internet.

5.2 Link Layer Protocols

Link layer protocols provide media access, and packet transmission and reception service to the upper layer protocols. In sensor neworks, link layers also provide information about the nodes and links in the neighborhood. Although high channel utilization is important, energy efficiency has been the focus of link layer research in wireless sensor networks.

Idle listening is the largest avoidable energy expenditure attributed to wireless communication in sensor networks. Idle listening is a phenomenon in which a node expends energy keeping its radio on even while no packet transmission or reception takes place. On radio chips such as the IEEE 802.15.4/ZigBee-compatible CC2420 radio, the radio consumes as much power in receive mode as it does while transmitting a packet. Fortunately, this cost is avoidable if we can duty-cycle the radio. However, duty cycling makes the media access problem harder: When a transmitter is ready to transmit a packet, the radio on the receiver might be off. Additional mechanisms are necessary to ensure both the transmitter and the intended receiver have their radios on at the time of packet transmission.

Two main radio duty-cycling approaches have been proposed in low-power MAC design for sensor networks. Coordinated or synchronous duty-cycling orchestrates radio duty cycles across the network in a predetermined schedule. This requires that nodes synchronize their schedules with each other, and may involve the use of a centralized scheduler, such as the base station. Uncoordinated or asynchronous approaches have no explicit coordination across the nodes. Some link layer designs combine these two approaches, but these are not typically used in practice. Uncoordinated schemes are far more common than coordinated schemes because they allow nodes to transmit packets at any time without requiring the overhead and complexity of explicit synchronization. We describe B-MAC and X-MAC, two examples of such protocols below.

5.2.1 B-MAC

B-MAC (Polastre et al. 2004) is a canonical example of a low-power MAC protocol for sensor networks. For coordination between multiple transmitters, B-MAC uses clear channel assessment and back-offs, as described further here. To enable low-power operation, B-MAC uses a technique called *low-power listening* to duty-cycle the radio. B-MAC also supports link-layer acknowledgments.

B-MAC uses *clear channel assessment* to determine the presence or absence of valid packet transmission in the channel. Clear channel assessment involves nodes determining the noise floor and comparing the level of signal with the noise floor to determine if the channel is clear. The noise floor can be different across different nodes, based on the ambient channel conditions. The noise floor can also change over time. B-MAC performs periodic estimation of the noise floor by sampling the ambient noise level in the environment. This sampling is done during times that are unlikely to have packet transmissions – the short wait time between a packet and its acknowledgment, for example. The median of these samples is used as the noise floor estimate for a given time. These estimates are also averaged over time using exponential averaging to yield a more stable estimate of the noise floor.

When B-MAC receives a request to transmit a packet, it must first determine if the channel is clear. Sampling the channel once and comparing against the noise floor results in a large number of false positives. The key insight used to increase robustness in B-MAC's clear channel assessment algorithm is that comparing the low outliers of multiple samples against the noise floor results in far greater accuracy than using threshold-based channel assessment. So, B-MAC samples the signal level in the channel a few times, computes the outlier, and compares against the noise floor. If the channel is clear, B-MAC transmits the packet. If the channel is not clear, the node backs off for a randomized interval of time. When the back-off timer expires, it performs clear channel assessment again. If acknowledgments are enabled, the transmitter waits for the acknowledgment before receiving or transmitting other packets.

B-MAC uses *low-power listening*, also called *preamble sampling*, to duty-cycle the radio. By default, the radio is left in a low-power sleep state, in which it is unable to receive incoming packets. Every *sleep interval*, each node turns on the radio and checks for an incoming packet from a transmitter. If a packet is detected, the node leaves its radio on to receive the packet. To avoid receiving a partial packet (e.g., if the receiver starts listening in the middle of a transmission), the transmitter must transmit a *preamble* before the actual packet, which is set to the length of the receiver's sleep interval. The long preamble ensures that a receiver will listen at least once during a preamble transmission.

In one of the evaluation experiments, the authors ran B-MAC on a testbed of 14 Mica2 sensor nodes. During the experiment, each node sent one data packet every three minutes to the base station using a collection routing protocol. The result showed that the nodes achieved a worst-case duty-cycle of 2.5 % and less than 1 s data delivery latency over six hops.

B-MAC allows applications to easily configure the MAC parameters, such as the sleep interval and whether link-layer acknowledgments are enabled. Allowing such interaction with the MAC enables applications to achieve desired duty-cycling and energy efficiency depending on the context and need of the application at a given time.

In low-power sensor networks, the sleep interval can be in the order of several hundred milliseconds. This requires transmitting an equally long preamble before each packet, which consumes a large amount of energy. In addition, long preambles increase packet transmission latency. They also cause nearby nodes that are not the intended recipient of a packet to wake up and attempt to decode the incoming packet, thereby wasting energy.

B-MAC makes the tradeoff of reducing receivers' idle listening power consumption by increasing the transmitters' power consumption when sending packets. This scheme is appropriate for low-data-rate applications in which transmissions occur infrequently. Reducing the overhead of packet transmissions has been the subject of much subsequent work, including the X-MAC protocol described below.

5.2.2 X-MAC

X-MAC (Buettner et al. 2006) is another uncoordinated duty-cycling MAC for sensor networks. X-MAC uses a series of short preambles and embeds the destination of the packet in the preamble. When a node checks for channel activity and receives the preamble, it can quickly determine if it is the intended recipient of the packet. X-MAC introduces a short wait time between the preambles. During these wait times, the intended receiver of the packet can signal the transmitter (through an ACK transmission) that the receiver's radio is on and ready to receive the packet. This signaling allows the transmitter to truncate the series of preambles and transmit the body of the packet, as shown in Figure 5.2.

These mechanisms together improve the performance of unicast data transmissions over the low-power listening scheme used by B-MAC. In an indoor testbed of TelosB nodes, with a sleep interval of 200ms and with five contending packet transmitters, each node sending one packet every second, the

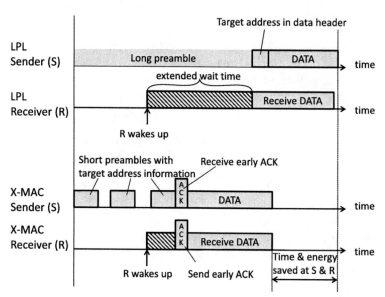

Figure 5.2. Comparison of transmitter-receiver coordination mechanism used in B-MAC (LPL) and X-MAC. Redrawn from Buettner et al. (2006).

authors report a duty-cycle of 20% compared to 65% with B-MAC's low-power listening.

5.2.3 Coordinated Duty Cycling

X-MAC still suffers from several limitations, such as that broadcast packet transmissions must still be transmitted for the full sleep interval. A great deal of ongoing research is investigating new protocols that attempt to reduce these overheads. Coordinated duty-cycling approaches can transmit broadcast packets more efficiently.

Coordinated duty-cycling MAC protocols explicitly compute the schedules for packet transmission and reception across the network. The nodes keep their radios on during these schedules and turn the radios off the rest of the time. Two types of schedule organization are common among the coordinated duty-cycling MAC protocols. MAC protocols such as S-MAC (Ye et al. 2002) and T-MAC (van Dam and Langendoen 2003) organize the schedules across the entire network (or a group of nodes) such that all the nodes (or a group of nodes) turn their radios on at the same time and start packet transmission and reception. The second approach organizes the schedules across the network such that the schedules reflect the routing topology or a specific data flow pattern. For example, the MAC protocol used by Dozer (Burri et al. 2007), described in Section 5.7, coordinates the radio schedules only between immediate neighbors, whereas DMAC (Lu et al. 2007) organizes the schedules along a routing path in a staggered pattern such that the packets can be forwarded along a path without interruption. Robust time synchronization and link dynamics, which often necessitate schedule revision at small timescale, are the key challenges to coordinated duty-cycling. Coordinated duty-cycling link layers seem less successful in practice than uncoordinated duty-cycling link layers.

5.2.4 Link Quality Estimation

Discovering nodes and links in the neighborhood provides valuable information to the routing and other protocols in sensor network. Most important function of neighborhood discovery is accurately and efficiently estimating the link qualities to the neighbors.

Most routing protocols in sensor networks use the ETX link metric (De Couto et al. 2003) to quantify the link reliability. ETX is a widely used link and path metric in wireless routing, and is defined as the *expected number of transmissions* for a packet to reach a destination node, assuming link-layer ARQ is in use. A perfect link has an ETX of 1; a larger ETX value implies a less reliable link.

In sensor network link quality estimation, the emphasis is on computing ETX of links efficiently, because each probe transmission costs valuable energy. Early proposals for computing ETX of links use a periodic link quality probes and compute average these estimates over time (Woo et al. 2003). Just relying on link quality probes or beacons that are sent at a low rate makes the estimate less agile and hence inaccurate over the short term. However, the short-term link reliability is what determines the success or failure of a packet transmission at a given instant.

As an alternative, the so-called *four-bit link quality estimator* (Fonseca et al. 2007) estimates the quality of wireless links using information from the physical, link, and network layers in combination.

The physical layer can provide immediate information based on the radio's ability to decode an incoming radio packet. Many radios report the Received Signal Strength Indicator (RSSI) for incoming packets. The CC2420 radio used by many sensor nodes additionally reports a Link Quality Indicator (LQI) for each packet. This information is available on each packet reception and tends to have high variation: Each packet received can have different RSSI or LQI value. Although physical layer information cannot provide complete description of link quality, it can often be used as a fast estimate of link quality that can be improved using information from the link layer and link quality probes.

The link layer can provide information regarding success or failure of packet transmission on a link. Most sensor network link layer protocols support acknowledgments of unicast packet transmission. If a packet acknowledgment is received on a link, then we can take that as an indication of a good forward link (data transmission) and a good reverse link (acknowledgment transmission). Unlike computing ETX using separate probe transmissions, use of acknowledgments allows the estimator to compute and update the ETX at the time scale of one packet transmission.

The network layer can provide information regarding the importance of a link, such as a link being on a path to the root. Although physical and link layer information can be used to compose an accurate and agile estimator, the network layer has information regarding which links are on the path to the root or on better paths to the root. The link estimator can use this information to evict links, and sometimes even high-quality links, from the link estimation table to make room for links that are likely to be used by the routing protocol.

The four-bit estimator uses narrow and portable interfaces to access these sets of information from the physical, link, and network layers: Only four bits of information need to be exchanged between the link estimator and these layers (Figure 5.3). The four-bit estimator can work on any radio platform that provides the proposed interface. These interfaces are designed to be easily implementable across different platforms and make few assumptions about the specific radio technology or hardware. The four-bit estimator has been implemented on sensor node platforms that use CC1000, CC1100, CC2420, TDA5250, and RF230 transceivers.

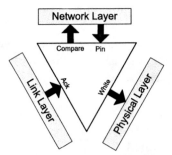

Figure 5.3. A link estimator, represented by the triangle in the center, uses four bits of information from the three layers. Outgoing arrows represent information the estimator requests on packets it receives. Incoming arrows represent information the layers actively provide.

In summary, the link layer is an active area of research in sensor networking. As energy efficiency becomes a focus on other mobile computing domains, some of the ideas from sensor networks link layer might inform designs in those spaces.

5.3 Tree-Based Routing

Collection trees are a core building block for sensor network applications and protocols. In their simplest use, collection trees provide an unreliable datagram routing layer that deployments use to gather data from the entire network to a small number of collection points, such as a single base station. Additionally, tree collection protocols provide the routing topology that underlies transport protocols such as RCRT and Flush, described in Section 5.5.

A collection protocol builds and maintains minimum-cost trees to nodes that advertise themselves as tree roots. Figure 5.4 shows an example of a routing tree formed in the network as the result of running a collection protocol. Collection is address-free: When there are multiple roots, collection sends the packets to root with the minimum cost without knowing its address.

The design goals for collection are:

Reliability: A collection protocol should deliver at least 90% of end-to-end packets when a route exists, even under challenging network conditions; 99.9% delivery should be achievable without end-to-end mechanisms.

Robustness: The protocol should be able to operate without tuning or configuration in a wide range of network conditions, topologies, workloads, and environments.

Efficiency: The protocol should achieve this reliability and robustness while incurring little overhead in terms of management traffic or framing overhead.

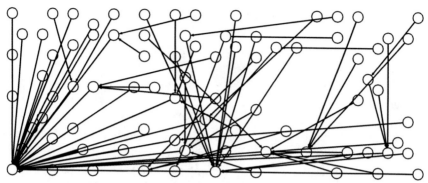

Figure 5.4. Result of running the Collection Tree Protocol on an indoor testbed of 85 nodes. All the nodes form a route to the collection point or root (node at the bottom left corner).

Many collection protocols have been proposed for sensor networks. Some collection protocols, such as MultihopLQI (MLQI 2009), are platform-dependent, whereas collection protocols such as MintRoute (Woo et al. 2003) and CTP (Gnawali et al. 2009) are platform-independent. We will use CTP as an example to study the design of a collection routing protocol.

5.3.1 CTP: Collection Tree Protocol

CTP (Gnawali et al. 2009) is a tree-routing protocol for sensor networks. Building on the functionality provided by the four-bit link estimator (Section 5.2) to accurately estimate link qualities, CTP incorporates two mechanisms to achieve high reliability, robustness, and efficiency. CTP uses an adaptive rate controller for routing beacons. As long as the routing gradient in the network is consistent, nodes reduce the frequency of control packets over time. When the network detects a loop or other inconsistency, nodes send control traffic more quickly to repair the topology. CTP also uses data traffic to actively probe the topology, detecting routing problems and repairing them as needed. This enables CTP to be highly agile and respond to broken links within a few packets.

All the nodes running CTP maintain an estimate of the cost (based on the ETX metric [De Couto et al. 2003] of its route to the root. A given node's cost is the cost of its next hop plus the cost of its link to the next hop: The cost of a route is the sum of the costs of its links. Roots advertise a cost of zero. Each data packet contains the transmitter's local cost estimate. When a node receives a packet to forward, it compares the transmitter's cost with its own. Since cost must always decrease, if a transmitter's advertised cost is not greater than the receiver's, then the transmitter's topology information is stale and there may be a routing loop. Using the data path to validate the topology in this way allows a protocol to detect possible loops on the first data packet after they occur.

Collection protocols typically broadcast control beacons at a fixed interval. This interval poses a basic tradeoff: A small interval reduces how stale information can get and how long a loop can persist, but uses more bandwidth and energy. A large interval uses less bandwidth and energy but can let topological problems persist for a long time. CTP's use of adaptive beaconing breaks this tradeoff, achieving both fast recovery and low cost. It does so by extending the Trickle algorithm (Levis et al. 2004) to maintain its routing topology.

The most relevant property of the Trickle algorithm is the adaptive timer interval. When the routing paths are reliable and stable, the timer interval is increased exponentially, thereby decreasing the rate at which beacons are transmitted. When events such as link dynamics or detected inconsistencies necessitate routing path repair, the Trickle timer is reset to a small interval, thereby increasing the rate at which the routing beacons are sent. More specifically, the routing layer resets the beacon interval to a small value on three events:

1. **It is asked to forward a data packet from a node whose ETX is not higher than its own.** The protocol interprets this as neighbors having a significantly out-of-date estimate and possibly a routing loop. It beacons to update its neighbors.
2. **Its routing cost decreases significantly.** The protocol advertises this event because it might provide lower-cost routes to nearby nodes. In this case, "significant" is an ETX of 1.5.
3. **It receives a packet with the pull bit set.** The "pull" bit advertises that a node wishes to hear beacons from its neighbors, for example, because it has just joined the network and needs to seed its routing table. The pull bit provides a mechanism for nodes to actively request topology information from neighbors.

In a network with stable links, both the first and second events are rare. As long as nodes do not set the pull bit, the beacon interval increases exponentially to one routing beacon every eight minutes. When the topology changes significantly, however, affected nodes reset their intervals to 128 ms and transmit to quickly reach consistency.

Gnawali et al. (2009) evaluated CTP on 12 different indoor testbeds with 20 to 310 nodes. They report achieving 90–100% data delivery reliability with duty-cycled and non-duty-cycled link layers.

5.3.2 In-network Aggregation

Apart from its use for path computation and as an underlying building block for higher-level protocols, tree-based routing is often used to guide the placement of in-network aggregation mechanism in sensor networks. In-network aggregation enables nodes in the network to combine the data received from multiple nodes into lossy or lossless summaries (depending on the application requirement) and transmit these summaries to the destination, thereby reducing communication overhead. In-network aggregation can be successfully used in sensor network applications that report summaries of data to the user: The user perceives no difference in the reported data, even though the summaries are computed in the network to minimize communication overhead.

TinyDB (Madden et al. 2002), Directed Diffusion (Intanagonwiwat et al. 2000), and Synopsis Diffusion (Nath et al. 2004) are representative examples of systems that employ in-network aggregation to minimize communication overhead. These systems define the operation a node performs to transform data received from multiple nodes into a single data item. Thus, each node transmits one packet to the destination as opposed to forwarding all the packets it receives.

As an example, consider computing the minimum value of all sensor readings in a network. To compute this aggregate, each node receives data from its children in the routing tree, computes the minimum of all those data items and its own sensor reading, and transmits only the computed minimum value to its parent in the tree. When all the nodes in the network run this algorithm along the routing tree, the root receives the minimum value of the entire network. With in-network aggregation, each node in the network transmits only one packet to its parent on each round.

The collection of data readings has been the dominant traffic profile of sensor network applications. Although most applications report all the data to the root, there are promising in-network aggregation algorithms that compute accurate or approximate aggregates, thereby avoiding the communication overhead of transmitting all data to the root.

5.4 Dissemination

Most sensor networks require the ability to send configuration parameters or control messages to the entire network. Dissemination protocols are used to reliably broadcast information from one or a small set of nodes to the entire network in an energy- and bandwidth-efficient manner.

Efficient dissemination protocols are challenging to build. A simple flooding approach, in which each node rebroadcasts the received dissemination packet, can lead to congestion and cause substantial packet loss. Avoiding congestion and collisions to improve efficiency requires moderating the pace at which information is flooded to the network. Of course, such pacing results in higher dissemination latency.

The dissemination protocols in sensor networks have these design goals:

- **Reliability.** A disseminated packet must eventually be received by all nodes in the network.
- **Efficiency.** Dissemination must use as few packet transmissions as possible.
- **Low latency.** Data must be disseminated to the network as quickly as possible.

We describe Drip (Levis and Tolle 2008) and DIP (Lin and Levis 2008) as two case studies of dissemination protocols. Both of these protocols use the Trickle algorithm (Levis et al. 2004) as their underlying mechanism to time the transmission of dissemination packets, so we describe that first.

5.4.1 Trickle

Trickle (Levis et al. 2004) is an eventual consistency protocol that efficiently bring data stored by multiple nodes in the network to a consistent state. In Trickle, nodes store some local state (such as a data object or software binary) with an associated *version number* that is monotonically increasing. When the state is updated by any node, the version number is incremented. Trickle is used to maintain consistency in the version number across the network, whereas the Drip protocol, described later in the chapter, is used to transfer the actual state between nodes.

Trickle is a gossip-based protocol: Each node periodically broadcasts its known version number to its neighbors. If a node hears a later (that is, higher) version number than its own, it pulls the new state from the node and updates its version number accordingly.

To limit overhead, Trickle adapts the rate at which version information is broadcast by sensor nodes. Each node maintains a local timer that determines the rate at which the version information is broadcast. The timer rate is adjusted based on the agreement between nodes in the local radio neighborhood. If all nodes in a neighborhood have the same version, there is no need to rebroadcast frequently. The timer interval is increased exponentially following each rebroadcast by a node, until it reaches a maximum interval duration. If a node receives information with a higher version number, the timer is reset to a small interval, which allows new information to be rapidly disseminated (Figure 5.5). To avoid congestion and redundant transmissions, if a node detects that another node has transmitted the new information before its timer expires, it cancels the scheduled transmission and increases its timer interval.

5.4.2 Drip

Drip (Levis and Tolle 2008) is a data dissemination protocol built on top of the Trickle algorithm. Whereas Trickle allows a node to discover the existence of new versions of data, Drip adds the mechanism to transfer the data object

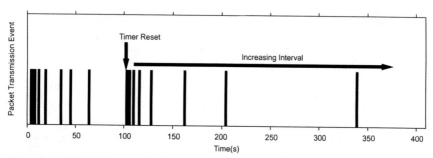

Figure 5.5. Trickle timer used to time the packet transmissions of a dissemination protocol. The Trickle timer starts with a small interval and doubles the interpacket interval after each transmission until the timer is reset.

itself. The canonical use case is to supply all nodes with a new set of configuration parameters. Trickle is used to inform nodes of the existence of the new parameters, whereas Drip ensures all nodes actually get a copy of them.

Once a node learns that its version is out of date, it broadcasts a request for the latest version. A node receiving this request replies by pushing the data object to the requesting node using a series of packet transfers with link-layer acknowledgment and retransmission. Because data transfers happen over a single radio hop, they can be done efficiently without the need for explicit packet routing; using Drip, the entire network can eventually receive the newest version of a data object injected by one node. Figure 5.6 shows the evolution of the communication overhead of Drip; the number of packet transmissions settles at the rate corresponding to the maximum Trickle interval despite the high overhead in the beginning.

5.4.3 DIP

Trickle and Drip are designed to support a single data object and associated version information. DIP (Lin and Levis 2008) is a protocol that can efficiently

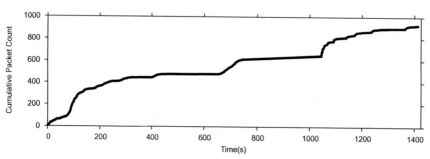

Figure 5.6. Drip's communication overhead grows slower over time, eventually settling at the rate corresponding to the maximum Trickle interval when there are no dissemination requests.

disseminate a large number of data objects. Although Trickle can be used for this purpose, the overhead of maintaining a separate timer and performing individual broadcasts for each data object becomes prohibitive as the number of data objects increases.

A simple approach is to transmit the complete set of known version numbers each round, using the Trickle algorithm. However, given the limited size of radio packets on typical sensor node platforms, it is typically not feasible to do this when the number of data objects is large. DIP uses an alternative approach, called the *scan* algorithm, that maintains a single timer and advertises a subset of the version numbers known by the node each time the timer fires. The challenge is deciding which subset of version numbers to advertise. DIP's scan algorithm collects estimates of data item version numbers based on neighbor advertisements. DIP then prioritizes the broadcast of data item versions that are likely to be newer than the ones in the neighborhood over rest of the data items. Although this approach has low communication overhead, it increases the latency for detecting the presence of new version numbers in the network.

DIP uses the *search* algorithm to locate the updated values using a hash tree. DIP advertises a summary of all the versions of data objects stored by a node, which is computed as a hash of the version numbers. When a node receives a hash summary different than its own, it knows that at least one item on the transmitter node is of a different version than its own. However, it cannot tell which item is different. The receiver then advertises its own summary: Instead of a single hash covering all its data version numbers, it transmits a set of summary hashes over smaller ranges, but covering the same range of data where the difference occurred. This allows the original transmitter to narrow down the changed data version to the size of the range of the hash summary in the packet. This iterative exchange of messages with smaller ranges results in the nodes traversing the hash tree to find the data item that has changed. When the nodes determine the exact data item that is different, the nodes exchange the data value corresponding to the new version number.

When a large number of new data items are introduced for dissemination, the search algorithm can require a large number of message exchanges to accurately identify the updated data item. In this scenario, it is more efficient to advertise each data item because new versions can be identified using just a single message. Also, switching to scanning after the search has narrowed the update to a small set of items is more efficient than continuing the message exchange between the nodes to search down the hash tree because lost packets will cause the search to go back up the hash tree and the constant factors of searching are higher than scanning with small ranges. The key insight behind DIP's efficiency is its switching between the search and scan algorithm depending on the network conditions and the number of updates.

DIP's communication overhead and new version detection latency stays asymptotically constant with the order of the number of data objects, but incurs logarithmic communication overhead in identification of the data item that has changed. The authors evaluated DIP on a testbed of 80 MicaZ motes with sixty-four data objects. In one experiment, the authors updated eight of the data items. DIP disseminated these updates to the entire network with 50% and 15% lower overhead and 80% and 60% less time than the search and scan protocols, respectively.

5.4.4 Other Dissemination Protocols

Dissemination protocols are used not only to disseminate commands and configuration parameters, but also to build higher-layer application-specific dissemination services. One interesting example is disseminating executable programs to the sensor networks. Deluge and Maté are good examples of such code dissemination protocols.

Deluge (Hui and Culler 2004) allows new executable binaries to be disseminated to the network, which might be on the size of tens of kilobytes, which is fairly large considering the limited radio bandwidth and high energy cost for packet transmission. The executable program is divided into pages and injected into a source node that acts as the seed for the dissemination protocol. When a new page is received by a node, it advertises the page number. As an optimization, a node advertises page i only if all previous pages $[0, i)$ are also available at the node. As in Drip, a node that learns of the availability of an executable page it does not yet have will transmit a request for the page to be transferred to it. Once the transfer of all the executable pages is complete, Deluge reprograms the node with the assembled program image and reboots the node.

Instead of transferring entire executable images, it is also possible to make use of a simple virtual machine on each sensor node, which can execute compact bytecode programs. Maté (Levis and Culler 2002) is an example of such a system. Maté programs are organized as a collection of capsules, each of which can fit within a single radio packet. When a node receives a capsule with a new version number, the node installs and runs the code in the capsule and forwards the capsule to its neighbors. This process continues until the program is disseminated and installed across the network. This greatly reduces the overhead for reprogramming the network, but restricts programs to using the limited instruction set defined by the Maté virtual machine. However, the Maté VM can be customized with application-specific bytecodes implementing native functions.

5.5 Reliable Transport

Many applications for WSNs require reliable transfer of data from sensor nodes to the base station. This is especially true in domains involving high-resolution signal collection, such as structural (Chintalapudi et al. 2006; Kim et al. 2007b), acoustic (Allen et al. 2008), or seismic (Werner-Allen et al. 2006) monitoring. Transferring a large volume of data from a sensor node over a multihop path to a base station is challenged by packet loss, radio channel contention, and lack of buffer space on nodes acting as routers.

Although TCP is widely used on the Internet for reliable stream-based transport, this approach is not suitable for lossy multihop wireless networks. This is primarily because TCP interprets packet loss as being due to buffer overflow in routers, and tunes the congestion window size accordingly. In WSNs, however, the *rate* at which a source node injects packets into the network is the critical parameter, since contention for the radio channel along a multihop path limits the effective rate at which packets can be relayed. Unlike wired networks, in wireless networks, packet loss is caused by transients in link conditions, collisions, and routing path churn, rather than by persistent congestion. In this section, we describe two approaches to reliable transport in sensor networks, Flush and RCRT.

5.5.1 Flush

Flush (Kim et al. 2007a) is a reliable bulk transfer protocol for multihop WSNs. The protocol operates over a routing tree such as MintRoute or CTP (Section 5.3) rooted at a base station. The base station transmits a request for data object stored on a given node using a broadcast flood (Section 5.4). The node hosting the data object breaks it into multiple packets, each with a corresponding sequence number, and streams those packets to the sink over the multihop path.

Flush relies on several techniques to ensure reliability and high throughput. First, both link-layer ACKs and end-to-end selective NACKs are used to recover from lost packets. Second, data is streamed along the routing path to the sink at a rate that is chosen carefully to avoid intrapath contention. Finally, Flush only supports a single bulk transfer in the network at a time, to avoid interflow interference.

Flush relies both on link-layer ARQ and end-to-end NACKs for reliability. At the link layer, a packet will be retransmitted up to four times before being dropped by the sender. Limiting the number of link-layer retransmissions is necessary to account for changes in the underlying routing topology or node failures. Link-layer retransmission paves over intermittent losses due to radio channel noise and significantly reduces the number of expensive end-to-end retransmissions.

The sink will send a selective NACK containing the sequence numbers of missing packets in the flow after it believes the last source packet has been transmitted, or after a timeout based on an estimate of the end-to-end RTT. Upon reception of the NACK, the source retransmits those packets up the routing tree in the manner described earlier. This process repeats until the sink has received the entire object.

The key contribution of Flush is its *rate control* algorithm that determines the peak rate at which the source node can inject new packets into the network without inducing loss due to intrapath collisions. In a multihop path, as nodes forward data to the root, those transmissions potentially collide with other transmissions both upstream and downstream along the path. A simple solution would be to avoid any pipelining, but this would eliminate spatial reuse of the radio channel, thereby reducing throughput.

Flush dynamically estimates the maximum sustainable sending rate to maximize throughput and avoid contention along the routing path. Nodes measure their own packet transmission delay and receive feedback from upstream nodes on their delays, and those of other nodes that might interfere with their transmissions. The set of potential interfering nodes is determined based on snooping the radio channel; Flush assumes that a node can overhear packet transmissions from the interfering nodes. Flush relies on the underlying MAC protocol to schedule individual packet transmissions, so this rate control is performed at the transport layer.

Combining these techniques, Flush achieves reliable transfer throughput that closely matches the best possible performance measured using a fixed transmission rate. The key is that Flush *automatically* determines the optimal rate, which depends on the node's depth in the routing tree and the overall network topology. Flush scales well with long routing paths: The protocol has been evaluated using a 48-hop linear chain of nodes deployed outdoors.

5.5.2 RCRT

RCRT (Paek and Govindan 2007) is another reliable transport protocol. Unlike Flush, RCRT handles multiple concurrent reliable flows. This is necessary in cases where all nodes are generating data simultaneously, requiring real-time streaming of the data back to the sink, owing to the limited buffer capacity. In RCRT, the base station performs centralized congestion control for all source nodes, using global knowledge of the performance of each flow. This approach permits global application of policies to drive allocation of network capacity.

Like Flush, RCRT uses end-to-end NACKs for repairing lost packets. The essential congestion control mechanism is based on measuring the *time to repair a loss*. If losses are repaired quickly enough, there is no need for rate adjustment. However, if the repair time exceeds the expected packet round-trip time, congestion is present in the network, and transmission rates are adjusted. RCRT adapts the *aggregate rate of all flows* using an AIMD adaptation scheme. Once the new aggregate rate is determined, individual flow rates are calculated based on the current policy in use. Example policies include a demand-proportional scheme, in which rates are allocated in proportion to each flow's desired rate, and a fair policy in which all flows receive an equal rate.

The authors evaluated RCRT on an indoor testbed of forty nodes, using tree routing with paths up to eight hops in length. They show that RCRT is able to sustain a per-node traffic demand of 0.8 packets/s, which is 88% of the optimal sustainable rate for their testbed. Further, RCRT does not suffer congestion collapse when load is increased. RCRT's goodput is more than double that of IFRC (Rangwala et al. 2006), another protocol designed for network-wide rate control.

Comparing Flush and RCRT, the main difference is that Flush supports a single reliable flow whereas RCRT supports multiple simultaneous flows and supports policies to balance the amount of bandwidth allocated to each. Flush computes the transmission rate at the sink using feedback from downstream nodes, whereas RCRT performs a centralized search for stable transmission rates at the base station. Unlike Flush, RCRT does not assume that nodes can overhear packet transmissions from other nodes. Both protocols rely on end-to-end selective NACKs for reliability, and both can work with any routing protocol that provides bidirectional paths to the sink.

5.6 Support Protocols

Apart from protocols for data communication, wireless sensor networks often employ a range of *support* protocols that provide services such as time synchronization and node localization. In this section, we will briefly discuss three representative protocols in this class: the Flooding Time Synchronization Protocol (FTSP) (Maroti et al. 2004), localization using acoustic beacons (Simon et al. 2004), and radio interferometric localization (Maróti et al. 2005).

5.6.1 The Flooding Time Synchronization Protocol

Time synchronization is an essential service for sensor networks in which the data acquired by nodes must be accurately timestamped against a global clock. Individual sensor nodes use oscillator crystals with a tolerance of around 40 ppm, causing the local clocks of each nodes to drift substantially. Moreover, this drift varies over time owing to fluctuations in temperature and voltage. Applications such as acoustic or seismic monitoring (Simon et al. 2004; Chintalapudi et al. 2006; Werner-Allen et al. 2006) require time accuracies in the millisecond or microsecond range, so it is inadequate to perform a one-time translation of each node's local clock to a global timebase. Time synchronization must be run periodically.

A simple approach would have a central node (such as the base station) advertise a global time to all nodes in the network, which would set their local clocks accordingly. However, propagating timebase information throughout a large network is challenging owing to the use of multihop paths and the timing uncertainty of radio communication. Transmitting and receiving radio messages incurs nondeterministic delays due to the MAC protocol, transmission and reception overheads, and interrupt processing delays. These delays must be accounted for when relaying timebase information through the network.

The Flooding Time Synchronization Protocol (FTSP) (Maroti et al. 2004) operates as follows. A single node acts as the root of a synchronization tree; the local clock of the root node is used as the global timebase. Failure of the root node leads to election of a new root. The root periodically beacons the global time value. Nodes within one hop of the root receive these beacons and compute a mapping from their local clock to the global timebase, as described later in this chapter. Each node rebroadcasts the beacon, allowing the global time to propagate throughout the network.

FTSP carefully accounts for the uncertain communication delays by timestamping outgoing beacon packets *after* the MAC delay, just prior to the actual transmission of the first byte of the packet. Likewise, a receiver timestamps the received packet on the arrival of the first reception interrupt. On the Chipcon CC1000 radio used in the study (Maroti et al. 2004), an interrupt is generated for each received byte of the packet. Receivers can determine the interrupt processing jitter by measuring the time between each successive interrupt (Figure 5.7).

To compensate for clock drift, each node measures the offset between the sender's timestamp and its local clock. Given that clock drift is expected to be linear over short time intervals, nodes perform a linear regression on these offsets to correct their local clocks for drift.

FTSP has been evaluated extensively on a testbed of 60 Mica2 nodes. All nodes were placed within radio range of each other, but a 6-hop multihop

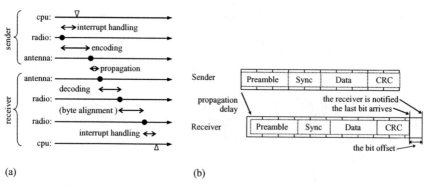

(a)   (b)

Figure 5.7. To minimize the impact of nondeterministic delays in packet timestamping on time synchronization accuracy, FTSP implementation on the CC1000 radio records timing for each byte-boundary in a packet and computes an aggregate over these times as the packet timestamp. *Source:* Maroti et al. (2004).

topology was induced in software. To measure time synchronization accuracy, a *reference broadcaster* node transmits periodic messages that all nodes in the testbed receive and timestamp using FTSP. Since it is assumed that all nodes should receive the message at the same time (module differences in RF propagation delay, which are negligible in the small testbed), comparing the timestamps of this global event across nodes allows one to assess timing accuracy. The authors show that FTSP exhibits an average pairwise error of 3 μs and an overall maximum error of less than 14 μs.

5.6.2 *Localization Using Acoustic Ranging*

Accurately determining the position of sensor nodes is another essential service in many deployments. Although many sensor networks involve static nodes that may be placed in the field by hand, it is inconvenient and error-prone to rely on manual surveying or GPS for determining a node's position. GPS only works outdoors and only provides an accuracy of a few meters. For sensor networks deployed in an ad hoc fashion (e.g., dropped from an airplane) or involving mobile nodes, it is critical that the network be able to self-localize.

Localization is a heavily studied topic, and a wide range of techniques have been proposed. One of the most common approaches involves ranging using acoustic time-of-flight (Simon et al. 2004). Sensor nodes are equipped with a sounder and a microphone. A source node broadcasts a radio message immediately followed by an acoustic chirp that can be detected by nearby nodes using their microphones. Receivers timestamp the arrival of the RF message and the chirp. Since it is assumed that the RF propagation delay is negligible, the time-of-flight of the chirp can be readily computed as the time between the reception of RF message and the acoustic chirp. Assuming the speed of sound in the environment is known (which can vary based on temperature and humidity), the approximate range to the source node can be computed.

This process is made more challenging by the limited signal-processing capability of motes, missed or incorrect chirp detections, reflections, and limited acoustic sensing range. To address these issues, the system transmits multiple chirps that are combined in postprocessing to enhance SNR. A digital bandpass filter is also used to suppress noise. In Simon et al. (2004), this ranging technique is shown to be accurate to within 10 cm over a range of up to 9 m.

Once a set of pairwise range estimates are known, they can be combined to determine the relative location of nodes in the network. One approach is to collect range estimates at a central node (e.g., the base station) and perform iterative optimization until a stable configuration is determined. Landmark nodes with known locations are used to anchor the network's orientation to geographic coordinates.

5.6.3 Radio Interferometric Localization

As an alternative to acoustic ranging, it is possible to use RF signals alone to obtain extremely accurate range estimates between nodes. The radio interferometric technique described in Maróti et al. (2005) involves two nodes transmitting sinusoid RF patterns at known frequency offsets. A pair of receiver nodes can determine the beat pattern induced by the interfering signals. This results in a series of equations that relates the phase offset of the beat patterns to the nodes' relative locations. Eight nodes are able to localize themselves in three dimensions. Using a network of 16 nodes, the authors demonstrate the average positional error to be 3 cm with a maximum of 6 cm. However, this technique is not yet appropriate for mobile networks because it involves extensive measurements and calibration. It also requires the use of radios that can be configured to transmit an unmodulated sine wave. Although more recent 802.15.4 radios provide a test mode with this capability, it is not possible to tune the frequency of the sine wave at fine enough granularity.

5.7 Cross-Layer Concerns

Sensor network protocol designs face a tension between the desire to exploit layering and the need for cross-layer optimizations to get the best efficiency and performance. Although this problem is evident in conventional networks as well, in sensor networks it is particularly pronounced due to the extreme resource limitations of sensor nodes.

Sensor network protocols have typically followed a layering principle that separates the physical, link, routing, and application layers, as we have presented in this chapter. However, many designs perforate the layer interface by providing control knobs and feedback to higher levels of the stack. B-MAC (Polastre et al. 2004) exposes a range of control parameters such as the listen interval and preamble length. SP (Polastre et al. 2005b) provides a neighbor table with link state and congestion information. The FTSP time synchronization protocol (Maroti et al. 2004) relies on link-layer packet timestamping well below the FTSP protocol layer itself. Each of these examples illustrates the need for cross-layer information in designing sensor network protocols. Given that sensor networks are not constrained by legacy application software and standards compliance, the community has had the opportunity to explore the protocol design space more broadly and experiment with vastly alternative designs.

Taking cross-layer design to the extreme, Dozer (Burri et al. 2007) is a system for low-power data collection for environmental monitoring. Unlike previous designs that largely separated the MAC, routing, and application layers, Dozer fully integrates these functions to achieve extremely efficient operation. The goal is to permit nodes to spend the maximum amount of time in an

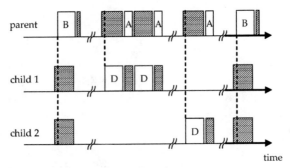

Figure 5.8. Packet transmission and reception time synchronization between a parent and its two children. The parent sends a beacon (B) to determine the transmission slots for the children. The children transmit the data messages (D), which are explicitly acknowledged (A). *Source:* Burri et al. (2007).

energy-efficient sleep mode. In Dozer, nodes make use of tree routing. TDMA is used for scheduling communication; each parent node in the tree defines the TDMA schedule for its children (Figure 5.8). This approach allows the parent to avoid idle listening by waking up only when a child is about to transmit. There is no explicit collision-avoidance mechanism apart from randomization of TDMA schedules across nearby nodes.

Dozer's tree formation protocol is based on periodic beacons that each node transmits to maintain the TDMA schedule for its children. A node wishing to join the tree first listens for beacons and selects a parent based on the potential parent's depth in the tree and the number of children it has (which can be determined from the beacon message). The node joins by sending a *connection request* to the chosen parent, which adds it to the TDMA schedule. Nodes periodically listen for beacons from other potential parents and cache this information so that a new parent can be selected quickly in the event of a parent failure. Nodes that are unable to join the network drop into a *suspend mode* to avoid polling the radio channel continuously.

Dozer incorporates a number of techniques to further improve efficiency. Each node pads its TDMA schedule by a random amount on each round to avoid schedule collisions between sibling nodes in the tree. This avoids the need for explicit coordination of TDMA schedules across nodes. Child nodes predict the length of each round's random padding using a pseudorandom number generator in which the seed is shared with the parent. Although there is no guarantee that collisions will not happen, the randomized schedule makes multiple collisions in successive rounds far less likely.

Dozer has been evaluated in an indoor network of forty nodes over the span of several weeks. The sensor sampling interval was set to be 120 s and the beacon interval was set to 30 s. The network obtained very low message losses (an average of 1.29%) with an overall average radio duty cycle of just 0.1679%. This yields a mean energy consumption of just 0.082 mW.

5.8 The Emergence of IP

In recent years, there has been increased interest in linking wireless sensor networks with conventional IP-based networks at the IP protocol layer. Initially, it was thought that the IP family of protocols would be too heavyweight to run on resource-limited sensor nodes, or that such protocols would be inappropriate for the new demands of sensor network applications. However, a substantial engineering effort by two IETF working groups – IPv6 over Low power WPAN (6LoWPAN) and Routing Over Low-Power and Lossy Networks (ROLL) – has resulted in efficient IP protocol implementations for sensor networks.

5.8.1 IP Packet Frame

The IETF 6LoWPAN working group has standardized the encoding of IPv6 packets on 802.15.4 networks with its publication of RFC 4944 (Montenegro et al. 2007). The IEEE 802.15.4 standard is likely to be the most pervasive link layer used in low-power wireless networks, such as sensor networks. Transmission of IPv6 packets over these links is challenging due to the limited packet size supported by 802.15.4. Whereas IPv6 packets can be as large as 1,280 bytes, 802.15.4 only supports a maximum packet size of 102 bytes, assuming maximum frame header overhead. By default, the IPv6 header occupies 40 bytes, leaving little room for application payload.

To reduce this overhead, the 6LoWPAN working group has defined a format based on header compression. In this scheme, a transmitter substantially compresses the header of outgoing packets by substituting or eliding header fields that have common values or that can be inferred from other header fields. The 40-byte IPv6 header can be compressed to as little as two bytes.

5.8.2 IP-Based Routing Protocols

Although no IP-based routing protocol for sensor networks has been standardized, there are several IP-based protocols in use in sensor networks both in commercial products and research.

An IP-based routing protocol must support unicast routing between arbitrary nodes in the network. In sensor networks, a combination of tree-based routing and source routing is typically used to enable unicast routing. In Blip, an IPv6 protocol stack implemented in TinyOS, all the nodes in the network first form a tree-routing topology. Then the nodes periodically send their topological information, such as their parent in the routing tree, to the root of the network. When a node needs to send a packet to an arbitrary node in the network, the packet is first forwarded to the root. The root, which has topological information for all the nodes in the network, then source-routes the packet to the destination.

The recently formed IETF ROLL working group is tasked with standardizing the routing protocols for the low-power networks such as wireless sensor networks. The working group, after a survey of wireless network protocols, concluded that no existing IETF protocol meets all of the requirements of these low power networks. At the time of this writing (October 2010), the ROLL working group has produced a proposed standard for routing protocols to be used in these lossy and low-power networks.

5.9 Sensor Networks and the Future Internet

Wireless sensor networks have broad implications for the future design of the Internet. The sheer number of sensor network nodes that may be deployed in the future raises significant challenges in terms of naming and addressing, communication protocols, resource management, and reliability. Of course, sensor networks differ substantially from conventional Internet hosts. They are extremely resource-limited; connected via wireless mesh networks; and often operate at low duty cycles. Further, the traffic produced by sensor networks is not typically dominated by unicast end-to-end flows; much traffic is multipoint-to-point (e.g., data aggregation up a spanning tree) or point-to-multipoint (dissemination to all nodes in the WSN).

Two opposing views have emerged with respect to the relationship between sensor networks and the Internet at large. At one extreme, sensor networks are treated as special-purpose appliances that would be connected to the Internet via a gateway. The gateway would translate between TCP/IP (and higher-layer protocols, such as HTTP or Web Services calls) and a low-level, possibly proprietary, protocol used within the WSN itself. ZigBee (The ZigBee Alliance 2009) is emerging as one contender for the back-end sensor network protocol and defines both routing and device profiles for a range of applications. The "smart gateway" approach presumes that sensor networks will evolve independently of the rest of the Internet, using specialized protocol implementations that are tailored for the resource-constrained, low-duty-cycle nature of sensor networks. The gateway can provide additional services such as storage and caching.

At the other extreme, sensor networks would be treated as first-class citizens on the Internet, communicating directly via TCP/IP to other Internet-based clients and applications. The ROLL working group of the IETF is developing routing solutions for sensor networks based on an end-to-end IP solution. In Section 5.8, we summarize the various technical directions being explored within this space. The upshot is that it is now possible to communicate with WSN nodes using a variant of IPv6 with special support for the limited memory, bandwidth, and energy capacity of sensor nodes. This approach significantly narrows the gap between WSNs and standard Internet applications.

The tension between these two competing approaches arises because sensor networks challenge the "end-to-end principle" (Saltzer et al. 1984) that has predominated the Internet architecture for decades. A smart gateway pushes complexity into the network (rather than the edges) to support the limited capabilities of a sensor network, whereas an IPv6-based solution upholds the end-to-end principle.

Regardless of the protocol stack used, bridging between WSNs and the Internet raises a number of special considerations that have yet to be fully resolved. The first is the nature of communication between user applications and sensor networks. End-hosts communicating with a sensor network must recognize that these devices cannot be treated like conventional Internet hosts in terms of the data rate, latency, and nature of traffic that WSNs can support. Although it is technically possible to run a lightweight Web server on a WSN node running an IPv6 stack, this may not be the best way for users to interface to the network. More likely, a programmatic interface (e.g., via RPC) will be required, as well as a portal to access a sensor network providing access to aggregate and historical data. Moreover, making WSNs directly accessible via the Internet will no doubt raise concerns over the impact of malware or buggy protocol implementations. Deployed WSNs may need to be "protected" from DoS attacks, port scanners, and viruses that would potentially disrupt their operation.

Moreover, transport and routing protocols present challenges when running across the boundary between the Internet and a resource-constrained sensor network. As described earlier, sensor network routing protocols are significantly different than their Internet counterparts. As a result, routing paths spanning the Internet and sensor networks will traverse segments with very different characteristics. From the transport protocol perspective, packet loss may occur for very different reasons on different segments of the path. As an example, TCP interprets packet loss as the result of congestion, which is often true in wired networks but less prevalent within a multihop wireless network. As a result, TCP buffers, windows, and timers will either stretch the limited sensor node resources or underutilize Internet end-host resources. Thus, transport protocols must evolve to work well under these asymmetric conditions.

Of course, link and routing protocols only provide the lower layers of a protocol stack. It is still necessary to provide protocol layers for accessing sensor data, tasking sensors, and administrative control. These protocols have yet to be defined, though it may be possible to leverage existing standards, such as IEEE 1451, which provides mechanisms for communicating, with a wide range of sensors and actuators.

Once sensor networks have been deployed in more widespread settings, discovery protocols will be a major concern. Applications must be able to query the operational characteristics of the sensor network, such as what types of sensors are installed; the locations of those sensors; and whether sensors have been recently calibrated or serviced. Likewise, the network should export metadata to report whether sensor nodes have failed. This is critical when requesting aggregate data from a large network, since the number and placement of failed nodes can have a substantial impact on the results that are returned.

Finally, conventional approaches to naming and addressing – such as DNS – seem to be ill-suited to sensor networks, which may consist of a large number of nodes whose population and capabilities change over time. Rather than addressing individual sensor nodes, a more appropriate paradigm may be to name the *data* using a semantic addressing scheme. A declarative query interface such as TinyDB (Madden et al. 2002) allows a user to request data with given filtering, aggregation, and periodicity parameters.

Although this chapter focuses on static sensor networks, a new class of networks involving mobile nodes is emerging. Examples include wildlife tracking using GPS collars (Zhang et al. 2004) and opportunistic collection of sensor data using cell phones carried by individuals (Mun et al. 2009). Hybrid networks, combining both static and mobile sensors, are another important future direction. Mobility makes it difficult to name nodes and requires a different approach to routing given that nodes may be disconnected from the network for significant periods of time. Furthermore, using sensors that are not necessarily "owned by" the sensor network in which they participate (such as cell phones) raises a number of issues in terms of accountability, security, and privacy.

If history is any guide, the increasing diversity of devices and applications connected to the Internet will soon encompass sensor networks as well. However, the traffic characteristics produced by sensor networks will be substantially different than conventional uses of the Internet, even with the increasing prevalence of other embedded and mobile devices, such as smartphones. Sensor networks will mostly be generators of traffic rather than sinks, so the focus will be on optimizing the outflow of data. Likewise, sensor networks may not be limited by the demand for human-tolerable access latencies, opening up the design space even further.

5.10 Conclusions

Wireless sensor networks are a fundamentally new kind of distributed computing system that present new opportunities and challenges for network protocol design. Their extreme constraints on energy, memory, bandwidth, and computational resources has led to new protocol designs at every layer of the stack, including link, routing, reliable transport, and application interfaces. Overall, sensor network protocol designs strive for low-power operation in the face of variable link conditions, node failures, and changing application requirements.

As a result, much of the research to date has focused on cross-layer approaches that highly specialize the protocol stack for a given application. The recent emergence of lightweight IPv6 implementations for sensor nodes has led to new questions about the role that sensor networks should play in the evolving Internet architecture. Although many open questions remain, we expect that further experience with this technology in both academic and commercial settings will lead to increased convergence with the Internet as a whole.

References

Allen, Michael, Girod, Lewis, Newton, Ryan, Madden, Samuel, Blumstein, Daniel T., and Estrin, Deborah. 2008. VoxNet: An Interactive, Rapid-Deployable Acoustic Monitoring Platform. *Proceedings of the 7th International Conference on Information Processing in Sensor Networks (IPSN '08)*.

Buettner, Michael, Yee, Gary V., Anderson, Eric, and Han, Richard. 2006. X-MAC: A Short Preamble MAC Protocol for Duty-Cycled Wireless Sensor Networks. *Proceedings of the 4th International Conference on Embedded Networked Sensor Systems (SenSys '06)*, pages 307–320.

Burri, Nicolas, von Rickenbach, Pascal, and Wattenhofer, Roger. 2007. Dozer: Ultra-Low Power Data Gathering in Sensor Networks. *Proceedings of the 6th International Conference on Information Processing in Sensor Networks (IPSN '07)*, pages 450–459.

Chintalapudi, Krishna, Paek, Jeongyeup, Kothari, Nupur, Rangwala, Sumit, Caffrey, John, Govindan, Ramesh, Johnson, Erik, and Masri, Sami. 2006. Monitoring Civil Structures with a Wireless Sensor Network. *IEEE Internet Computing*.

De Couto, Douglas S. J., Aguayo, Daniel, Bicket, John, and Morris, Robert. 2003. A High-Throughput Path Metric for Multi-Hop Wireless Routing. *Proceedings of the 9th ACM International Conference on Mobile Computing and Networking (MobiCom '03)*.

Fonseca, Rodrigo, Gnawali, Omprakash, Jamieson, Kyle, and Levis, Philip. 2007. Four Bit Wireless Link Estimation. *Proceedings of the Sixth Workshop on Hot Topics in Networks (HotNets VI)*.

Gnawali, Omprakash, Fonseca, Rodrigo, Jamieson, Kyle, Moss, David, and Levis, Philip. 2009. Collection Tree Protocol. *Proceedings of the 7th ACM Conference on Embedded Networked Sensor Systems (SenSys'09)*.

Hui, Jonathan W., and Culler, David. 2004. The Dynamic Behavior of a Data Dissemination Protocol for Network Programming at Scale. *Proceedings of the 2nd International Conference on Embedded Networked Sensor Systems (SenSys '04)*. ACM Press, pages 81–94.

Intanagonwiwat, Chalermek, Govindan, Ramesh, and Estrin, Deborah. 2000. Directed Diffusion: A Scalable and Robust Communication Paradigm for Sensor Networks. *Proceedings of the International Conference on Mobile Computing and Networking (MobiCom '00)*.

Kim, Sukun, Fonseca, Rodrigo, Dutta, Prabal, Tavakoli, Arsalan, Culler, David, Levis, Philip, Shenker, Scott, and Stoica, Ion. 2007a. Flush: A Reliable Bulk Transport Protocol for Multihop Wireless Networks. *Proceedings of the 5th International Conference on Embedded Networked Sensor Systems (SenSys '07)*.

Kim, Sukun, Pakzad, Shamim, Culler, David, Demmel, James, Fenves, Gregory, Glaser, Steve, and Turon, Martin. 2007b. Health Monitoring of Civil Infrastructures Using Wireless Sensor Networks. *Proceedings of the International Conference on Information Processing in Sensor Networks (IPSN '07)*.

Levis, Philip, and Culler, David. 2002. Maté: A Tiny Virtual Machine for Sensor Networks. *Proceedings of the 10th International Conference on Architectural Support for Programming Languages and Operating Systems (ASPLOS X).*

Levis, Philip, and Tolle, Gilman. 2008. *Dissemination of Small Values.* TinyOS Extension Proposal TEP-118.

Levis, Philip, Patel, Neil, Shenker, Scott, and Culler, David. 2004. Trickle: A Self-Regulating Algorithm for Code Propagation and Maintenance in Wireless Sensor Networks. *Proceedings of the First USENIX/ACM Symposium on Networked Systems Design and Implementation (NSDI '04).*

Lin, Kaisen, and Levis, Philip. 2008. Data Discovery and Dissemination with DIP. *Proceedings of the 7th International Conference on Information Processing in Sensor Networks (IPSN '08),* pages 433–444.

Lu, Gang, Krishnamachari, Bhaskar, and Raghavendra, Cauligi S. 2007. An Adaptive Energy-Efficient and Low-Latency MAC for Tree-Based Data Gathering in Sensor Networks. *Wirel. Commun. Mob. Comput.,* 7(7), 863–875.

Madden, Samuel, Franklin, Michael J., Hellerstein, Joseph M., and Hong, Wei. 2002. TAG: A Tiny AGgregation Service for Ad-Hoc Sensor Networks. *Proceedings of the 5th USENIX Symposium on Operating Systems Design and Implementation (OSDI '02).*

Maroti, M., Kusy, B., Simon, G., and Ledeczi, A. 2004. The Flooding Time Synchronization Protocol. *Proceedings of the Second ACM Conference on Embedded Networked Sensor Systems (SenSys '04).*

Maróti, Miklós, Völgyesi, Péter, Dóra, Sebestyén, Kusý, Branislav, Nádas, András, Lédeczi, Ákos, Balogh, György, and Molnár, Károly. 2005. Radio Interferometric Geolocation. *Proceedings of the 3rd International Conference on Embedded Networked Sensor Systems (SenSys '05),* pages 1–12.

MLQI. 2009. *The MultiHopLQI protocol.* http://www.tinyos.net/tinyos-2.x/tos/lib/net/lqi

Montenegro, G., Kushalnagar, N., Hui, J., and Culler, D. 2007. *Transmission of IPv6 Packets over IEEE 802.15.4 Networks.* Network Working Group RFC4944.

Mun, M., Reddy, S., Shilton, K., Yau, N., Boda, P., Burke, J., Estrin, D., Hansen, M., Howard, E., and West, R. 2009. PEIR, the Personal Environmental Impact Report, as a Platform for Participatory Sensing Systems Research. *Proceedings of the 7th Annual International Conference on Mobile Systems, Applications and Services (MobiSys '09).*

Nath, Suman, Gibbons, Phillip B., Seshan, Srinivasan, and Anderson, Zachary R. 2004. Synopsis Diffusion for Robust Aggregation in Sensor Networks. *Proceedings of the 2nd International Conference on Embedded Networked Sensor Systems (SenSys '04),* pages 250–262.

Paek, Jeongyeup, and Govindan, Ramesh. 2007. RCRT: Rate-Controlled Reliable Transport for Wireless Sensor Networks. *Proceedings of the 5th International Conference on Embedded Networked Sensor Systems (SenSys '07),* pages 305–319.

Polastre, Joseph, Hill, Jason, and Culler, David. 2004. Versatile Low Power Media Access for Wireless Sensor Networks. *Proceedings of the Second ACM Conference on Embedded Networked Sensor Systems (SenSys '04).*

Polastre, Joseph, Szewczyk, Robert, and Culler, David. 2005a. Telos: Enabling Ultra-Low Power Wireless Research. *Proceedings of the Fourth International Conference on Information Processing in Sensor Networks: Special track on Platform Tools and Design Methods for Network Embedded Sensors (IPSN/SPOTS '05).*

Polastre, Joseph, Hui, Jonathan, Levis, Philip, Zhao, Jerry, Culler, David, Shenker, Scott, and Stoica, Ion. 2005b. A Unifying Link Abstraction for Wireless Sensor Networks. *Proceedings of the Third ACM Conference on Embedded Networked Sensor Systems (SenSys '05).*

Rangwala, Sumit, Gummadi, Ramakrishna, Govindan, Ramesh, and Psounis, Konstantinos. 2006. Interference-Aware Fair Rate Control in Wireless Sensor Networks. *SIGCOMM Comput. Commun. Rev.,* 36(4), 63–74.

Saltzer, J. H., Reed, D. P., and Clark, D. D. 1984. End-to-End Arguments in System Design. *ACM Trans. Comput. Syst.*, **2**(4), 277–288.

Simon, Gyula, Maróti, Miklós, Lédeczi, Ákos, Balogh, György, Kusy, Branislav, Nádas, András, Pap, Gábor, Sallai, János, and Frampton, Ken. 2004. Sensor Network-Based Countersniper System. *Proceedings of the 2nd International Conference on Embedded Networked Sensor Systems (SenSys '04)*, pages 1–12.

The ZigBee Alliance. 2009. *The ZigBee Alliance*. http://www.zigbee.org

Tolle, Gillman, Polastre, Joseph, Szewczyk, Robert, Culler, David, Turner, Neil, Tu, Kevin, Burgess, Stephen, Dawson, Todd, Buonadonna, Phil, Gay, David, and Hong, Wei. 2005. A Macroscope in the Redwoods. *Proceedings of the Third ACM Conference on Embedded Networked Sensor Systems (SenSys '05)*.

van Dam, T., and Langendoen, K. 2003. An Adaptive Energy-Efficient MAC Protocol for Wireless Sensor Networks. *Proceedings of the 1st ACM Conference on Embedded Networked Sensor Systems (SenSys '03)*.

Werner-Allen, Geoff, Lorincz, Konrad, Johnson, Jeff, Lees, Jonathan, and Welsh, Matt. 2006. Fidelity and Yield in a Volcano Monitoring Sensor Network. *Proceedings of the 7th USENIX Symposium on Operating Systems Design and Implementation (OSDI '06)*.

Woo, Alec, Tong, Terence, and Culler, David. 2003. Taming the Underlying Challenges of Reliable Multihop Routing in Sensor Networks. *Proceedings of the First ACM Conference on Embedded Networked Sensor Systems (SenSys '03)*.

Ye, Wei, Heidemann, John, and Estrin, Deborah. 2002. An Energy-Efficient MAC protocol for Wireless Sensor Networks. *Proceedings of the IEEE Conference on Computer Communications (Infocom '02)*, pages 1567–1576.

Zhang, Pei, Sadler, Christopher M., Lyon, Stephen A., and Martonosi, Margaret. 2004. Hardware Design Experiences in ZebraNet. *Proceedings of the 2nd International Conference on Embedded Networked Sensor Systems (SenSys '04)*, pages 227–238.

6

Network Services for Mobile Participatory Sensing

Sasank Reddy, Deborah Estrin, and Mani Srivastava

Abstract

The rapid explosion of mobile phones over the last decade has enabled a new sensing paradigm – participatory sensing – where individuals act as sensors by using their mobile phones for data collection. Participatory sensing relies on the sensing capabilities of mobile phones, many of which have the ability to detect location, capture images and audio, the networking support provided by cellular and WiFi infrastructure, and the spatial and temporal coverage along with interpretive abilities provided by the individuals that carry and operate mobile phones. If successfully coordinated, participants involved in data collection using their mobile phones can open up new possibilities uniquely relevant to the interests of individuals, groups, and communities as they seek to understand the social and physical processes of the world around them. Responsibly realizing a vision of sensing that is widespread and participatory poses critical technology challenges. To support mobile participatory sensing applications, the future Internet architecture must provide network services that enable applications to select, task, and coordinate mobile users based on measures of coverage, capabilities, and participation and performance patterns; attestation mechanisms that enable sensor data consumers to assess trustworthiness of the data they access; and privacy and auditing mechanisms that enable sensor sources to control sharing and disclosure of data.

6.1 Mobile Participatory Sensing Vision

6.1.1 Individuals Carrying Mobile Phones as Sensors

Embedded wireless sensing provides scientists and engineers unique insights into the physical and biological processes of the natural and "built" environments. Here we consider a shift into the public realm, a move that anticipates sensing's use by the general public and suggests new possibilities for understanding social, political, or, more generally, "urban" processes. In this expanded view, sensing can serve as a technological platform for advocacy – "making a case" through distributed documentation of some need. Or it can be a tool for introspection into the habits and situations of individuals and commu-nities – self-discovery through private or social data analysis. A key distinction between this use of sensing and traditional embedded scientific applications is its reliance on individuals' participation in data collection and analysis. Perhaps more important, however, is an accompanying proliferation of purpose; that is, the applications of publicly deployed sensing actively emerge from the inter-ests of the public. Traditional approaches to networked sensing cannot achieve this because embedding the necessary sensors in real-world environments is too costly, requires broad deployments that are likely to be either aesthetically or politically unacceptable, and ultimately proves to be inflexible in the face of diverse users' needs. As an alternative to this sensing of the public, we consider sensing by the public (Burke et al. 2006; Eisenman et al. 2006; Paulos et al. 2007). We take as our starting point the cellular and WiFi networks that currently support billions of mobile phone users. Most phones are already equipped with acoustic, image, and location sensors – in the form of microphones, cameras, and GPS, WiFi, or cellular positioning – and a Bluetooth interface that can be used to connect external sensors. They also provide text and graphics entry for the manual description of events. These devices can be tools for sensing, and we focus on what it would take to establish their role in a participatory sensing network. In this context, mobile devices are the sources of digital con-tent, network services provide higher-level understanding and organization of personally contributed data streams, and mobile users are active in defining, participating in, and analyzing data from coordinated observing "campaigns." The resulting platform is parsimonious, introducing very little new equipment into the environment and requiring from participants only the data necessary to achieve the impacts they desire, and yet is uniquely able respond to diverse need and interest.

6.1.2 Types of Participatory Sensing

6.1.2.1 Authored versus Ad Hoc Data Collection

In the simplest model of participatory sensing, individual participants gather sensor data about social and environmental processes, publish, and share it in an ad hoc fashion. "On the scene" citizen reporting, like CNN's I-Report, where an individual's serendipity is an asset, is an example that has emerged in popular culture. The relatively uncoordinated nature of this approach as a sensing paradigm, however, limits its utility for campaigns that have stricter requirements in what, where, and how data should be collected. To address this deficiency, we introduce the notion of "authored" campaigns. In this model, mobile phone-based data gathering is coordinated across a potentially large number of participants over large spans of space and time. Such coordinated sensing could be initiated by individuals, groups, or institutions and might involve dynamic decisions about the data being gathered, the spatial extent and temporal frequency of sampling, and the overall level and character of the participation needed. Network services are necessary to support the critical element of human participation.

6.1.2.2 Opportunistic versus Guided Sensing

In authored data collections, the level of coordination can range from being opportunistic to being guided. In the opportunistic case, participants are involved in an autonomous manner in which the sensing on the mobile phone occurs without the participant's direct involvement. The main goal is to obtain necessary sensor values without putting a burden on the participant, and thus the system infers situations when sensing should occur and activates the appropriate sensing on behalf of the participant. Examples of opportunistic involvement include taking pictures from the camera automatically every twenty seconds while the phone is exposed externally or sampling the microphone when the phone is held. At the opposite end of the spectrum is guided sensing where network services work in coordination with the participant to inform them of specific campaign needs, such as where spatial or temporal gaps exist in the data collection. The system can provide suggestions to participants of sensing needs in the field, as well and incentivize them to fill sensing gaps. For instance, a service can provide a route plan that maximizes sensing utility or inform the user of nearby sensing opportunities as they walk through an area of interest.

6.1.3 Application Space

Participatory sensing enables data-collection campaigns that can make an impact in a wide variety of application spaces including urban planning, environmental monitoring, and cultural exploration. Here we give scenarios of how mobile phone sensing can be used for such "make a case" sensing deployments. These campaigns show the need for network services that account for individuals' geospatial coverage, availability, and reputation for delivering useful campaign data while respecting participant privacy concerns.

6.1.3.1 Truck Traffic Assessment

Our first example is inspired by T. S. Lena and colleagues' work with community documentation of diesel truck traffic in the Hunts Point peninsula of New York's South Bronx, home to a primarily low-income population and a hub in the tri-state freight transportation system (Lena et al. 2002). Consider a community with higher-than-average asthma hospitalizations and deaths, which is concerned about diesel truck exhaust in their neighborhoods, a primary source of airborne particulate matter. Documenting average diesel truck traffic counts through many streets in the neighborhood would create a valuable resources for community members to (1) assess the amount of traffic relative to zoning and regulatory requirements; (2) find unexpected "hot spots" of traffic; (3) coordinate with a university or public health organization to supplement more specialized monitoring; all to (4) generate material necessary to advocate for further study, legislation, or research. In this application, there could be many willing community participants, but with minimal free time and a need to obtain high data credibility. Coordinated, participatory sensing campaigns could be employed by the community and university to best organize their willing and intelligent human resources to capture truck traffic counts through both directed sensing ("please go to this corner and make some recordings") and participatory interaction ("you're already near a place we need data, please take a few photos or enter the number of trucks you see"). This coordination is made more challenging because the time and spatial variations in truck traffic could be initially informed, perhaps under the guidance of the participating university based on existing environmental models, traffic studies, legislation, or other data.

6.1.3.2 Citywide Resource Survey

The second campaign example is inspired by the Getty Conservation Institute's Historic Resource Survey Project [GCI08] and the USC GeoDec (Shahabi et al. 2006) group's "social image mapping" project. The Getty is collaborating with the City of Los Angeles to develop professional survey methodology to document the city's historic resources (primarily buildings). Once this is done, the city will face the challenge of actually implementing the data collection. Even though professional survey is not possible using mobile phones with untrained participants, a citywide participatory project that involves everyone in deciding what is historic and why, and building up a secondary library of media documentation, is very exciting and can be implemented. A coordinated participatory sensing campaign can contribute geo-tagged historic resource images, audio, and other data to augment the Getty's data collection. This enables never-before-possible documentation of our built environment, which involves people in decision making about what to document, when, and with what thematic keywording, but coordinated by intelligent network services.

6.1.4 Network Services for Coordination, Feedback, and Privacy

Overall, participatory sensing's challenges and promises come from the same characteristics. First, the approach's tremendous potential emerges from leveraging existing technology and infrastructure to actively consider human concerns. It is precisely this reliance on systems designed for other purposes that challenges us to create more human-aware network services to coordinate participation. Second, as coordinated and model-assisted sensing scales up, it requires network management of credibility and reputation on behalf of participants and their self-expressed goals. Finally, given a network with coordination functions, people will participate in sensing things that matter to them, but it is their intimate involvement in the sensing process that must be respected and responsibly designed for, in a secure, flexible, and transparent approach to participation, data control, and privacy regulation.

6.2 Context Inference and Coordination

For participatory sensing campaigns to succeed, network services need to exist that operate continuously on behalf of all involved in a sensing campaign to (1) identify the potential participants who are suited to the goals of the campaign; (2) negotiate with the participants the constraints on their involvement, and incentivize them to participate; (3) opportunistically exploit sensing and data-sharing opportunities that present themselves as people move around; and (4) optimize the sampling coverage while assuring credibility of sensor data and conforming to constraints negotiated with the participants. The network service architecture, shown in Figure 6.1, will embody these functions through the Recruiter, Coordinator, and Guardian modules whose designs will be impacted by various human factors. As their names suggest, the three modules represent network services that respectively select participants for a campaign, coordinate them to perform the sensing task, and monitor their performance throughout the data-collection effort. Their roles in the architecture are described in detail in the next subsection. The system has to select from and manage a diverse population of potential but uncommitted participants with different availability, mobility and activity patterns, history of participation, diligence, predisposition, skills, timeliness, phone capability, and privacy constraints. Further complexity arises because humans are self-willed, intelligent, and creative. These human factors will be captured for each participant through models of context-annotated mobility profiles, reputation, and privacy constraints.

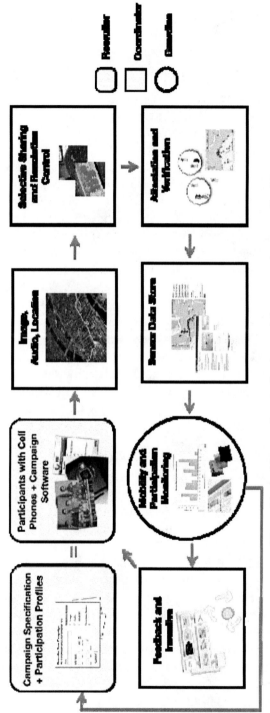

Figure 6.1. Architecture for coordinating participatory sensing data collections.

159

6.2.1 Architecture Components

6.2.1.1 Recruiter

Like many Web 2.0 applications that rely on contributions from individuals, a campaign seeks concerned participants willing to volunteer their time to help with data collection and analysis. We envision a Recruiter that takes campaign specifications and participant profiles as input and recommends potential participants, much like the friend-finding features of social networking sites. This engine should be useful for the group initiating a campaign as well as those who would like to volunteer, allowing the latter group to judge the feasibility of their participation. The campaign specifications might include the sensing modalities needed, the regions of space and time over which to conduct the campaign, the overall campaign budget (which may consider all human and material resources needed to run the campaign, not just cases where participants are compensated for their time), the demographic diversity of participants, and so on.

Multipart profile information for each potential participant is the other crucial input to the Recruiter. The profile contains information in regards to the incentives needed by the participant, along with interest vectors. Furthermore, participants are evaluated in terms of capabilities, availability, and performance. Capability information captures relevant characteristics of the cell phone carried by that individual, such as the set of sensing modalities and their quality. Availability information would indicate when and where the participant is likely able to gather and contribute data. This would include models of the participant's mobility and activity in space and time, as well as constraints due to privacy rules. Performance information would indicate how this individual performed on previous sensing campaigns in terms of metrics such as quality and timeliness of contributed data, consistency relative to their commitments, and responsiveness to data collection requests. The campaign designer could compare released performance information with what they consider minimum qualifications for their campaign.

The designer would then use participant recommendations to selectively recruit a participant list that achieves the highest utility while adhering to the campaign resource budget. In keeping with a participatory approach, at the time of recruitment, the system could negotiate with the participant a level of commitment to the campaign that would be part of the basis for valuation, incentives, and reputation in the system. At a technical level, the recruitment problem is similar to that of sensor selection in traditional embedded applications like those studied in Ganesan et al. (2004), Krause et al. (2006), and Krause et al. (2008), with the obvious distinction that these papers do not consider direct human involvement at all or only consider certain aspects of it in the measurement process. In the case of campaign recruitment, recommending participants requires the more difficult task of modeling factors tied to human behavior.

6.2.1.2 Coordinator

The Coordinator orchestrates data collection by remotely tasking and configuring participants' cell phones, keeping owners in the loop of the coordinated sensing process, and attesting the authenticity of collected data through verification techniques that check that samples represent the phenomenon that occurred at that time and space and were taken by a human (as opposed to an automated bot). Furthermore, the verification methods could be used to assign reputation scores to participants involved in the campaign. The Coordinator is supported by software that runs on the mobile phone that facilitates in-situ data collection, remote configuration, and interactive feedback (Burke et al. 2006; Froehlich et al. 2007).

One of the main objectives of the Coordinator is to promote participation. It can do this by a feedback system that is informed by persuasive computing – prompts, an incentive system, and social validation are employed (Fogg 1998). Prompts can be visual and auditory aids that remind participants to take samples. The objective is to deliver these prompts in a simple, clear, and nonobtrusive fashion based on the participants' current location and the sensing uncertainty at that location relative to campaign requirements (Intille 2004). An incentive system based on "credits" that can be redeemed for monetary rewards or for additional capabilities in the campaign system will be used to encourage high-quality participation (Pryor 2002). Credits can be removed from participants as well if they deliberately deliver wrong information (for the purpose of collusion or some other type of self-gain). Finally, the Coordinator may use social validation – the concept that people determine what is correct based on what other people consider is correct – by showing a particular participants contribution level compared to other individuals involved in the campaign system (Cialdini 2001). By providing this relative comparison, we hope to encourage individuals to compete to achieve or keep a high participation level.

The Coordinator is also involved in the verifying whether data contributed by a participant actually took place at a particular time and location and that it was contributed by a human as opposed to an automated program (bot). More details about these verification and attestation mechanisms are given in the next section.

6.2.1.3 Guardian

Working in close connection with the Coordinator module in our architecture is the campaign Guardian module. The Guardian observes overall campaign performance and how each participant is doing relative to their negotiated commitment and the campaign's needs. It must assess participant activity/mobility patterns, and their sensing performance in terms of quality and utility of their contributions. In soft real time, the Guardian provides campaign status to the Coordinator in case participants join or leave, must be added or removed, or need incentives to collect better data or data at desired locations. Over longer time scales, the Guardian updates the profiles associated with each participant based on data about their performance relative to their commitments and mobility/activity pattern during the campaign. Although performance tracking during the execution of a system is not new, monitoring and assessing data collection by humans, especially for coordination and execution purposes, is something we consider novel and challenging. This updated information provided by the Guardian can then be used by the Recruiter to adapt the participant list based on the current behavior of participants in the campaign.

6.2.2 Multipart Profiles

Multipart profile information for each potential participant is crucial to both recruitment and execution. Capability information captures relevant characteristics of the cell phone carried by that individual, such as the set of sensing modalities and their quality. Availability information would indicate when and where the user is likely able to gather and contribute data. This would include models of user's mobility and activity in space and time, as well as constraints due to privacy rules. Performance information would indicate how this individual performed on previous sensing campaigns in terms of metrics such as quality and timeliness of contributed data, consistency relative to their commitments, and responsiveness to data collection requests. Commitment weights each of these relative to the negotiation made with the participant for each campaign.

The annotated mobility profiles used to model participants' behavior over time and geography could be quite fine-grained, such as whether one is outside, walking, eating, or on the phone with a colleague. In general, annotated mobility profiles are quite difficult and inconvenient to sense, and may even require additional sensor hardware. However, even the coarse-grained macro notions of context that can be gathered with existing infrastructure are of great utility in coordinating participating sensing. Good examples are whether the participant is indoor or outdoors, and mode of transportation, whether one is stationary, walking, running, biking, or traveling in a vehicle. Figure 6.2 shows an example of such traces for an individual over a period of eight days. Different portions of each daily trace are coded in different shades to indicate the inferred activity state of the individual at various locations and times: still, walking, or in a vehicle. The activity state was inferred in real time by software running

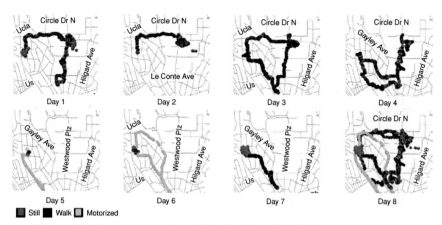

Figure 6.2. Context annotated mobility data for an individual over several days.

the GPS-equipped mobile phone and employing machine-learning algorithms. Moreover, although the traces are quite diverse, they do show several locations and paths that occur frequently, corresponding to repetitive mobility patterns that may be useful in deciding the suitability of a participant for a sensing task.

6.2.2.1 Inferring Context Using Mobile Phones

We envision a system that gathers <location, time, context> traces for participants according to their privacy constraints and then represents it as a compact evolving model used to assess how an individual could contribute to a specific campaign's needs. Location and time can be obtained via GPS embedded in many cell phones or from network infrastructure. Capturing mobility or activity context is more difficult, especially given our goal not to rely on hardware beyond the mobile phone. The challenge lies in identifying the context using sensors likely to be embedded on the cell phones, without requiring its owner to significantly change their habits of use.

Recent work has had good success in identifying significant locations, indoor versus outdoor status, and mode of transportation. For instance, density-based clustering of GPS location points has been employed with distance and time boundaries to divide mobility traces of individuals into locations and routes (Kang et al. 2005; Reddy et al. 2008a; Zhou et al. 2004). Likewise, GPS on the cell phone has been used as an indoor versus outdoor sensor by using a vector of features including the number of satellites available for GPS, geometric dilution of precision, accuracy, and speed variance. Finally, transportation mode of an individual has been inferred using both GPS and accelerometer features. Specifically, coarse-grained transportation mode classification, such as whether a user is stationary, walking, or in motorized travel, has been inferred by using only GPS data by dividing routes based on changes points (speed close to zero, loss of GPS signal) and then calculating features based on speeds for these segments, such as average, maximum, and minimum speeds, along with distance information (Zheng et al. 2008). Also, fine-grained transportation mode inference, which includes differentiating between running and biking along with other modes, is possible with high accuracy based on analyzing the GPS speed along with accelerometer features such as mean, variance, and certain frequencies of the magnitude force vector (Reddy et al. 2008b).

6.2.2.2 Mobility Profiles for Coverage Assessment

To assess a participant's suitability for the space-time coverage needs of a campaign, the context-annotated mobility information must be represented in a participant's profile in a compact manner, adaptable to variations in their behavior, and efficiently query-able by the Recruiter. Mobility modeling for coordinating participatory sensing requires predicting the statistics of movement patterns at fine granularity over longer period of times than is currently done in cellular networks. It differs from prior work on mobility models for network simulation, which focuses on generating mobility traces (Bai et al. 2003; Hong et al. 2001; Jardosh et al. 2003; Tian et al. 2002). Also, it differs from traffic aids and resource allocation for wireless hand-off systems, which focus on short-term prediction of location (Bhattacharya and Das 1999; Choi and Shin 1998; Hariharan and Toyama 2004; Krumm and Horvitz 2006; Simmons et al. 2006).

Based on both anecdotal evidence and our exploratory data gathering, we know that human mobility has common patterns (Eagle and Pentland 2006), such as repeating routes and frequent locations. But human mobility also exhibits significant temporal jitter on a day-to-day basis. For example, one can imagine taking the same route from home to work but departing at different times in the morning or going to the grocery store every week but during different days of the week. Also, schedules of individuals might change over time, and the mobility model needs to be able to adapt. College students often have dramatic schedule shifts from one semester to another, so the mobility model will need to adapt to these changes quickly. But the updating scheme should also be aware of outliers (vacations, conference visits, etc.). Figure 6.3 shows an example of mobility profiles as they change over time for an individual where both a natural variation and dramatic shift is shown. Natural variations occur as the individual visits different stores, restaurants, and so forth, whereas the dramatic shift occurs due to an event such as change in place of work or residence. The algorithms and models used for mobility profiling must capture the natural variations in mobility patterns and also be able to detect dramatic shifts to allow trigger retraining.

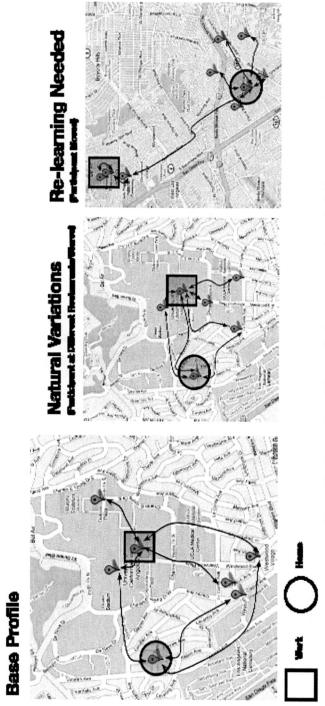

Figure 6.3. Changes in mobility profile for several weeks for an individual.

165

Overall, the models will require a mix of data mining to identify and cluster patterns and statistical representation to identify the patterns and use them as higher-level building blocks for expressing the overall mobility behavior. One such technique organizes mobility information into a profile that consists of an "association matrix" that captures the amount of time spent in a particular context during a time period. This association matrix is used to infer which individuals would be the "best fit" for coverage. Individuals' profiles can be compared over a time period for consistency by performing Singular Value Decomposition to obtain the eigenbehaviors (main column signatures in the association matrix) and then comparing consecutive time periods of eigenbehaviors by calculating similarity (Eagle and Pentland 2006; Gonzalez et al. 2008; Reddy et al. 2008a).

6.3 Data Attestation and Credibility

Participatory sensing systems must establish credibility of collected information, considering that contributions come from participants with varying skill, intent, and understanding of the campaign's needs. They must do this while allowing participants to regulate their own privacy and participation. For example, in some cases, participants may be anonymous from either the perspective of the human campaign organizers or even the system itself. To increase confidence in contributed data, the system could verify samples as (1) taken at a particular location and time, (2) capturing the phenomenon of interest that occurred, and (3) contributed by an authorized participant and not an automated bot. Like other security measures, such verification results will not be absolute, and we thus consider them as input to a participant's reputation. Even with a tamper-proof trusted platform running trusted software (Aissi et al. 2004), there is sufficient variation in each measure that verification would not be binary. We focus on the compelling and more immediately scalable scenario, in which participants can use available mobile platforms. In the following section, we describe each verification task in more detail and suggest mechanisms to achieve them.

6.3.1 Verifying Participant Context (Location)

High confidence in where and when samples were taken increases their credibility. Providing direct assurance of location and time of capture is very difficult without having trusted platform components in the mobile device (Aissi et al. 2004). We approach this problem in a simpler but more immediately practical sense by verifying the location and time of the contributor. Thus, when a sample is uploaded, the system is at least able to verify that the sample was taken at a location and a time at or before the upload time. We propose creating a location and time attestation service that a participant's mobile device can query. A related approach has been presented in (Lenders et al. 2008). Location attestation can be implemented using trusted infrastructure or with location fingerprints. In the first case, a trusted infrastructure exists, such as a WiFi network, which can be queried to attest location and time. Essentially, the participant's device sends its location and time according to its own measure, obtained either through GPS or the cellular network, to the verification service. The service compares the submitted location and time with its network-observed location using wireless signal strength and triangulation techniques (Letchner et al. 2005) and time, and sends back a certificate if it matches. This certificate can be then uploaded by the participant when data is submitted. The second method does not rely on infrastructure but instead works by having the participant's device advertise and listen for WiFi and Bluetooth beacons. It then uploads the identifiers of scanned beacons along with its location obtained through GPS or the cellular network to the verification service. As regulated by the personal privacy decisions of participants, the verification service maintains a database of all devices, their time, location, and fingerprints defined by the set of devices they see. It uses this database to compare and verify the locations of participants who enable this feature to increase the credibility of their data.

6.3.2 Verifying Validity of Sampled Information

The system could also assess whether contributed data represents an occurrence of the phenomenon of interest. For instance, in our second campaign example, if a participant contributes resource images from another city, creates a digitally altered image, or includes an image from the past as if it was the present – how can we identify such misrepresentations? To address this challenge, the system could coordinate other participants in the campaign to cross-check the validity of each other's contributions. For instance, if a person is located within a certain area near where a sample was previously contributed, the system would request them to also take a sample at that location as well. This same request could be issued to other participants involved in the campaign and used to validate the contribution. For cases where the phenomenon exhibits dynamic behavior over time, constructing such opportunistic verification requests becomes more challenging but still important given the value of additional samples in building a model that establishes the validity of the documentation.

6.3.3 Verifying Human Contributions

Verifying that a human is in the loop during data collection is crucial to the respectability of participatory systems and to the creation of an equitable and relevant reputation framework. When reputation is meant to reflect some measure of engagement, the system is open to attack by automated "spam" processes that simulate participation. To counter this, we propose an in situ challenge-response system (Naor 1997). While a participant is sampling and uploading data, the Coordinator provides a challenge that, with high probability, only a human, and not a computer process, can respond to in a timely manner. Specifically, we propose the use of a sensing CAPTCHA, the Completely Automated Public Turing Test to Tell Computers and Humans Apart, introduced by Ahn (2003). In participatory sensing, we can request tagging or annotation of other participants' samples as a challenge, and thus have the added value of "crowd-sourced" data verification as well (Gentry et al. 2005). Similarly, the challenge could use a randomized sequence of data captures done under a small set of known conditions (e.g., camera orientations, newspaper inserted into the image, simultaneous playing of audio signals that are picked up during audio capture, etc.). If the tags or data series match, then the system confirms that the participant collected the sample at the claimed place and time. In addition, a participant can be asked to rank the quality of people's samples, and in this way the system is able to score data. Finally, by sending the same challenge to several participants, the system can more reliably score and verify contributions (Chew and Tygar 2005).

6.3.4 Reputation Measure for Contributors

Beyond providing a measure of confidence in a given sample, the verification mechanisms described previously could also contribute to a measure of overall reputation for campaign participants, some of which could leverage trusted information, such as that from augmented handset hardware, without directly revealing it to the campaign organizers. The system can keep a reputation score associated with a participant based on how that participant performs as a data contributor relative to their commitments and campaign needs. Tracking user reputation is not new – in fact, it has been widely employed on the Internet to track whether to trust a user for transactions (Ebay) or in providing good input to a system (Yahoo Answers, Amazon MTurk) (Jøsang et al. 2007; Resnick et al. 2000; Resnick and Zeckhauser 2002). In the default operation of our system, providing verified data will result in a high reputation, whereas contributing samples suspected as invalid or contrived will result in a lower reputation score. The reputation score of the individual can also be associated with their sample to give data users a sense of the system's assessment of sample credibility to consider when they employ shared data. To achieve this, we face challenges that include how necessary privacy mechanisms affect verification and reputation calculations. For example, reputation may not be a scalar value but instead be a vector of performance and participation assessment metrics. Whereas performance reputation may be affected by privacy mechanisms lowering the verifiability of samples (blurring or adding noise to data, say), participation reputation would not be affected by these controls. Additionally, reputation mechanisms should be customizable by the campaign creator, given their knowledge of what success, reliability, and credibility mean for their campaign and its participants.

An additional challenge for maintaining reputation scores comes from the idea of identity. In participatory sensing, we envision a range of identity options for campaign designers and participants: For some, there is strong protection of anonymity even from the managing system; for others, pseudonyms are created for particular campaigns but a consistent identity is known by the system; and for still others, authentication is based on time spent in a location, or device identity is decoupled from user identity. Mapping reputation services to these identities is challenging. It is tempting to follow the lead of many Internet services and reward users for maintaining a single, trackable identity across campaigns. Certainly this makes reputation management more immediately effective and its corresponding credibility metrics easier to understand. But this may be counterproductive or unnecessary in many cases where other forces exist "offline" from the system to regulate use – for example, through preexisting social mechanisms for authenticating participants. In these cases, intracampaign reputation based on community norms will be sufficient. We believe that participatory sensing systems can be created with sufficiently configurability to address this variety, as long as the network services do not fix a single concept of identity.

6.4 Privacy

Privacy is a long-standing topic in mobile computing, especially with respect to the delivery of location-based services. In mobile participatory sensing, privacy becomes a first-order challenge because the sensing is enabled as a fine-grained resolution and is directly associated with the individuals performing the sampling. For instance, the data collected can be used to quantify habits, routines, associations, and the data (especially location traces) is easy to mine to obtain this personnel information. Furthermore, there are a host of negative impacts (location-based discrimination, safety and security threats) that could result if privacy is not considered seriously. The approach that the research community has taken to tackle this challenging issue has focused on two fronts: creating design principles that network services must meet to help balance data collection and privacy concerns of participating individuals; and designing system architectures that support core data services for privacy, such as audit of sampled data and the ability to enable filtering and resampling of information for sharing purposes.

6.4.1 Privacy Principles

There are many software architectures emerging to support participatory sensing in a privacy-preserving fashion. These systems typically focus their design around three underlying principles, as outlined by Shilton et al. (2009): participant primacy, data legibility, and longitudinal engagement. Participant primacy is the concept that participants should own the data they collect and have the ultimate control on how the data is used, whom the data is shared, and how long the data is retained. Services need to exist to support these privacy-based data-collection decisions. Data legibility encourages systems to create visualizations so that participants can make sound decisions about their privacy needs – specifically providing intuitive interfaces to export processing, sharing, and retention details by components interacting with the collected data. Longitudinal engagement is the ideal that systems should strive to keep participants involved in the complete data-collection cycle, from initial sampling to processing, usage, and deletion, encouraging them to be active privacy stewards of their data.

6.4.1.1 Participant Primacy

Participants should have the ultimate control over how their data is used. For this to occur, however, certain tools need to exist in regards to data transformation, storage, and access control to help participants make sound privacy-enhanced decisions (Caceres et al. 2009; Hong and Landay 2004; Shilton et al. 2009). Transformation deals with how the data should be presented to different data sinks. For instance, in the case of a location stream, the sampling frequency of provided data could be changed (instead of providing a service with a location update every second, the sampling rate provided can be changed to five minutes), or the resolution could be adjusted (the level of uncertainty could be adjusted from a few feet to miles, if necessary). Currently, many mobile systems involve data simply being sent to an end-point service and no intermediate storage. By having a tool that enables the backup of all information sent, both audits and future dissemination is possible. An end-point service might perform a certain type of inference on data, but unless the participant has access to the raw data sent initially, there is no way to check whether the inference is being performed correctly or even adheres to the statement of service. Furthermore, the data collected for one particular data collection campaign might be useful to another in the future. By having a backed-up copy, the data is available for future use. Finally, by incorporating access control as part of a toolset for data collectors, participants have fine-grained control over who gets to access their data, how long it should be retained, and whether access should be revoked. The usefulness of this access control mechanism can be seen especially when unwanted information that gets uploaded to a service. Without having access control services in place, there would be no way for the user to revoke rights to the mistakenly uploaded data.

6.4.1.2 Data Legibility

Privacy is a negotiation between participants and sensing organizers of what information to share or withhold. But in order for individuals to make sound decisions about their sharing policies, system legibility is key. They must understand who is asking for the data (identity); what the data will reveal about them (granularity); what the organizer wants to use the data for (purpose); and how long data will be retained by a requesting organizer (retention) (Reddy et al. 2008a). The system must communicate to participants the nature of the processing that is occurring with their sampled data, along with whom the data is shared with and for how long. Communication between the system and the participant is essential to the system's legibility: the ways in which a system enables people of all technical backgrounds to make informed disclosure decisions. Concerns of visibility and accessibility lead to considerations of data visualization and interpretation in relation to privacy. Work has emerged that faces the challenges associated with visualizing information obtained through data collection. Tools exist to enable "mashups" of collected data on a geo-spatial frame. The idea or providing a platform to enable "social data analysis" and "casual visualization" is becoming important (Wattenberg et al. 2007). To this end, these same visualization techniques are being employed to enhance system legibility (Mun et al. 2009).

6.4.1.3 Longitudinal Engagement

A key ideal that must exist with privacy tools for participatory sensing data collectors is continued engagement. Specifically, privacy should not be a one-time operation that occurs when data collection is first initiated, but instead should be an ongoing engagement with the participant throughout the data-collection life-cycle. Systems should be designed so that participants are reminded of their privacy settings and actively confirm that their settings are still valid. This can come in the form of regularly scheduled reminders that require active acknowledgment. For instance, FireEagle, a location-sharing service, sends out an email alert every three months to encourage participants to check their privacy settings (FireEagle 2009). This feedback can be simply alerts, as FireEagle does currently, or more detailed summaries of privacy policies that are in place. Furthermore, if changes on how data are being used occur, feedback should be given to participants so that they can make a sound, timely decision on whether their data should continue to be exported or if changes need to be made to the resolution or retention policies associated with the data.

6.4.2 Personal Data Vault

Several research groups have been working on experimental architectures to enable the privacy principles to be enacted in implementation form. Most of these architectures revolve around the idea of a personal data vault (PDV) that acts as a proxy for the participant in interacting with various applications that can use the data being collected. This PDV is designed to have a number of intelligent features including the ability to perform backup and republishing, enable access control and auditing of data usage, and perform adaptive filters on the actual data in terms of resolution control and sampling frequency.

6.4.2.1 Storage and Republishing

Similar to existing backup services that exist for personal computers, one essential task that a PDV can perform is the backup of all data sent to it. Furthermore, since backups exist, the PDV can also act as an intelligent republishing tool so that information can be disseminated to other applications both in real time and in a delayed fashion. The storage of data is important because the ecosystem for end-point services is changing rapidly, and a service that is popular today might not be the one used in the future (Caceres et al. 2009). Easily being able to take existing data that was backed up and sent to another service at a future time is invaluable. Furthermore, the republishing strategy helps with efficiency on the phone. If a participant is running multiple data-collection campaigns, then it is more energy-efficient to send the data to a PDV and have it republish to other services on behalf of the participant (Caceres et al. 2009).

6.4.2.2 Access Control of Data Usage and Audit Trails

Another important feature of the PDV is the ability to perform access control and audit the usage of data. Access control works similar to current systems that exist on the desktop for sharing files with various individuals and groups. But the PDV would also incorporate features specifically available on the mobile phone, such as context information. Participants might set access control policies for data based on location (data is shared only in certain zip codes or other spatial regions), time (information is collected during certain parts of the day only), and activity (data is collected only if the user is performing a certain transportation mode) (Shilton et al. 2009). In addition to this access control mechanism, the PDV would also incorporate a trace audit tool that records access, use, inference, and manipulation of data by corresponding end-point services. Having this ability to audit data usage would require external services to log transformations and sharing back to the PDV, and a signing system could exist to verify that certain services perform their advertised tasks (Shilton et al. 2009).

6.4.3 Resolution Control and Resampling

In accordance with the ideal of participant primacy, the PDV should provide tools to enable participants the ability to have fine-grained control of the resolution and sampling exposed to external services. Although this can apply to any modality, one in which this is especially important is location data. Instead of streaming the raw location field at the finest-grained level to external applications, the participant can instruct the PDV to share certain resolution levels to particular end points. For instance, the resolution can range from having a resolution of a few meters to one that is at a zip code or city level (Parker et al. 2006). In addition to resolution control, the PDV can also be used to change the sampling rate associated with sensor readings. Even if the mobile phone is publishing information at a very high rate (i.e., location updates every second), the PDV can instead share this data at a lower sample rate (every five minutes, one hour, or more) (Caceres et al. 2009; Parker et al. 2006). Furthermore, the PDV can delay when the data is actually sent to external applications as well, enabling a "lag" between when data is collected and when it can be used for external inference. This lag can also be used as a buffer for participants to make sharing decisions.

6.5 Implications for the Future Internet

The creation over the past decade of unanticipated applications of the Internet, such as web services, peer-to-peer (P2P) file sharing, networked gaming, IP telephony, and mobile application, has motivated researchers to revisit the core Internet infrastructure and the original architecture choices. The emerging class of mobile participatory sensing applications described in this chapter carries similarly significant implications for the Internet. While many prototypes of mobile participatory sensing applications are being realized over the current Internet infrastructure, experience also suggests that for these applications to scale, certain essential services will need to be incorporated in the fabric of the Internet.

The primary impact of mobile participatory sensing applications on the Internet architecture is not at the lower-layer protocols for routing, transport, and so on. Rather, these applications motivate the need for the network to provide primitives for privacy-aware sharing of personal sensory data, and for handling of certain physical context as a first-class entity.

Sharing of personal sensory data poses conflicting demands from producers and consumers. To the former, the network has to provide control over the quality of information disclosed to different consumers. To the latter, the network has to provide information attributes permitting its quality, provenance, and overall trustworthiness to be assessed. Doing these would require automated and cryptographically secure components in the network.

The handling of physical context as a first-class entity by the network would be limited to contextual information that has universal use, namely location, direction, and speed. Beside the need to formalize representation and dissemination of such information, the challenge is in ensuring that the information is trustworthy and that the client is also provided with an assessment of its quality.

Most techniques to estimate physical context are prone to cheating and adversarial manipulation, and network can take proactive measures to verify context information.

We anticipate the emergence of specialized mediator entities providing these context-handling and data-sharing services as becoming an integral part of the network fabric.

6.6 Conclusions

The challenge and the promise of participatory sensing come from the same characteristics. First, the systems' tremendous potential emerges from the use of existing mobile phone technology and cellular wireless infrastructure. But it is precisely this reliance on systems designed for other purposes that challenges us to create network services to coordinate participation. Second, this coordinated sensing scales down as well as up. It can bring value to even a few people, but increases in accuracy, scope, and worth as more participate – as long as credibility and reputation can be managed. Finally, having technology and coordination, people will participate in top-down, bottom-up, and personally reflective sensing about things that matter to them, but it is their intimate involvement in the sensing process that must be respected and responsibly designed for in a secure, flexible, and transparent approach to participation, data control, and privacy regulation. The future Internet can support such applications at large scale by incorporating as an integral part of its fabric certain critical services, such as sharing of data streams while ensuring trustworthiness and respecting privacy, and first-class handling of verifiable contextual information.

6.7 Acknowledgments

The work described in this chapter is part of a collective effort by several members of the Center for Embedded Networked Sensing, including Jeff Burke, Mark Hansen, Min Mun, Vids Samanta, and Katie Shilton. This research is funded by the NSF under grant CNS-0627084 and by the Center for Embedded Networked Sensing. Any opinions, findings, and conclusions or recommendations expressed in this material are those of the author(s) and do not necessarily reflect the views of the funding agencies.

References

Ahn, L. V., Blum, M., Hopper, N. J., and Langford, J. 2003. CAPTCHA: Using Hard AI Problems for Security. *Advances in Cryptology – Eurocrypt*, pages 294–311.
Aissi, S., Maruyama, H., Miura, F., Nakamura, T., Saito, D., Takeshita, A., Wheeler, D. and Yoshihama, S. 2004. Trusted Mobile Platform Protocol Specification. *OASIS*. http://xml.coverpages.org/TMP-ProtocolV10.pdf

Bai, F., Narayanan, S., and Helmy, A. 2003. IMPORTANT: A Framework to Systematically Analyze the Impact of Mobility on Performance of Routing Protocols for Ad hoc Networks. *INFOCOM*.

Bhattacharya, A., and Das, S. K. 1999. LeZi-Update: An Information-Theoretic Approach to Track Mobile Users in PCS Networks. *Mobile Computing and Networking (Mobicom)*.

Burke, J., Estrin, D., Hansen, M., Parker, A., Ramanathan, N., Reddy, S., and Srivastava, M. B. 2006. Participatory Sensing. ACM SenSys Workshop on World-Sensor-Web (WSW'2006).

Caceres, R., Cox, L., Lim, H., Shakimov, A., and Varshavsky, A. 2009. Virtual Individual Servers as Privacy-Preserving Proxies for Mobile Devices. *ACM SIGCOMM Workshop on Networking, Systems, and Applications on Mobile Handhelds (MobiHeld)*, pages 37–42.

Cialdini, R. 2001. *Influence: Science and Practice*. Allyn & Bacon.

Chew, M., and Tygar, J. 2005. Collaborative Filtering CAPTCHAs. *Proceedings of the Second International Workshop in Human Interactive Proofs*, pages 66–81.

Choi, S., and Shin, K. G. 1998. Predictive and Adaptive Bandwidth Reservation for Hand-Offs in QoS-Sensitive Cellular Networks. *SIGCOMM Comput. Commun. Rev.* 28(4).

Eagle, N., and Pentland, A. 2006. Reality Mining: Sensing Complex Social Systems. *Personal Ubiquitous Comput.*, pp. 255–268.

Eisenman, S. B., Lane, N. D., Miluzzo, E., Peterson, R. A., Ahn, G. S., and Campbell, A. T. 2006. MetroSense Project: People-Centric Sensing at Scale, *ACM SenSys Workshop on World-Sensor-Web (WSW'2006)*.

FireEagle. 2009. Yahoo, http://fireeagle.com.

Fogg, B. J. 1998. Persuasive Computer: Perspectives and Research Directions. *Conference on Human Factors in Computing Systems (SIGCHI)*.

Froehlich, J., Chen, M. Y., Consolvo, S., Harrison, B., and Landay, J. A. 2007. MyExperience: A System for In Situ Tracing and Capturing of User Feedback on Mobile Phones. *Proceedings of the 5th International Conference on Mobile Systems, Applications and Services (Mobisys)*.

Ganesan, D., Cristescu, R., and Beferull-Lozano, B. 2004. Power-Efficient Sensor Placement and Transmission Structure for Data Gathering under Distortion Constraints. *Information Processing in Sensor Networks, Third International Symposium*, pages 142–150.

Gentry, C., Ramzan, Z., and Stubblebine, S. 2005. Secure Distributed Human Computation. *Proceedings of the 6th ACM Conference on Electronic Commerce*, pages 155–164.

Getty Conservation Institute. 2008. *Los Angeles Historic Resource Survey*, http://www.getty.edu/conservation/field_projects/lasurvey/

Gonzalez, M. C., Hidalgo, C. A., and Barabasi, A. L. 2008. Understanding Individual Human Mobility Patterns. *Nature*, 453, 779–782.

Hariharan, R., and Toyama, K. 2004. Project Lachesis: Parsing and Modeling Location Histories. *Geographic Information Science*.

Hong, J. I., and Landay, J. A. 2004. An Architecture for Privacy-Sensitive Ubiquitous Computing. *Conference on Mobile Systems, Applications, and Services (Mobisys)*.

Hong, X., Kwon, T. J., Gerla, M., Gu, D. L., and Pei, G. 2001. A Mobility Framework for Ad Hoc Wireless Networks. *Proceedings of the Second International Conference on Mobile Data Management (MDM)*, pages 185–196.

Intille, S. S. 2004. A New Research Challenge: Persuasive Technology to Motivate Healthy Aging. *Information Technology in Biomedicine, IEEE Transactions*, 8(3), 235–237.

Jardosh, A., Belding-Rover, E. M., Almeroth, K. C., Suri, S., 2003. Towards Realistic Mobility Models for Mobile Ad hoc Networks. *Proceedings of Mobile Computing and Networking (Mobicom)*, pages 217–229.

Jøsang, A., Ismail, R., and Boyd, C. 2007. A Survey of Trust and Reputation Systems for Online Service Provision. *Decis. Support Systems*, 43(2), 618–644.

Kang, J., Welbourne, W., Stewart, B., and Borriello, G. 2005. Extracting Places from Traces of Locations. *Mobile Computing and Communications Review*.

Krause, A., Guestrin, C., Gupta, A., and Kleinberg, J. 2006. Near-Optimal Sensor Placements: Maximizing Information while Minimizing Communication Cost. *Information Processing in Sensor Networks*.

Krause, A., Horvitz, E., Kansal, A., and Zhao, F. 2008. Toward Community Sensing. *Proc. of Information Processing in Sensor Networks (IPSN)*.

Krumm, J., and Horvitz, E., 2006. Predestination: Inferring Destinations from Partial Trajectories. *Proceedings of Ubiquitous Computing (Ubicomp)*, pages 243–260.

Lena, T. S., Ochieng, V., Carter, M., Holguin-Veras, J., and Kinney, P. L. 2002. Elemental Carbon and PM2.5 in an Urban Community Heavily Impacted by Diesel Truck Traffic. *Environmental Health Perspectives*, 110(10), 1009–1015.

Lenders, V., Koukoumidis, E., Zhang, P., and Martonosi, M. 2008. Location-Based Trust for Mobile User-Generated Contents: Applications, Challenges and Implementations. *Proceedings of the 9th IEEE Workshop on Mobile Computing Systems and Applications (HotMobile 2008)*, pages 60–64.

Letchner, J., Fox, D., and LaMarca, A. 2005. Large-Scale Localization from Wireless Signal Strength. *Proc. of the National Conference on Artificial Intelligence (AAAI)*, pages 15–20.

Mun, M., Reddy, S., Shilton, K., Yau, N., Boda, P., Burke, J., Estrin, D., Hansen, M, Howard, E., and West, R. 2009. PEIR, the Personal Environmental Impact Report, as a Platform for Participatory Sensing Systems Research. *Conference on Mobile Systems, Applications and Services (Mobisys)*, pages 55–68.

Naor, M. 1997. Verification of a Human in the Loop or Identification via the Turing Test. Unpublished Manuscript. http://www.wisdom.weizmann.ac.il/~naor/PAPERS/human.ps

Parker, A., Reddy, S., Schmid, T., Chang, K., Saurabh, G., Srivastava, M., Hansen, M., Burke, J., Estrin, D., Allman, M., and Paxon, V. 2006. Network System Challenges in Selective Sharing and Verification for Personal, Social, and Urban-Scale Sensing Applications. *IEEE Workshop on Mobile Computing Systems and Applications (HotMobile)*, pages 37–42.

Paulos, E., Honicky, R., and Goodman, E. 2007. "Sensing Atmosphere," Workshop on Sensing on Everyday Mobile Phones. *ACM Conference on Embedded Networked Sensor Systems (SenSys 2007)*.

Pryor, K. 2002. Don't Shoot the Dog!: The New Art of Teaching and Training. *Interpet*.

Reddy S., Burke, J., Estrin, D., Hansen, M., and Srivastava, M. 2008a. Determining Transportation Modes on Mobile Devices. *IEEE International Symposium on Wearable Computing (ISWC)*.

Reddy S., Shilton, K., Burke, J., Estrin, D., Hansen, M., and Srivastava, M. 2008b. Using Context Annotated Mobility Profiles to Recruit Data Collectors in Participatory Sensing. *International Symposium on Location and Context Awareness (LoCA)*.

Resnick, P., Kuwabara, K., Zeckhauser, R., and Friedman, E. 2000. Reputation Systems. *Communications of the ACM*, 43(12), 45–48.

Resnick, P., and Zeckhauser, R. 2002. Trust Among Strangers in Internet Transactions: Empirical Analysis of eBay Reputation System. *The Economics of the Internet and E-Commerce*, 11, 127–157.

Shahabi, C., Yao-Yi Chiang, Chung, K., Kai-Chen Huang, Khoshgozaran-Haghighi, J., Knoblock, C., Sung Chun Lee, Neumann, U., Nevatia, R., Rihan, A., Thakkar, S., and You, S. 2006. Geodec: Enabling Geospatial Decision Making. *Multimedia and Expo, 2006 IEEE International Conference*, pages 93–96.

Shilton, K., Burke, J., Estrin, D., Hansen, M., Govindan, R., and Kang, J. 2009. Designing the Personal Data Stream: Enabling Participatory Privacy in Mobile Personal Sensing. *Conference on Communication, Information and Internet Policy (TPRC)*.

Simmons, R., Browning, B., Yilu Zhang, and Sadekar, V. 2006. Learning to Predict Driver Route and Destination Intent. *Intelligent Transportation Systems Conference (ITSC)*, pages 127–132.

Tian, J., Hahner, J., Becker, C., Stepanov, I., Rothermel, K. 2002. Graph-based Mobility Model for Mobile Ad Hoc Network Simulation. *Proceedings of the 35th Annual Simulation Symposium,* pages 337–344.

Wattenberg, M., J. Kriss, J., and McKeon, M. 2007. ManyEyes: A Site for Visualization at Internet Scale. *IEEE Transactions on Visualization and Computer Graphics.*

Zheng, Y., Liu, L., Wang, L., and Xie, X., 2008. Learning Transportation Mode from Raw GPS Data for Geographic Applications on the Web. *ACM WWW Conference.*

Zhou, C., Frankowski, D., Ludford, P., Shekhar, S., and Terveen, L. 2004. Discovering Personal Gazetters: An Interactive Clustering Approach. *ACM International Conference on Advances in Geographic Information Systems (GIS).*

7

Supporting Cognitive Radio Network Protocols on Software-Defined Radios

George Nychis, Srinivasan Seshan, and Peter Steenkiste

7.1 Introduction

Over the past few years, an increasingly diverse and ever-changing wireless spectrum has created a need for cognitive radio networks. Such networks leverage spectrum sensing and information from each layer in the protocol stack to overcome spectrum diversity by adapting all layers (e.g., the MAC and PHY) on the fly. By doing so, cognitive radios can achieve the greatest level of performance, given the current networking conditions. For example, in areas where access to the spectrum is highly contended, the radio can switch from using a carrier sense multiple access (CSMA) MAC protocol, to a time division multiple access protocol that reduces overhead in accessing the spectrum to increase capacity and reduce collisions. Despite the increased recent activity in cognitive radio networks, supporting the development of protocols at the MAC and PHY layers, as well as cross-layer optimizations for such networks, has been extremely challenging. Commodity wireless hardware does not facilitate such development, because the majority of MAC functionality is placed on the network interface (NIC) hardware, where programmability is limited and access to the software that runs on the NIC is often restricted.

The limited programmability of wireless NICs makes Software-Defined Radios (SDRs) an attractive alternative for building cognitive radio network protocols. SDRs implement the majority of functionality, including the physical and link layers, in software running on commodity hardware, making all layers of the protocol stack easy to modify. The SDR hardware simply translates the signals between the RF and digital domains, and the software does the majority of the processing. The processing of the digitized samples in SDR architectures[5,8,14,16,17] is commonly distributed across various processing units – including FPGAs and CPUs located on the SDR device and the host machine. Unfortunately, the high degree of flexibility offered by SDRs does not automatically lead to flexibility in cognitive radio network protocol implementations. The heterogeneous processing units and interconnecting buses in such architectures often contribute large latencies and jitter in processing. These latencies can severely cripple the ability of the MAC layer, for example, to effectively respond to channel conditions, time transmissions, and communicate in a timely manner, which reduces the performance of the radio. These are important functions in a cognitive radio network, because being able to respond timely to spectrum conditions is the essence of a cognitive protocol. Placing the functionality on the SDR radio hardware to avoid these latencies would again make it difficult to develop, which is what made SDRs an attractive platform over NICs.

In this chapter, we explore an API enabled through the addition of a control channel and metadata that enables rich information and control between the radio hardware and the host, allowing adaptation at all layers on a per-packet basis. Additionally, we present a novel *split-functionality* approach to implementing core cognitive radio network functions that enables high-performance MAC (a common layer for adaptation) and cross-layer implementations. In this approach, a part of the cognitive radio function is placed on the SDR radio hardware for performance reasons, and a part on the host CPU to maintain easy customization of the protocol (at any layer) and radio. A set of novel techniques is presented to achieve the split of the functions, and how to properly distribute the functionality between the processing units on the hardware. Finally, we present a design and implementation of a cognitive switching layer that would allow control of the radio, such as the MAC protocol it is running, in the future Internet. Such a layer could be accessible by the future Internet through a global controller, or allow for local coordination within a single LAN.

This chapter makes the following contributions: Given that cognitive radio networks strongly leverage adaptation at the MAC layer, we place a major focus of our work at this layer. We identify a set of core protocol functions, from which many MAC and cognitive protocol layers are built, as well as cross-layer implementations that must be implemented close to the radio for performance and efficiency reasons. We define a *split-functionality* architecture that allows the functions to be implemented near the radio hardware while maintaining control on the host CPU through an API. We present an implementation of our architecture using the GNU Radio[5] and USRP[14] SDR platform. Using our implementation, we characterize the performance-flexibility tradeoffs for key protocol features. For example, our results show three orders of magnitude greater response time of the radio to spectrum conditions. Finally, we use our implementation for an end-to-end evaluation of the split-functionality architecture. We show how the system can support high-performance cognitive network protocols by first implementing 802.11-like and Bluetooth-like protocols for experimentation over the air, and then a cognitive protocol that can switch between the two MAC protocols based on current network conditions. The rest of the chapter is organized as

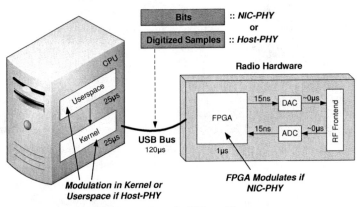

Figure 7.1. Generic SDR architecture.

follows. We discuss current radio architecture and its impact on MAC protocol development in Section 7.2. In Sections 7.3 and 7.4, we explore the core wireless MAC and cognitive protocol requirements and introduce our *split-functionality* architecture. Section 7.5 provides details for each component implementation with evaluation results. Finally, we present end-to-end evaluation results, related work, and a summary of our results in Sections 7.6 through 7.8.

7.2 Software-Defined Radio Architecture and Challenges

Software-defined radio architecture varies in the exact nature of the processing units and interconnecting buses; however, a common architecture which will be the focus of this work is shown in Figure 7.1. The frontend is responsible for converting the signal between the RF domain and an intermediate frequency, and the A/D and D/A components convert the signal between the analog and the digital domain. Physical and higher-layer processing of the digitized signal are executed on one or more processing units. Typically, there is at least an FPGA or DSP close to the frontend. The frontend, D/A, A/D, and FPGA are usually placed on a network card that is connected to the host CPU by a standard bus. In the next section, we quantify the delays between each of these components. In a *Host-PHY* architecture, the majority of the signal processing (e.g., modulation) would be done on the host-machine in userspace or in the kernel. On ther other hand, in a *NIC-PHY* architecture, this signal processing is done on the FPGA or a similar processing unit on the radio hardware.

Unfortunately, SDRs have fallen short in providing rich physical layer information to the protocol stack and have failed to provide high-performance flexible protocol (e.g., MAC and cross-layer) implementations. Functionality implemented on the radio hardware (e.g., modulation) will have good performance but lack flexibility and will be hard to modify. We refer to this architecture as

NIC-PHY, exemplified by WARP.[17] The opposite is true of functionality in a *host-PHY* architecture where the functionality is placed on the host CPU (e.g., GNU Radio[5] and the USRP[14]). A *host-PHY* architecture has been considered incapable of supporting even core protocol techniques (e.g., carrier sense) due to the large processing delays inherent to the architecture.[12,15] However, the goal of our work is to support high-performance flexible protocol implementations in a *host-PHY* architecture to enable many MAC protocols, cognitive protocol techniques, and cross-layer optimizations such as those proposed between the MAC and PHY layers.[4,6,7] Given that cognitive radio network protocols often heavily adapt at the MAC layer, we focus the majority of our work on optimizing performance and development at this layer while increasing the reactiveness of the radio to the spectrum.

In the next section, we explore delay and jitter measurements in the *host-PHY* architecture, which are the major limiting factor on the performance of MAC implementations[12,15] and the responsiveness of the radio to cognitive protocol techniques. By understanding the sources of the delay and quantifying them, we can explore a *split-functionality* approach (Section 7.4) that places pieces of techniques (e.g., carrier sense) before specific sources of delay or jitter to achieve greater performance. Therefore, it is important to not only understand the total delay in the system, but the delay between each major component in Figure 7.1.

7.2.1 Delay in Software-Defined Radios

In this section, we explore sources of delay in a *host-PHY* architecture for the purpose of understanding why MAC and cognitive network protocol implementations have suffered in performance, such that we can overcome the delay. We use GNU Radio and the USRP for this work, which[12] present as a delay measurement for the platform; however, they focus on user-level measurements, largely ignoring precise measurement of delays between the kernel and userspace, and kernel and the radio hardware. Such measurements are important because they can provide insight into whether implementing protocol techniques in the kernel is sufficient to overcome the performance problems associated with user level implementations.

To obtain precise user and kernel-level measurements, we modified the Linux kernel to record nanosecond precision timestamps on the USB data between the host and radio hardware at various times in the transmission and receive process. All user-level timestamps are taken in user space right before data is submitted to the kernel, or when the data is first read in user space. To measure as close to the USB bus as possible, timestamps in the kernel are recorded at the last point in the kernel's USB driver before the DMA write request is generated, or after a DMA read request. To measure the roundtrip time between GNU Radio

Table 7.1. *SDR* host-PHY *Architectural Delay Measurements*

Units: μs	Avg	SDev	Min	Max
User→Kernel	24	10	22	213
Kernel→User	27	89	13	7000
4096 Kernel→FPGA	291	62	204	360
512 Kernel→FPGA	148	35	90	193
GNU Radio→FPGA	612	789	289	9000

(in user space) and the FPGA, we introduce a ping command on a control channel that we implement (Section 7.4.2). Using the measurements described earlier, we are also able to identify the sources of the delay by calculating the user-to-kernel space delay, kernel-to-user space delay, and roundtrip time between the kernel and FPGA based on ping. We ran the user process at the highest priority to minimize scheduling delay. We used the default 4,096 byte USB transfer block size for all experiments, and then perform an additional kernel to FPGA RTT experiment using a 512 byte transfer block size, the minimum possible, in an attempt to minimize queueing delay.

Averaged over 1,000 experiments, the delay results are presented in Table 7.1. The results show that the roundtrip time is dominated by the kernel-FPGA roundtrip time (291 out of 612 μs), whereas the user-kernel and kernel-user times are relatively modest (24 and 27 μs). The remaining time (270 μs) is spent in the GNU Radio chain. The high latency of the kernel-FPGA roundtrip time is somewhat surprising, given that the effective measured rate of the USB with the USRP is 32 MB/s. Focusing on the latencies between 4KB and 512B, the difference is only a factor of two, suggesting that the setup cost for transfers contributes significantly to the delay. The kernel-FPGA time also includes the time it takes for the data to pass through the USRP USB FX2 controller buffers, and to be copied into the FPGA for parsing. The time taken for the data to pass through the USRP USB FX2 controller buffers and copied into the FPGA for parsing also contributes to the kernel-FPGA RTT. The standard deviations and the min/max values show that the user-kernel and kernel-FPGA times are not highly variable, therefore contributing only a small amount of jitter. On the other hand, the kernel-user times are extremely variable, resulting in a high standard deviation for the GNU Radio ping delays. This is clearly the result of process scheduling.

7.2.2 Implications of SDR Latency on Cognitive Protocol Implementations

The delays shown in Section 7.2.1 have strong implications on cognitive protocol development, especially on protocols running at the host. Although the host CPU is easy to program, the significant delay and jitter shown between the radio hardware and host CPU will impact a host-based cognitive radio network protocol's ability to react quickly to the spectrum, and for the MAC layer to precisely control packet timing, or implement small, precise interframe spacings. We conclude that *time-critical radio or MAC functions should not be placed on the host CPU.* On the other hand, processing close to the radio hardware on the FPGA has the opposite properties, making it attractive for implementing delay-sensitive functions and adapting quickly to the spectrum. Unfortunately, code running on the radio hardware is often closed-source and much harder to change because it is often hardware-specific and requires a more complex development environment. Therefore, we conclude that *in order to be widely applicable, the control of flexible MAC implementations and cognitive techniques should reside on the host.*

When distributing functionality between the host and radio hardware, three key properties of the SDR will be affected: network performance, flexibility in protocol implementations, and reprogrammability. Unfortunately, as discussed earlier, these properties are in conflict with each other and achieving the highest level for each is not possible. In this chapter, we present a split-functionality architecture that implements part of key protocol functions on the radio hardware for performance, and an additional part on the host that provides full control. As we will show, this allows us to simultaneously score very high on all four metrics, and it also allows developers and users to make tradeoffs across the metrics. Even though developers will always have to make tradeoffs, the negatives associated with specific design choices are significantly reduced in our design. Note that this does not imply that our design can support any arbitrary or even all existing MAC designs and cognitive network techniques. However, we believe that it is capable of supporting most of the critical features of modern designs that can be quickly adapted for cognitive techniques using a control channel we introduce. Therefore, we must first identify core functions that the design must support high performance and flexible implementations of, from which modern MAC designs and cognitive techniques can be built.

7.3 Core Cognitive Radio and MAC Functions

An ideal platform for cognitive radio network development, with a focus on a highly adaptable MAC layer, should support well-known MAC protocols, novel designs, and various cognitive techniques. A study of current wireless protocols including WiFi (both Distributed and Point Coordination Function), Zigbee, Bluetooth, and various research protocols shows that they are based on a common, core set of techniques such as contention-based access (CSMA), TDMA, CDMA, and polling. In this section, we identify the subset of functions that a platform must implement efficiently in order to support a wide range of MAC protocols and cognitive radio techniques. In further sections, we focus on splitting these core functions in the architecture in such a way that we achieve high performance of each, while maintaining flexibility for development and fine-grained control over the functions to adapt the radio to the spectrum for cognitive techniques.

- **Precise Scheduling in Time**: TDMA-based protocols require precise scheduling to ensure that transmissions occur during time slots. Imprecise timing can be tolerated by using long guard periods; however, this degrades performance. Surprisingly, modern contention-based protocols also require precise scheduling to implement interframe spacing (i.e., DIFS, SIFS, PIFS), contention windows, back-off periods, and so on.
- **Spectrum Sensing/Carrier Sense**: Contention-based and cognitive radio network protocols often use spectrum sensing and carrier sense to detect other transmissions and available spectrum. Carrier sense may use simple power detection (e.g., using signal strength) or may use actual bit decoding. Network interfaces need to transmit shortly after the channel is detected to be idle. Additional delay increases both the frequency of collision and also the minimum packet size required by the network.
- **Backoff**: When a transmission fails in a contention-based protocol, a backoff mechanism is used to reschedule the transmission under the assumption that the loss was caused by a collision. Backoff is related to precise scheduling, but focuses more closely on fast-rescheduling of a transmission without the full packet transmission process (e.g., modulation).
- **Fast Packet Recognition**: Many MAC performance optimizations could use the ability to quickly detect an incoming packet and identify that it is relevant to the local node in a timely and accurate manner. For example, detecting and identifying an incoming packet before the demodulation procedure can reduce resource use on the processing units and on the bus.
- **Dependent Packets**: Dependent packets are explicit responses to received packets. A typical example is control packets that are associated with data packets, for example, for error control (e.g., ACKs) or for improved channel access (e.g., RTS/CTS). Network interfaces need to generate these packets quickly and transmit them with precise time scheduling relative to the previous packet.
- **Fine-grained Radio Control**: Cognitive radio networks need to adapt to the spectrum quickly, therefore the radio should also be able to switch and adapt all layers on the fly in a timely manner. Frequency-hopping spread spectrum protocols such as Bluetooth and the recently proposed MAXchop algorithm[9] require fine-grained radio control to rapidly change channels according to a pseudorandom sequence. Recent designs[1] for minimizing interference require the ability to control transmission power on a per-packet basis.

- **Access to physical layer information**: Many MAC protocol optimizations and cognitive network techniques could benefit from access to radio-level packet information. Examples include using a received signal strength indicator (RSSI) to improve access point handoff decisions or to locate unused spectrum, and using information on the confidence of each decoded bit to implement partial packet recovery.[6]

It is difficult to argue that this (or any) list of core functions is the correct one and is complete, but we believe that it is sufficient to implement a broad range of interesting MAC protocols and cognitive radio techniques. To provide some degree of confidence in this statement, we describe our implementation of an 802.11-like CSMA protocol and a Bluetooth-like TDMA protocol using our framework in Section 7.6, as well as a cognitive technique that switches between the two layers. As such, this is a reasonable first "toolbox" that protocol developers can extend over time.

7.4 Split Functionality Architecture

Having derived a set of core functions in Section 7.3, we can now determine the types of delay that can affect the performance of each function and discuss how to overcome them. For example, most cognitive radio network protocols need spectrum sensing and need to react quickly to the spectrum; however, the delays inherent in a host-based implementation in the given SDR architecture would make these functions inefficient or ineffective. We first introduce limitations that prevent high-performance implementations of the core functions, and then discuss how to overcome these limitations.

7.4.1 SDR Architectural Limitations

Enabling high-performance implementations of the core functions from Section 7.3 is prevented by three major factors in SDR architecture:

- **Bus delay:** A constant delay introduced by bus transmission is relatively easy to accommodate in supporting *precision scheduling*. However, large delays impact *spectrum sensing and carrier sense*, *dependent packets*, and *fast packet recognition*, as they require information, which is significantly delayed, to perform some task.
- **Queuing delay:** Although queueing delay can be smaller than the bus-transmission delay, it increases the amount of jitter in the system, which makes precision scheduling difficult, if not impossible, at the microsecond level (common in current protocols). In related work,[11] it is shown that this compression can be so significant in the given architecture that spacing

transmissions by less than 1 ms cannot be achieved reliably using host-CPU based scheduling.

- **Stream-based architecture of SDRs**: The frontend operates on streams of samples that can make *fine-grained radio control* and *access to physical layer information* from the host ineffective. The reason is that it adds complexity to the interaction between a MAC layer executing on a host CPU (or NIC CPU) and the radio frontend, because it is difficult to associate control information or radio information with particular groups of samples (e.g., those belonging to a packet). This problem consists of two components: (1) how to propagate information within the software environment that performs physical and MAC layer processing; and (2) how to propagate the information between the host and the frontend, across the bus and SDR hardware. This first issue is being addressed in the GNU Radio design with the introduction of m-blocks,[2] which is briefly discussed in Section 7.7, but we must address the second issue.

7.4.2 Overcoming the Limitations

We now present an architecture that overcomes the above limitations. The goal is to allow as much of the protocol to execute on the host as possible to achieve the flexibility and ease of development goals, both of which are important to a wireless platform for protocol development, as identified in Section 7.2. However, we must ensure that the high latency and jitter between the host and radio frontend does not result in poor performance and limited control, the other two criteria in Section 7.2. This is done by introducing two architectural features, *per-block meta-data* and a *control channel*, shown in Figure 7.2. The novelty is not in the two new architectural features, but in how we use them to implement the core MAC functions (Section 7.3) in such a way that we maintain flexibility while increasing performance (Section 7.5). We first discuss both features in more detail.

Per-block meta-data: Enabling the association of information with a packet is crucial to the support of nearly all of the core requirements in Section 7.3. Each packet is modulated into blocks of samples, for which we introduce per-block meta-data. The meta-data stored in the header includes a timestamp (inbound and outbound), a channel flag (data/control), a payload length, and single-bit flags to mark events such as overrun, underrun, or to request specific functions that we implement on the radio hardware. We limit the scope of the meta-data to the minimum needed to support the core requirements, thus minimizing the overhead on the bus.

Control channel: The *control channel* allows us to implement a rich API between the host and radio hardware and allows for less frequent information to

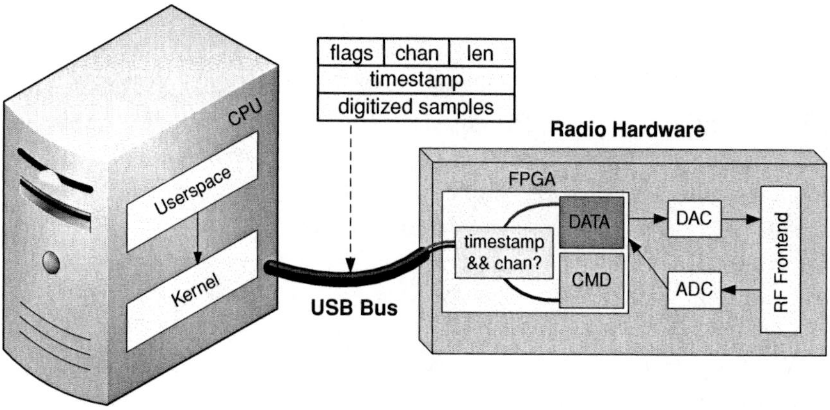

Figure 7.2. Split SDR architecture.

be passed. It consists of control blocks that are interleaved with the data blocks over the same bus. Control blocks carry the same meta-data header as data blocks but have the channel field in the header set to *CONTROL*. The control block payload contains one or more command subblocks. Each subblock specifies the command type, the length of the subblock, and information relevant to the specific command (e.g., a register number). Examples of commands include reading or writing configuration registers on the SDR device, changing the carrier frequency, and setting the signal sampling rate.

With these two features, we can effectively partition the core functions into a part that runs on the radio hardware close to the radio frontend, and a control part that runs on the host. Of course, meta-data and control channels are used in many contexts. The contribution lies in how we use them to partition the core functions, which is the focus of the next section.

7.5 Evaluating the Split-Functionality Approach for Cognitive Radio Networks

We now examine how the split-functionality approach can be used to implement the core functions described in Section 7.3, and just as importantly, how the split-functionality architecture can enable protocols that can react more quickly to the spectrum without sacrificing flexibility. We only present a subset of the functions that are crucial in supporting cognitive network protocols, which illustrate the split-functionality approach, and refer the reader for the details of the remaining functions in related work.[11] We focus our discussion on the GNU Radio and USRP platform.

7.5.1 Spectrum Sensing and Carrier Sense

The ability of a cognitive radio to react to the current state of the spectrum is extremely important to the performance of the radio and the effectiveness of the cognitive techniques. Given that cognitive techniques require information about the current state of the network to adapt, if this information is stale, the network will adapt inappropriately. The quicker the radio can adapt, the greater the performance will be. A perfect example of such a technique that requires physical layer information in a timely manner to react properly is carrier sense. The performance of carrier sense is crucial to CSMA protocols: The longer it takes to transmit a packet after the channel goes idle, the greater the chance of collision. Measuring the reactiveness of carrier sense is a raw measurement of the reactiveness of the radio. This timing, which we will refer to as *reactiveness*, is shown in Figure 7.3. Reactiveness is crucial to cognitive radio networks: It is the time it takes for the radio to adapt to a change in the spectrum. We therefore present the split-functionality design of carrier sense to demonstrate how we can achieve greater performance of the core function, and how the split-functionality design increases the reactiveness of the radio.

7.5.1.1 Carrier Sense Design and Evaluation

To significantly increase the reactiveness of the radio to the spectrum, and therefore the performance of carrier sense, we must avoid the associated delays by placing the decision at the radio hardware. However, the decision process should be controlled by software running on the host CPU to maintain flexibility. The first assumption we can make is that when a host wishes to perform carrier sense, it can modulate a packet and pass the computed samples to the radio hardware to wait for the carrier to be idle. The per-block meta-data for the transmission has a single bit flag set to indicate that the block should be held until there is no carrier using a locally computed RSSI value. The host can

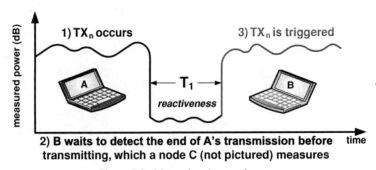

Figure 7.3. Measuring the reactiveness.

control the carrier sense threshold via the control channel. We use an RSSI value recorded in the radio hardware to implement a simple RSSI threshold carrier sense mechanism. Therefore, we split the carrier sense implementation in our split-functionality design by placing the carrier sense triggering mechanism on the radio hardware, and full control over the RSSI threshold and carrier sense algorithm on the host. As our evaluation will show, this allows us to achieve greater performance (a smaller reaction time), without sacrificing flexibility.

We now present an evaluation of this carrier sense design in comparison to performing carrier sense at the host. This compares the reactiveness of the radio if the core functions were implemented solely at the host, to the reactiveness of using a split-functionality approach. In the host implementation, the host estimates the received signal strength from the incoming sample stream and uses thresholds to control outgoing transmissions. We use the evaluation setup illustrated in Figure 7.3, where a USRP (node C) monitors two node's transmissions by measuring the magnitude of received complex samples. At 8 megasamples per second, the monitoring node (C) achieves a precision of 125 nanoseconds for measuring the reactiveness of the radios. The two contending nodes (A and B) exchange the channel using carrier sense 100 times, and we measure the spacing between each transmission as the reactiveness, as illustrated in Figure 7.3. The first contending node, A, finishes transmission TX_n, and B takes T_1 time to detect the channel as idle and begin transmission TX_{n+1}. T_1 represents the reactiveness.

As shown in Figure 7.4, taking the average gap observed across 100 exchanges, the results were 1.5 μs and 1.98 ms for the split-functionality and host implementations, respectively. The host-based latency could be reduced closer to 1 ms, or on the order of tens of microseconds, by moving the functionality to the USRP device driver, or the kernel, respectively. In our evaluation, the times were recorded at an application-level block in GNU Radio where a MAC protocol would reside. These measurements illustrate our design's ability

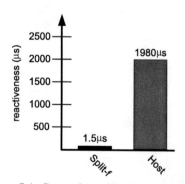

Figure 7.4. Comparing achieved reactiveness.

to reduce the carrier sense blind spot by *three orders of magnitude* while maintaining host control on a per-packet basis. This can significantly increase the capacity in the channel by reducing the time it takes to detect it is idle. The host can even control the threshold on a per-packet basis by placing a control packet with a new threshold on the bus before the data packet.

7.5.2 Fast Packet Recognition

Cognitive radio network protocols not only need to react quickly to changes in the spectrum, but also to incoming packets. For example, cognitive radios may exchange packets to inform each other of protocol parameters, or even of complete switches of a layer (e.g., changing from TDMA to CSMA at the MAC layer). Therefore, the radio must be able to identify incoming packets to the node in a timely manner. Additionally, traditional software-defined radios in the receive state will stream captured samples at some decimated rate between the radio hardware and the host. For many MAC protocols, such as CSMA-style designs, the radio cannot determine when packets for the attached node will arrive. As a result, the radio must remain in the receiving state. The downside to this is that the demodulation process uses significant memory and processor resources despite the fact that incoming packets destined for the radio are infrequent. As such radios become more ubiquitous and common for implementation, resource usage will become increasingly important, especially for energy-constrained devices such as the battery-powered Kansas University Agile Radio.[8]

One simple solution would be to send samples when the RSSI is above some threshold. However, this does not filter out transmissions destined to other hosts and external signals. A better solution would be to have the radio hardware look for the packet preamble and the destination address, then transfer a maximum packet size worth of samples to the host after any match. As we also describe in related work,[11] the ability to identify packets and process them partially on the SDR hardware is also critical to supporting low-latency MAC interactions (e.g., packet/ACK exchanges or RTS/CTS) in a high-latency architecture.

7.5.2.1 Fast Packet Recognition Design

Our goal is to accurately detect packets at the radio hardware without demodulating the signal (to keep flexibility). To achieve this goal, we perform signal detection. The most relevant work in signal detection comes from the area of radar and sonar system design. From this area, we borrow a well-known technique, called a *matched filter*, to detect incoming packets at the radio hardware without performing demodulation.

Matched filter: A matched filter is the optimal linear filter that maximizes the output signal-to-noise ratio for use in correlating a known signal to the

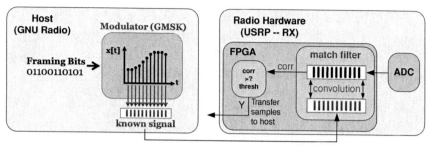

Figure 7.5. Matched filter and dependent packet design.

unknown received signal. For use in packet detection, the known signal would be the time-reversed complex conjugate of the modulated framing bits. These known signal's samples, which are referred to as coefficients, are stored in the matched filter's memory bank (Figure 7.5). The received sample stream is convolved with the coefficients. The result can be treated as a correlation score between the unknown and known signals. The correlation score is then compared with a threshold to trigger the transfer of samples to the host. The matched filter is flexible to different modulation schemes (e.g., GMSK, PSK, QAM) but requires a Fast Fourier transform for OFDM, given that the symbols are in the frequency domain. This would require an FFT implementation on the radio hardware.

To also detect that the frame is destined to the particular host, two different methods that have mathematically different properties can be used. *Single Stage*: Use a frame format where the destination address is the first field after the framing bits, and use this complete modulated sequence as the matched filter coefficients. *Dual Stages*: Detect the framing bits first, then change the coefficients to the modulated destination address. Our implementation uses the single-stage approach for simplification. However, a dual stage is more appropriate for monitoring multiple addresses such as a local address and a broadcast address.

7.5.2.2 Fast Packet Recognition Evaluation

We evaluate the effectiveness of the matched filter at detecting incoming sequences using simulations where we can control the noise level. Results are presented from over-the-air experiments with the presence of interference, multipath, and fading in related work.[11]

To evaluate the effectiveness of the matched filter with varying signal quality, we first run experiments with controlled signal-to-noise ratios (SNR) using the GNU Radio software. We introduce additive white Gaussian noise (AWGN) to control the SNR in terms of dB: $SNR(dB) = 10 * log_{10} * \frac{Power_{signal}}{Power_{noise}}$. To introduce the noise, we compute the signal power: $Power_{signal} = |Signal_{ampl}|^2$,

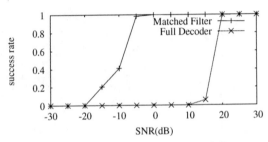

Figure 7.6. Success rate of the matched filter.

and then the noise power: $Power_{noise} = \frac{Power_{signal}}{SNR}$, based on the specified *snr*: $SNR = 10^{(snr/10)}$. For evaluation, 1,000 frames of 1,500 bytes are encoded using the Gaussian minimum-shift keying (GMSK) modulation scheme. These frames are used as the ground truth and mixed with the noise. We require that the matched filter detect the framing bits *and* that the transmission is destined for the attached host using the single-stage scheme (Section 7.5.2). The success rate is defined as the number of detected frames over the total number of frames in the dataset (1,000). For comparison, we also include the success rate of the full GMSK decoder. At a high noise level, even the full decoder will fail at detecting the frames. The success rate, as a function of the SNR, is shown in Figure 7.6. The results show that the matched filter can detect the frames at a much higher success rate than the decoder can, even at low SNR levels where the noise power is greater than the signal power.

Given these results, and further real-world results presented in related work,[11] we conclude that using the matched filter for detecting relevant packets is accurate enough that the host will never miss an actual frame due to the filter. In fact, the filter triggering samples to the host can be seen from a different perspective as providing further confidence to the host that there is actually a frame within the sample stream. The host could then perform additional processing in an attempt to decode the frame successfully.

7.5.3 Access to Physical Layer Information and Fine-Grained Radio Control

The underlying radio hardware in an SDR platform has many controls that are not configured by the transmitted sample stream (e.g., transmission frequency and power), and can make many observations that are not easily derived from the input sample stream (e.g., RSSI). We use our control channel between the SDR hardware and host to expose these controls and physical layer information to the MAC protocol implementation. Many existing network interfaces use similar

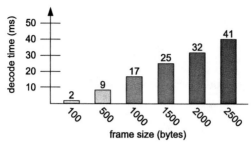

Figure 7.7. Decode times for various frame sizes.

designs for setting the transmission channel and obtaining RSSI measurements. One key difference is that our interface operates on blocks of samples instead of packets.

7.5.3.1 Physical Layer Information

Access to physical layer information at all other layers in the processing chain is important for supporting common cross-layer optimizations and extremely crucial to cognitive radio network protocols to adapt to the spectrum. This can be seen through recent work where per-bit confidence levels are used to perform partial packet recovery.[6] We enable this functionality in our architecture through the control channel and per-block meta-data. In our design, information from the SDR can be sent to the host using either the control channel or per-block meta-data. We use this mechanism to report RSSI to the host. Note that the host could calculate RSSI using the raw samples, but an RSSI value that takes into account the gain or attenuation in the RF stages is only available at the radio hardware. The control protocol is easily modified to support reporting additional properties; however, developers must reprogram the FPGA to report the desired values.

7.5.3.2 Radio Control

We also implement a set of radio hardware control messages on the control channel that can be synchronized with packet transmissions using the timestamp. For example, by placing a control block with a timestamp T before a data packet on the bus, which uses a *NOW* timestamp, the radio will be reconfigured at time T and the data packet will be transmitted immediately after the reconfiguration. This can be used to implement common techniques such as rapid frequency hopping or to reconfigure parts of the core functions that reside on the radio hardware. Unfortunately on the USRP, the daughterboards are tuned directly from the FX2 USB controller using the I^2C bus, which has no connection to the

FPGA. Therefore, we cannot issue daughterboard commands from the FPGA using the control channel and hardware clock to implement rapid frequency hopping. The USRP2 tunes the daughterboards directly from the FPGA. Therefore, if our design was implemented on the USRP2, unavailable at the time, rapid frequency hopping could be achieved.

7.6 MAC-Layer Evaluation

We now provide end-to-end results for a Bluetooth-like TDMA protocol and 802.11-like CSMA protocol. The protocols use the *split-functionality* design described in Section 7.5, and we compare their performance with that of full host-based implementations. This demonstrates the increased performance possible at the MAC-layer of the cognitive radio network. In the next section, we describe a design that enables a cognitive radio network protocol that switches between the MAC layers on the fly using information from the radio hardware (Section 7.6.3).

7.6.1 Bluetooth-Like TDMA Protocol

To illustrate the effectiveness of the overall system design, we implement a tightly timed Bluetooth-like TDMA protocol. Like Bluetooth, the network (piconet) consists of a master and a maximum of seven slaves. The slaves communicate with the master in a round-robin fashion within a slot time of 625 μs. Unlike Bluetooth, our protocol fixes its frequency instead of hopping (a limitation of the USRP discussed in Section 7.5.3), uses slightly simpler synchronization (bypasses *pairing*), and we also vary the slot guard time for evaluation.

   Each slave in the network synchronizes with the start of a round by listening for the master's beacon, and calculates the start of transmission as the logical synchronization time T. The beacon frame also carries the total number of registered slaves (N) and the guard time (T_g). The slave can then compute the total round time, which must account for the master: $T_r = N + 1 * (T_s + T_g)$, where T_s is the slot time (625 μs). The start of round k is computed as: $T_k = T + T_r * k$. We remind the reader that this is a logical time kept at each node, taken from the beacon frame that is a global reference point. Finally, each slave's slot offset is computed from its node ID (n), $\delta_n = n * (T_s + T_g)$, which is then used to compute the local start time of slave n's slot in round k: $T_{n(k)} = R_k + \delta_n$.

7.6.1.1 TDMA Results

We use two metrics in our evaluation: ability to maintain tight synchronization and overall throughput. The synchronization error at the master is 15 ns, computed by measuring the actual spacing of 1,000 beacons using a monitoring node (discussed in Section 7.5.1). This illustrates the tight timing of the master's beacon transmissions. To measure the synchronization error at the slaves, we record the calculated timestamps of 1,000 beacons at 4 slaves. Each timestamp should be exactly T_r apart from the next. The absolute error in spacing represents shifts in the slave's calculation of the start of the round. We find the maximum error of the 1,000 beacons at all 4 slaves to be 312 ns, with an average of 140 ns. This answers the question of our platform's ability to obtain tight synchronization at both transmitters (master) and receivers (slaves).

We compare a split-functionality implementation to a host implementation, which differ in their guard times. A guard time of 1 μs is used for the split-functionality implementation, which is nearly three times the maximum error. We use our roundtrip host and radio hardware delay measurements from Section 7.2.1, which accounts for both transmissions and reception timing variability, to estimate the host guard time needed. A guard time of 9 ms would be needed to account for the maximum error; however, this delay occurs rarely and we, therefore, present results using a guard time of 3 ms (approximately 3 * sdev) and a more realistic guard time of 6 ms based on our recorded delay distribution.

We perform 100 KB file transfers, varying the number of registered slaves and presenting averaged results across 100 transfers in Figure 7.8. The *split-functionality* implementation is able to achieve an average of four times the throughput of the host-based implementation. While we had only been able to answer the question of obtaining synchronization, we find that throughout the full transfers, no slave drifts into another slot period using only the initial beacon for synchronization, illustrating the ability to *maintain* tight synchronization. These results are promising for the development of TDMA protocols on the platform.

7.6.2 802.11-Like CSMA Protocol

We implemented two 802.11-like CSMA MAC protocols, one fully on the host CPU and one using our *split-functionality* optimizations including on-board carrier sense (Section 7.5.1), dependent packet ACK generation, and backoff

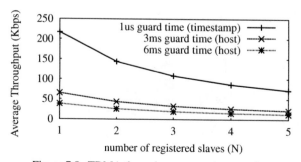

Figure 7.8. TDMA throughput comparison results.

(both found in related work[11]). The MAC implements 802.11's clear channel assessment (CCA), exponential backoff, and ACKing. Our protocol does not implement SIFS and DIFS periods; this work is in progress. For space reasons, we focus our description on how the 802.11-like protocol uses our architecture.

The host-based implementation places all functionality on the host CPU, including carrier sense, ACK generation, and the backoff. The optimized implementation uses the matched filter and SNR monitoring for ACK generation, and performs carrier sense and backoff on the radio hardware. We configure the USRPs for a target rate of 0.5 Mbps, and run 100 1 MB file transfers for each implementation using a center frequency of 2.485 GHz in an attempt to avoid 802.11 interference. This allows us to present results that highlight the differences in the implementation without the effect of uncontrolled interference. We also vary the number of nodes in the network, where each pair of nodes performs a transfer.

The results for the two implementations are shown in Figure 7.9. We see significant performance increases from the use of the split-functionality implementation. This nearly doubles the throughput on average, likely due to the time saved in decoding to generate the ACK, and the delays associated with carrier sense and backoff. We note that the matched filter detected every framing sequence, and the fast-ACK generation technique only failed two times over the total number of runs. To recover from these failures, we implemented a feedback mechanism on the host that checks the SNR monitoring technique's decision and retransmits. This is needed because we did not use a higher-layer recover mechanism like TCP.

7.6.3 Supporting Cognitive Switching of the MAC Layer

As discussed throughout this chapter, a cognitive radio network monitors current network conditions and adapts at all layers to achieve the greatest level

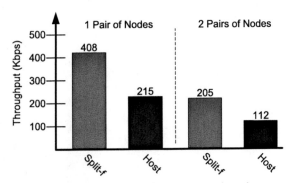

Figure 7.9. 802.11-like CSMA protocol results.

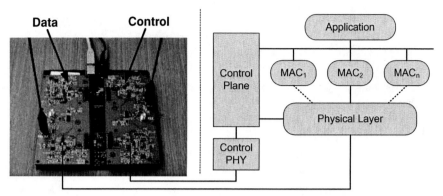

Figure 7.10. Design of a cognitive architecture for MAC layer switching.

of performance, given the current spectrum. For example, the radio can use spectrum-sensing information to find a less congested channel, or use noise/loss information from the PHY and MAC layers to change the protocols running to achieve a higher throughput. Under low loss rates and low congestion, the radio could use the CSMA protocol discussed in Section 7.6.2. Under high loss and congestion, the radio could use the TDMA protocol discussed in Section 7.6.1, which reduces the overhead involved with accessing the spectrum and reduces the chances of collision. While we have briefly presented a general design for a switching protocol, it is not the contribution of this section. The contribution is a general architecture and design of the components on a host-PHY radio, which allows for the switching of the layers, shown in Figure 7.10, as well as suggestions on how to better support layer switching in a host-PHY architecture. From this design, novel cognitive radio network protocols that govern the switching of the layers can be built. Additionally, it is a layer that can be accessible via a global controller in the future Internet to control the protocols (e.g., the MAC layer) in use by the cognitive radio. Through such a controller, new protocols could be designed and propagated to the radios for use.

As mentioned, the radio uses information from the protocol layers, such as the loss rate, to change the MAC that is operating. This logic cannot reside in the MAC layer because it needs to run independent of this layer. Additionally, we do not want to place it at the PHY layer because it violates the general hierarchy of networking protocols, and we would like the design to be general enough to allow switching of the physical layer. Therefore, we need to construct a new layer that monitors information from both the MAC and PHY layers, shown in Figure 7.10. We refer to this layer as a *control plane*, which has a connection to all possible MAC layers for control and status information. The control plane communicates to the MAC layers, activating the appropriate MAC protocol such that it communicates with the physical layer. There is also a connection from the control plane to the physical layer for control and status information such that it can adapt based on physical layer information. As it is important that all radios in the network operate using the same MAC layer, we construct a control channel using a second frontend in a different frequency band on the USRP, which uses a separate physical layer at the host. The control channel is used to communicate a switch of the MAC layer, which all other radios will also switch to. All radios in the network use the control channel to agree on a MAC layer, sharing current channel information such as congestion and loss. The radios use a basic consensus mechanism to choose the layer, based on beacons of the current MAC layer sent by each radio on the control channel. Once consensus is reached to change the MAC layer, the control plane uses the control-and-status channel to the MAC layers to disable or enable the appropriate protocols.

Using this general architecture, we are able to change the active MAC layer based on the current state of the network. However, we are not able to change the connections between the MAC layers and the physical layer. To our knowledge, current host-PHY architectures do not directly support switching of the MAC layer; in fact, their general architecture inhibits it. In a host-PHY software-defined radio architecture, such as GNU Radio, it is common that multiple modular processing blocks are connected to create a layer such as the MAC or PHY layers. At runtime, instances of the blocks are created, and the application specifies how the blocks are connected. To our knowledge, there is no host-PHY SDR that allows for the connection of these blocks, and what blocks are instantiated, to be changed at runtime. This prevents a cognitive radio network protocol from making "extreme" changes to the processing, such as removing a series of blocks to completely replace the MAC layer. Therefore, as future work, we propose host-PHY architectures take into consideration a growing need to dynamically change the processing of the radio in an optimal manner, such that all blocks do not need to be instantiated at runtime, which requires additional memory and logic such that all blocks are connected.

7.7 Related Work

We review related work in the area of MAC development. Existing platforms mostly use the extremes of the design space where either the majority of functionality is fixed on the network card (*Traditional NICs*) or performs all processing at the host (*Software-Defined Radios*).

Traditional NICs: Several efforts[3,10,13] have built new MAC protocols on top of existing commercial NICs (e.g., 802.11 cards). Unfortunately, commercial 802.11 cards implement the bulk of the MAC functionality in proprietary microcode on the card, limiting what functions can be changed by researchers. As a result, this approach is not very satisfactory: The range of MAC protocols that can be implemented is limited and performance (e.g., throughput, capacity) is often poor from the MAC needing to be implemented on the host. For example, past efforts have mostly implemented TDMA-based schemes.

Software-defined Radios: Software-Defined radios (SDRs) provide a compelling architecture for flexible wireless protocol development, considering that most aspects of both the MAC and physical layer are, by design, implemented in software and thus, in principle, easy to modify. However, so far, SDR efforts have focused on implementing the physical layer[16] whereas MAC and higher-layer protocol development has received little attention. Recent work by Schmid et al.[12] examines the impact of increased latency in SDRs using GNU Radio and the USRP. The authors address how the bus latency creates "blind spots" that increase collision rates when carrier sense is performed at the host, and how precomputation of packets is not possible without fully demodulating (at the host), resulting in larger interframe spacing. Our design provides solutions for both of these issues in Sections 7.5.1 and 7.5.2, respectively. Bus delay measurements were also taken by Valentin et al.[15]

A number of groups have developed software radios with architectures that differ from the current GNU Radio and USRP design by including a CPU on the radio hardware (NC-CPU), either as a separate component or as a core on the FPGA. Examples include the Rice University Wireless Open-Access Research Platform (WARP)[17] and USRP2. These designs are more expensive, but they offer additional flexibility for partitioning the MAC. However, there is still a nontrivial delay (compared with traditional radios) owing to physical layer processing and queueing. The NC-CPU is also likely to be slower than the host CPU, increasing the processing delay. Finally, in deployed products based on this architecture, the NC-CPU is likely to be off-limit to users, similar to the current situation with commercial wireless cards. As a result, we expect that our architecture will be useful for this type of platform as well.

7.8 Conclusions

In this chapter, we presented a set of techniques that support the implementation of diverse, high-performance cognitive radio network protocols on software radios. The work is motivated by an increasing diverse and ever-changing wireless spectrum, such that to achieve the greatest level of performance, the radio must adapt at all layers in the wireless networking stack. Software radios offer flexibility, but their architecture, specifically the delay between the host and the radio frontend, has traditionally been a problem for protocols. We introduce a split-functionally approach, which addresses this problem, and show that it enables the implementation of a set of core MAC functions and cognitive functions that can react to the spectrum more quickly for greater performance. An implementation for the USRP and GNU Radio, along with the implementation of an 802.11-like and Bluetooth-like protocol, shows the approach is effective.

Additionally, we presented a basic design to support cognitive switching of the MAC layer on the fly, which can be extended. To our best knowledge, these protocol implementations are the first high-speed, bidirectional MAC implementations for the GNU software radio platform. For future work, we plan to implement a more diverse set of protocols to further evaluate our design and implement the architecture on different SDR platforms to evaluate its generality.

References

[1] A. Akella, G. Judd, S. Seshan, and P. Steenkiste. Self-Management in Chaotic Wireless Deployments. *ACM MobiCom*, pages 185–199, 2005.

[2] BBN Technologies Corperation, GNU Radio Architectural Changes (m-block). http://acert.ir.bbn.com/downloads/adroit/gnuradio-architectural-enhancements-3.pdf

[3] C. Doerr, M. Neufeld, J. Fifield, T. Weingart, D. C. Sicker, and D. Grunwald. Multi-MAC – An Adaptive MAC Framework for Dynamic Radio Networking. *IEEE DySPAN*, 2005.

[4] S. Gollakota and D. Katabi. Zigzag Decoding: Combating Hidden Terminals in Wireless Networks. *ACM SIGCOMM*, 2008. ACM Press.

[5] Gnu radio. http://www.gnu.org/software/gnuradio/

[6] K. Jamieson and H. Balakrishnan. PPR: Partial Packet Recovery for Wireless Networks. *SIGCOMM Comput. Commun. Rev.*, 37(4): 409–420, 2007.

[7] S. Katti, D. Katabi, H. Balakrishnan, and M. Medard. Symbol-Level Network Coding for Wireless Mesh Networks. *ACM SIGCOMM*, 2008. ACM Press.

[8] Kansas university agile radio. https://agileradio.ittc.ku.edu/

[9] A. Mishra, V. Shrivastava, D. Agrawal, S. Banerjee, and S. Ganguly. Distributed Channel Management in Uncoordinated Wireless Environments. *ACM MobiCom*, pages 170–181, 2006.

[10] M. Neufeld, J. Fifield, C. Doerr, A. Sheth, and D. Grunwald. SoftMAC – Flexible Wireless Research Platform. *Fourth Workshop on Hot Topics in Networks (HotNets)*, 2005.

[11] G. Nychis, T. Hottelier, Z. Yang, S. Seshan, and P. Steenkiste. Enabling MAC Protocol Implementations on Software-Defined Radios. *NSDI*, 2009.

[12] T. Schmid, O. Sekkat, and M. B. Srivastava. An Experimental Study of Network Performance Impact of Increased Latency in Software Defined Radios. *WiNTECH'07*, 2007.

[13] A. Sharma, M. Tiwari, and H. Zheng. MadMAC: Building a Reconfigurable Radio Testbed Using Commodity 802.11 Hardware. *IEEE Workshop on Networking Technologies for Software Defined Radio Networks*, Reston, 2006.

[14] The Universal Software Radio Peripheral. http://www.ettus.com/

[15] S. Valentin, H. von Malm, and H. Karl. Evaluating the GNU Software Radio Platform for Wireless Testbeds. *Technical Report TR-RT-06-273*, 2006.

[16] Vanu Software Radio Systems. http://www.vanu.com

[17] Rice University Wireless Open-Access Research Platform (WARP). http://warp.rice.edu

8

Vehicular Networks: Applications, Protocols, and Testbeds

Mario Gerla and Marco Gruteser

Abstract

Vehicular networks are expected to be one of the major new application areas for wireless and Internet services. There are more than 600 million vehicles worldwide and these will be networked to achieve improvements to safety, traffic management, navigation, and user convenience. Vehicular networks (VANETs) have several elements in common with ad hoc mesh networks, but also have unique new requirements including high mobility, rapidly changing topology, multiple usage modes (vehicle-to-infrastructure [V2I] and vehicle-to-vehicle [V2V]), and the central importance of geo-location.

In the first part of this chapter, emerging VANETs are shown to be unique in the broad family of MANETs (Mobile Ad Hoc Networks). VANET services are reviewed and classified. A location-aware content distribution ("car-torrent") is then presented. Next, vehicle urban sensing is showcased for applications that range from traffic congestion/pollution measurements to distributed civilian surveillance. MobEyes, an urban surveillance application that supports forensic investigations, is then described and contrasted to other urban sensing projects.

In the second part of the chapter, the enabling VANET protocols are reviewed. First, physical and MAC layer standards for vehicular communications (DSRC, WAVE, and IEEE 802.11p) are reviewed. Then, new VANET network level protocol requirements are identified and solutions are discussed. Geo-location-based protocol architectures are introduced and briefly touch on complementary techniques such as geo-based handoff and geo-based beam adaptation for smart antennas. Security and privacy issues are addresses, with particular attention to location privacy. These protocols are illustrated with urban sensing applications.

The third part describes the role of the infrastructure in VANETs, and introduces the notion of MobiMESH, the wireless mesh architecture consisting of roadside (Access Points) APs. Functions such as Mobility Management (e.g., Geo Location Service) are supported by MobiMESH.

In the fourth part, experimentation of VANET protocols and applications is discussed, and two emerging VANET vehicular testbeds – C-VeT and ORBIT – are reviewed.

8.1 Introduction

Vehicular communications have been receiving increasing attention over the last ten years as a viable means of augmenting road safety and travel efficiency. The field has consequently attracted consistent investments from auto manufacturers and public transport authorities, further stimulating academic research. We have reached now a situation where the essential building blocks of vehicular networks (On Board Radios, Road Side APs, Reserved 5.9 Ghz spectrum, and dedicated communication standards [Standard Specification for Telecommunications and Information Exchange between Roadside and Vehicle Systems 2003]) are (almost) available, thus opening up interesting opportunities for a wealth of car-to-car applications.

On the one side, security-oriented applications are still the top priority for auto industry and transport authorities, and recent testbed experiments have proven the effectiveness of vehicular communications in preventing intersection crashes (ElBatt et al. 2006). On the other side, the availability of the technology is stimulating interesting debates on new and challenging applications to be supported by vehicular communication systems, and visionaries are looking beyond safety applications. Automatic and efficient traffic control services (using "Intelligent Transport" techniques) can greatly benefit from vehicular communications by reducing traffic congestion, possibly keeping under control the associated chemical pollution. Imagine a comprehensive urban traffic planning system that receives inputs from vehicles (e.g., route plans, destinations, sensor readings, positions, driver's preferences, etc.), processes such information to generate an "urban routing" plan, and implements the plan through the careful control of traffic lights. The control may be extended to actual vehicle routes, possibly rerouting the vehicle to alternate, less congested routes with the assistance of "navigator" companies.

The aforementioned traffic planning system also can be equipped with entertainment-oriented functionalities providing information on locally available resources (e.g., restaurants, movie theaters, museums, etc.) and supporting content distribution, sharing, and file streaming through peer-to-peer systems (e.g., Car-torrent [Nandan et al. 2005]) and e-commerce applications, as well as mobile Internet gaming. Moreover, a new paradigm of applications arises from the observation that vehicles can actually behave as collectors (i.e., "sensors") of information from the surrounding environment. Indeed, vehicles can be easily equipped with several sensing devices monitoring specific physical processes/ phenomena (cameras, microphones, pollution sensors, humidity, temperature, etc). Such sensing devices can be used to build up a distributed and enriched awareness of the vehicular environment, which, in turn, can boost the creation of "environment-aware" applications. As an example, vehicular surveillance systems can be built to support crime investigation, homeland protection, and suspicious activities monitoring. Further, massive distributed databases can be created and maintained storing commercial, entertainment, and cultural information.

From a network architecture point of view, we argue that to support all the aforementioned applications/services, vehicle-to-vehicle communications need to be supported and integrated into *roadside infrastructure*, which in turn must provide Internet connectivity and communication resiliency. As an example, crash prevention and intelligent transport applications would not be feasible or effective if they relied only on pure car-to-car communications under sparse vehicle distributions. Similarly, content distribution (via CarTorrent, say) services most likely must retrieve the original content in the Internet, thus calling for a fixed infrastructure to bridge the vehicles to the Internet. Thus, roadside infrastructure must be ubiquitous and instantly available to support all the above functions.

Roadside APs providing the contact point between the vehicular realm and the infrastructure are to be placed in special locations, to best serve the fast-moving vehicles, as opposed to the APs designed to support pedestrians, which are generally placed in shopping malls, popular bars, restaurants, bus/train stations, and other public places. To this extent, ideal places to install the roadside APs are traffic lights and more generally light poles, overpasses, and other public structures. Traffic lights in particular are perfectly positioned to act as traffic routers: They are ubiquitously distributed throughout urban centers in precisely the locations where traffic management is most required; they are equipped with power and directly maintained by local municipalities; and they have the best "view" of approaching vehicles and crossing pedestrians. Traffic lights and other roadside access points form neighborhood *wireless meshes* that are interconnected with each other via the infrastructure. Not all the roadside AP's have wired access to the Internet, due to cost and physical limitations. The wireless mesh will provide this interconnection in a simple and cost-effective way.

Vehicular protocols and applications can be adequately evaluated and validated only in an experimental setting. Various vehicular testbeds have recently been announced, many of them offering open access to experimenters. Given the difficulty to create test environments that capture the scale of an urban grid with millions of vehicles, there must be provision for powerful emulation platforms and rigorous validation tools that help bridge the gap between small-scale testbeds and large-scale simulators.

This chapter will introduce VANET architectures using a top-down approach. Requirements and applications are introduced first, followed by enabling protocols, supporting infrastructure functions, and testebds. The chapter is organized as follows. First, in Section 8.2, the VANET is compared and contrasted to closely related MANETs, and VANET unique properties are highlighted. Next, emerging VANET applications are described, including content delivery (CarTorrent/CodeTorrent) and "urban sensing." Section 8.3 follows, with the protocols that make such applications possible. The main focus is safety messaging/broadcast standards; mobility models/generators; routing, including emerging geolocation-based protocol architectures; DTN routing; and vehicular security and privacy. Section 8.4 identifies the role of the infrastructure and introduces the notion of a wireless mesh network and its role in support of mobility management. Section 8.5 will cover the emerging VANET testbeds (UCLA C-VeT; Rutgers Vehicle Testbed + ORBIT).

8.2 Vehicular Network and Application

8.2.1 VANET vs. MANET: What Is the Difference?

The first MANET (Mobile Ad Hoc Network) was borne about forty years ago, on the wake of the ARPANET successful debut. It was called Packet Radio Network (Kahn 1977) and was mainly viewed as a portable (at the light weight of 40 lb) radio for packet radio communications among soldiers in the battlefield. In the past forty years, the MANET has received enormous attention by wireless network researchers in academia as well as in the aerospace and military industry. Supported by steady funding from government and defense agencies, it has evolved to be an extremely sophisticated system both in radio and protocol designs. The most important application is tactical networking, followed by emergency and civilian protection scenarios. Excluding a few sensor networks (which are fixed anyway), commercial MANET applications are still in their infancy. The VANET is the prominent example of emerging MANET. In fact, it is the researchers' dream because it enables a number of exciting and compelling applications that have commercial potential. However, if researchers expect to extend mature MANET protocols to the VANET, they are going to be quickly disappointed. That is because the VANET is anything but an ordinary MANET.

To start, the conventional MANET is instantly deployable and reconfigurable in areas without infrastructure. Figure 8.1 contrasts the multi-hop, instantly deployable MANET with the wireless infrastructure network. The urban VANET

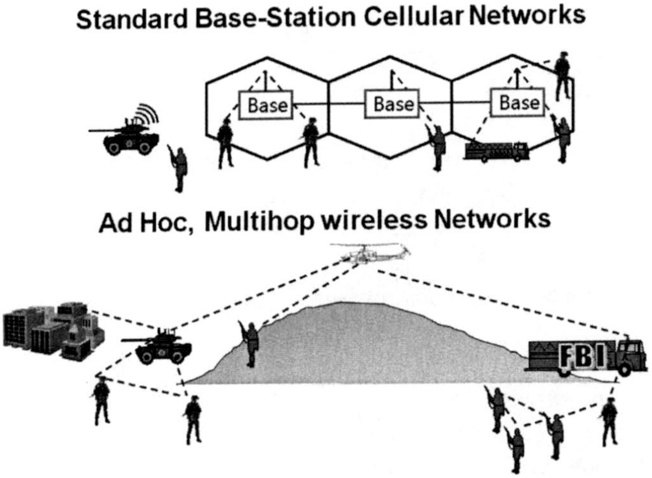

Standard Base-Station Cellular Networks

Ad Hoc, Multihop wireless Networks

Figure 8.1. Ad hoc multihop instantly deployable MANET versus wireless infrastructure network.

is also dynamically reconfigurable; however, in normal operating conditions, it can tap one or more different "infrastructures (3G cellular, WiFi or IEEE 802.11p, WiMAX). MANETs are typically deployed to satisfy a "temporary" need (e.g., battlefield, emergency, etc) – as qualified by the term "ad hoc." VANETs run welldefined, permanent applications (like safe navigation, crash prevention, road congestion monitoring, etc.).

Mobility is the key attribute of a MANET and is characterized by a motion pattern. In battle and emergency scenarios, the motion pattern is generally not well known in advance; routing architectures are compared under various semiarbitrary assumptions, like random way-point, group motion, coordinated motion (say, follow the leader, gather/scatter, etc.) depending on the specific applications. In the VANET, there is a much better understanding of the motion pattern (say, commuting traffic to/from work; business traffic to/from train station/ airport/convention center; shopping expeditions to malls, etc.). In fact, most of the VANET architecture evaluations are based on traces or on traffic patterns that have been validated by traces. Mobility in MANETs also implies battery constraints – like in a scouting team equipped with portable radios and exploring the forest for a few days. Thus, low energy protocols are a must. In a VANET, battery power can be assumed infinite for the purpose of communications and computing.

Mobile-to-mobile multi-hop routing on dynamically changing paths has been the trademark of MANETs. In fact, in MANETs, the lead application so far has been reliable data delivery (uni or multicast) to remote destinations. Most of the challenges in MANETs design stem from designing stable routing protocols and data transfer sessions (UDP and TCP) over such multi-hop paths. In VANETs, the delivery of data to remote destinations is not the lead application. Besides, the Internet infrastructure takes care of that. At most, data will travel a few vehicle-to-vehicle hops until a roadside AP is reached. Typical VANET applications require neighbor interactions like broadcasting alarms, P2P sharing of content, exchanging sensor information, and so on. So VANET routing is "proximity" driven rather than multi-hop to far destinations.

Multi-hop routing in VANETs still plays a role to get to a roadside AP a few hops away. More important, efficient V2V routing is required in special situations – for example, when the entire infrastructure has failed because of a disaster (e.g., Hurricane Katrina scenario), or when the infrastructure cannot be used for covert operations (e.g., homeland defense or peacekeeping operations in an unfriendly city). When V2V multi-hop routing is required, the preferred routing scheme is geographic routing, considering that virtually all vehicles will soon be equipped with GPS, and there are efficient techniques to fill in the gap in tunnels of urban canyons where the GPS signal is weak. Moreover GPS jamming is not as critical in VANET applications as it is in tactical MANET scenarios.

From the preceding discussion emerges the picture of a VANET that is quite different from the conventional MANET. In fact, vehicles will connect in most cases single-hop to the infrastructure like in a WLAN. V2V ad hoc networking will occur only it is necessary because of lack of nearby APs or applications latency constraints – say crash prevention, or emergencies and covert operations. We may describe the VANET as an Opportunistic Ad Hoc Network, where direct access to Internet (via WiFi, WiMAX, or 3G) is readily available but is opportunistically "bypassed" using the "ad hoc" if too costly or inadequate. For example, V2V is preferred for the exchange of navigation safety beacons and alarms among cars as shown in Figure 8.2. This drastic difference between the VANET and the conventional MANET is in part a loss, in the sense that it precludes the use of much of the classic MANET research generated over the last forty years. It does, however, open a tremendous opportunity of new research on this very novel environment in many areas including:

Physical and MAC layers:

- Radios (MIMO, multichannel, cognitive, SDR)
- Positioning in GPS deprived areas

Network Layer & Routing:

- Mobility models
- Network Coding

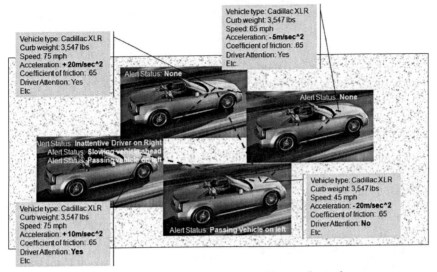

Figure 8.2. V2V navigation safety beacons and alarms exchanged among cars.

Routing:

• Geo routing, Content based routing, DTNs

Security and privacy
New Applications:

• Content distribution, mobile sensing, safety, etc.

8.2.2 Emerging Vehicular Applications

8.2.2.1 Classification and Requirements

As mentioned earlier, VANETs offer the opportunity to deploy, in addition to traditional MANET type applications, a broad range of innovative, peer-to-peer content sharing and dissemination applications. Although P2P sharing has been so far confined to the wired Internet (e.g., BitTorrent), the much increased storage and processing capacity of VANETs with respect to personal or sensor-based ad hoc networks make such applications now feasible in the mobile domain as well. Moreover, the fact that car passengers are a captive audience provides incentive for content distribution and sharing at a scale that would be unsuitable to other ad hoc network contexts. We describe a representative set of VANET P2P applications and classify them by the vehicle's role in managing data: as a data source, data consumer, source and consumer, or intermediary.

First, the vehicle is a unique *source* of data. It provides an ideal platform for mobile data gathering, especially in the context of monitoring urban environments (i.e., vehicular sensor networks) (Eriksson et al. 2008; Hull et al. 2006; Lee et al. 2006; Lee et al. 2008a; Lee et al. 2008b). Each vehicle can sense events (e.g., images from streets or the presence of toxic chemicals), process sensed data (e.g., recognizing license plates), and route messages to other vehicles (e.g., forwarding notifications to other drivers or police officers). As the vehicle removes processing power and storage space constraints, these sensors can generate and handle data at a rate not imaginable for traditional sensor networks. Vehicular sensor applications require persistent and reliable storage of data for later retrieval. Namely, they require networking protocols (including sophisticated query processing) to efficiently locate/retrieve data of interests (e.g., finding all the vehicles at a certain time and location).

Second, vehicles can be significant *consumers* of content. The on-board equipment is capable of supporting high-fidelity data retrieval and playback. For the duration of each trip, drivers and passengers make up a captive audience for large quantities of data. Examples include locality-aware information (map-based directions) and content for entertainment (streaming movies, music, and ads) (Nandan et al. 2005; Lee et al. 2006b; Nandan et al. 2006; Caliskan 2006). These applications require high network data rates and fast access to stored data.

In a third class of compelling applications, vehicles are both the *producers and consumers* of content. Examples include services that report on road conditions and accidents, traffic congestion monitoring, and emergency neighbor alerts – for example, my brakes are malfunctioning (Dikaiakos et al. 2005; Guo et al. 2005; Lee et al. 2006a; Nadeem et al. 2003; Park et al. 2006). Also, interactive applications (e.g., voice-over-V2V and online gaming) belong to this category. These applications require location-aware data gathering/dissemination and retrieval. In particular, interactive applications require real-time communication among vehicles.

Finally, all of the previously mentioned applications will need to rely on vehicles in an *intermediary role*. Individual vehicles in a mobile group setting must cooperate to improve the quality of the applicant experience for the entire network. Specifically, vehicles will provide temporary storage (caching) for others, as well as forwarding of both data and queries. In this capacity, they require reliable storage as well as efficient location of and routing to data sources and consumers.

The demands of these applications give us a list of requirements and challenges for vehicular applications.

Time sensitivity – Time-sensitive data must be retrieved or disseminated to the desired location within a given time window. Failure to do so renders the data useless. This mirrors the needs of multimedia streaming across traditional networks, and one can leverage relevant research results from the related areas.

Location awareness – Both data gathered from vehicles and data consumed by vehicles are highly location-dependent. This property has direct implications on the design of data management and security components. Data caching and indexing should focus on location as a first-order property, whereas data dissemination must be location-aware in order to maintain privacy and prevent tampering.

Most applications require methods of storing/retrieving such location/time sensitive information. As in MANETs, we can use structured approaches such as geographic hashing (Ratnasamy et al. 2002) and DHT (Caesar et al. 2006), or structureless approaches such as epidemic dissemination (Vahdat and Becker 2000). However, it is nontrivial to maintain structure in VANETs due to the high mobility, nonuniform distribution of vehicles and intermittent connectivity. Thus, most application protocols rely on variants of epidemic data dissemination such that the produced information is disseminated to nodes in an area where the information is produced (Caliskan et al. 2006; Dikaiakos et al. 2005; Lee et al. 2006a; Nadeem et al. 2003; Zhou et al. 2005).

8.2.2.2 Vehicles as Data Consumers: Content Distribution

Content distribution to vehicles ranges from multimedia files to road condition data and to updates/patches of software installed in the vehicle. Nandan et al. (2005) proposed SPAWN, a BitTorrent-like file swarming protocol in a VANET. In SPAWN, a file is divided into pieces and is uploaded into an Internet server. Each file has a unique ID (e.g., hash value of the file content), and each piece has a unique sequence number. Users passing by the APs download parts of the file. Once out of the range of APs, they cooperatively exchange missing pieces.

SPAWN is composed of the following components: peer/content discovery and peer/content selection. Due to intermittent presence of APs, SPAWN cannot use a centralized server as in BitTorrent that keeps track of all the peers. Instead, SPAWN uses a decentralized "gossiping" mechanism for peer/content discovery that leverages the broadcast medium of the wireless networks. A gossip message of a node contains a file ID, a list of pieces that the node has, a hop-count, and so on. For efficient gossiping, SPAWN uses gossiping methods, namely probabilistic spawn and rate-limited spawn. In the probabilistic spawn, nodes forward gossip messages with a certain probability, whereas in rate-limited spawn, nodes forward gossip messages in their buffer with a certain rate; for example, forwarding a random gossip message in the buffer every two seconds. The hop-count of a gossip message is incremented whenever a gossip message

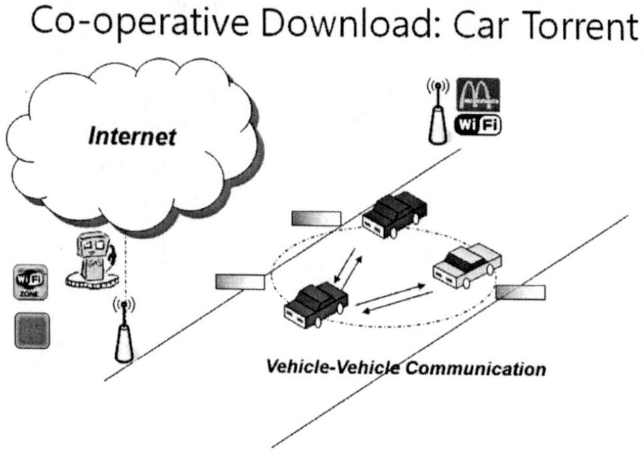

Co-operative Download: Car Torrent

Figure 8.3. Cooperative file downloading in a VANET.

is forwarded. For a given file, there are three types of users in the network: those who are interested in downloading the files, those who are uninterested in downloading the files, and those who do not understand the SPAWN protocol. These roles are considered in the gossiping. For instance, interested users may have a higher probability of packet forwarding than uninterested users.

After the peer/content discovery, a node has to select a peer to download a piece. Given that TCP connections spanning fewer hops perform better in multi-hop wireless networks, SPAWN uses proximity-driven piece selection strategies where the proximity is estimated by the hop-count in the gossip messages: (1) Rarest-Closest First chooses the rarest piece among all the peers in one's peer list, and breaks the tie based on proximity; (2) Closest-Rarest First selects the rarest piece among all the closest peers. Recall that BitTorrent uses a rarest piece first-selection strategy where the rarest piece among all the peers in its list is selected. After peer selection, the node finally downloads pieces by setting up a TCP connection. Any routing protocols such as AODV and DSR can be used for this purpose.

By simplifying SPAWN, Lee et al. (2007) proposed CarTorrent (Figure 8.3). Given that proximity is the key factor of peer selection, CarTorrent uses k-hop limited probabilistic gossiping, and Closest-Rarest First is used for peer selection. CarTorrent uses a cross-layer approach in that route discovery of underlying on-demand protocols is utilized for gossiping. Lee et al. (2006a) proposed CodeTorrent, a network coding-based content distribution protocol. Recall that BitTorrent-like protocols suffer from a coupon collection problem – that is, as a node collects more pieces, it will take progressively longer time to collect a new piece. It is known that network coding can mitigate this problem (Gkantsidis and Rodriguez 2005; Chiu et al. 2006). Figure 8.4 shows that CodeTorrent improves

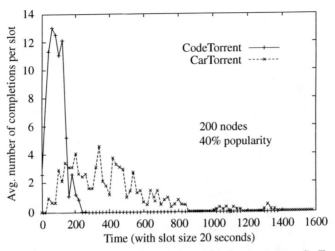

Figure 8.4. CodeTorrent improves download completion time versus CarTorrent.

download completion time versus CarTorrent by almost tenfold in a simulation experiment with 200 vehicles because it eliminates the "coupon collection" problem.

Eriksson et al. (2008a) proposed techniques to improve data delivery throughput. Quick – a streamlined WiFi client – reduces the end-to-end link establishment delay to a WiFi AP, and Cabernet Transport Protocol (CTP) improves the data throughput by differentiating congestion in wired links and packet loss in wireless links. Recently, Yoon et al. (2008) proposed Mobile Opportunistic Video-on-demand (MOVi), a mobile peer-to-peer (P2P) video-on-demand application. Since switching WiFi modes (between infrastructure and ad hoc modes) takes time, MOVi exploits the opportunistic mixed usage of roadside WiFi APs and direct P2P communications using Direct Link Service (DLS) in 802.11 standards that enables direction communications between nodes within a single BSS.

8.2.2.3 Vehicles as Data Sources: Vehicular Sensor Platforms

Vehicular networks are emerging as a new network paradigm of primary relevance, for example for proactive urban monitoring using sensors and for sharing and disseminating data of common interest. In particular, we are interested in urban sensing for effective monitoring of environmental conditions and social activities in urban areas using vehicular sensor networks (VSNs). Differently from traditional wireless sensor nodes, vehicles can easily be equipped with powerful processing units, wireless communication devices, GPS, and sensing

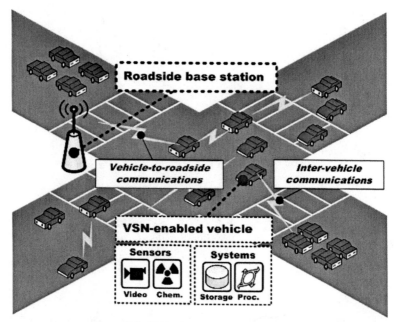

Figure 8.5. Vehicular Sensor Network (VSN).

devices such as chemical detectors, still/video cameras, and vibration/acoustic sensors. Figure 8.5 shows an application scenario.

MobEyes: Proactive Urban Monitoring Services
MobEyes aims to provide proactive urban monitoring services where vehicles continuously monitor events from urban streets, maintain sensed data in their local storage, process them (e.g., recognizing license plate numbers), and route messages to vehicles in their vicinity to achieve a common goal (e.g., to allow police agents to pursue the movements of specific cars). However, this requires the collection, storage, and retrieval of massive amounts of sensed data. In conventional sensor networks, sensed data is dispatched to "sinks" and is processed for further use (e.g., Direct Diffusion [Intanagonwiwat et al. 2000]), but that is not practical in VSNs due to the sheer size of generated data. Moreover, it is impossible to filter data a priori because it is usually unknown which data will be of use for future investigations. Thus, the challenge is to find a completely decentralized VSN solution, with low interference to other services, good scalability, and tolerance to disruption caused by mobility and attacks.

MobEyes is a novel middleware that supports VSN-based proactive urban monitoring applications (Lee et al. 2006a; Lee et al. 2008a; Lee et al. 2008b). Each sensor node performs event sensing, processing/classification of sensed data, and periodically generates data summaries with extracted features and context information such as timestamps and positioning coordinates. Summaries are then disseminated to other regular vehicles, making it possible for patrol cars to move to the scene of an accident, say, and opportunistically harvest from neighbor vehicles the summaries relative to that accident.

Summary Diffusion: Any regular node periodically advertises a packet with newly generated summaries to its current neighbors. Each packet is uniquely identified (generator ID + locally unique sequence number). This advertisement to neighbors provides more opportunities to the agents to harvest the summaries and thus reduces the delay to collect the desired data. The advertise period is set to optimize the tradeoff between harvesting latency and data channel load.

Neighbors receiving a packet store it in their local summary database. Therefore, depending on node mobility and encounters, packets are opportunistically diffused into the network. MobEyes is usually configured to perform "passive" diffusion: Only the packet source can advertise its packets. Two different types of passive diffusion are implemented in MobEyes: single-hop passive diffusion (packet advertisements only to single-hop neighbors) and k-hop passive diffusion (advertisements travel up to k-hop as they are forwarded by j-hop neighbors with $j < k$). MobEyes can also adopt other diffusion strategies – single-hop active diffusion, for instance – where any node periodically advertises all packets (generated and received) in its local database at the expense of a greater traffic overhead. As detailed in the following section, in a usual urban VANET, it is sufficient for MobEyes to exploit the lightweight k-hop passive diffusion strategy, with very small k values, to achieve an efficient level of harvesting.

Figure 8.6 depicts the case of a VSN node C1 encountering other VSN nodes while moving (for the sake of readability, only C2 is explicitly represented).

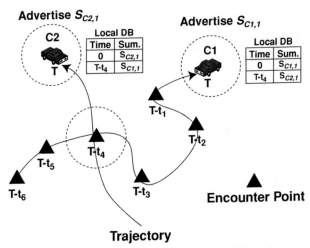

Figure 8.6. MobEyes single-hop passive diffusion.

Encounters occur when two nodes exchange summaries, that is, when they are within their radio ranges and have a new summary packet to advertise. In the figure, dotted circles and timestamped triangles represent, respectively, radio ranges and C1 encounters. In particular, the figure shows that C1 (while advertising $S_{C1,1}$) encounters C2 (advertising $S_{C2,1}$) at time T-t_4. As a result, after T-t_4, C1 includes $S_{C2,1}$ in its storage, and C2 includes $S_{C1,1}$.

Summary Harvesting: A MobEyes police agent harvests diffused summaries that meet particular criteria (typically, time and place where the data was collected) by proactively querying its neighbors. The ultimate goal is to collect all the relevant summaries generated in a given area. To focus only on missing summaries, a MobEyes agent compares its already collected set with the summary list at each neighbor (set difference problem) by exploiting a space-efficient data structure for membership checking, namely a Bloom filter. A Bloom filter for representing a set of n elements consists of m bits, initially set to 0. The filter applies k independent random hash functions $h_1, \ldots, h_k$ to MobEyes summary identifiers and records the presence of each element into the m bits by setting k corresponding bits. To check the membership of the element x, it is sufficient to verify whether all $h_i(x)$ are set.

In summary, the MobEyes harvesting procedure consists of the following steps:

- The police agent broadcasts a "harvest" request with its Bloom filter.
- Each neighbor prepares a list of "missing" summaries from the received Bloom filter.
- One of the neighbors returns missing summaries to the agent.
- The agent sends back an acknowledgment with a piggybacked list of just-received summaries. Upon listening or overhearing this, neighbors update their missing summary lists for the agent.
- Steps 3 and 4 are repeated until there are no missing summaries.

Note that Bloom filter membership checking is probabilistic. In particular, false positives may occur and induce MobEyes regular nodes not to send summaries still missing to the agent. The probability of a false positive depends on m and n (Fan et al. 1998). Nevertheless, in MobEyes, the agent can obtain a missing summary with high probability, because it is highly probable that other nodes have the summaries as time passes, and the harvesting procedure is repeated as the agent moves. For example, in usual VSN deployment scenarios (e.g., with ten neighbors on average), we can show that the probability of missing one summary due to false positives after repeating the procedure multiple times is very low.

Data Retrieval: Harvesting leads to summaries that only contain metadata. From the metadata the agent gets the ID of the actual vehicle that owns the data. The data must then be obtained from that vehicle. This entails finding the current vehicle location (via a *Location Server*) and routing a request (via *Geographic Routing*) to said vehicle to upload the data at its earliest convenience to the Internet. Both Geographic Routing and Location Server implementation are described in later sections of this chapter.

Related Urban Mobile Sensor Platform Projects

Recently, there have been several projects that addressed the sensing of urban data (traffic, pollution, road conditions, etc.) using mobile platforms (cell phones or vehicles). In CarTel (Hull et al. 2006), users submit their queries about sensed data on a portal hosted on the wired Internet. Then, an intermittently connected database is in charge of dispatching queries to vehicles and of receiving replies when vehicles move in the proximity of open access points to the Internet. Eriksson et al. (2008) proposed a system called Pothole Patrol that uses mobility of vehicles, opportunistically gathering data from vibration and GPS sensors, and processing the data to access road surface conditions. Yoon et al. (2007) proposed a method of identifying traffic conditions on surface streets using the GPS location traces collected from vehicles. Eisenman et al. (2006) proposed a three-tier architecture called MetroSense: Servers in the wired Internet are in charge of storing/processing data sensed by cell phones; Internet-connected stationary Sensor Access Points (SAP) act as gateways between servers and cell phone users viewed as mobile sensors (MS); MS move in the field opportunistically delegating tasks to each other and "muling" (Shah et al. 2003) data to SAP. MetroSense requires infrastructure support, including Internet-connected servers and remotely deployed SAP. Wang et al. (2006) proposed data delivery schemes in Delay/Fault-Tolerant Mobile Sensor Network (DFT-MSN) for cell phone-based pervasive information gathering. The trade-off between data delivery ratio/delay and replication overhead is mainly investigated in terms of buffer and energy resource constraints. CENS Urban Sensing project (Burke et al. 2006) addresses "participatory" sensing where cell phone-equipped agents of the same interest participate in an urban monitoring campaign. The data is uploaded to Internet servers via WiFi access points or the 3G network.

As it may have been noted, the previously mentioned urban sensing applications upload the data to the Internet at the earliest opportunity. This is because the data is of immediate need (e.g., traffic information in CarTel or pollution measurements in the CENS Participatory Sensing project). MobEyes differs from the previously described applications in that it collects indiscriminately a massive amount of information (like a security video camera in a shopping mall). Only a fraction of this information will be needed for forensic investigations in case of an accident. Thus MobEyes uploads no data or metadata to the Internet. Rather it opportunistically "disseminates" the data in the urban grid, making it easier for future investigator to search.

8.3 Enabling Protocols

Whereas vehicular networks enable a broad spectrum of applications, automotive safety is still a key motivating application that drives most of the protocol development activities. Initial safety applications (Robinson et al. 2007) such as lane change assistance (LCA), emergency electronic brake lights (EEBL) (Zang et al. 2008), and cooperative collision warning (CCW) (ElBatt et al. 2006) will steer the driver's attention through indicator light, as well as auditory and haptic signals, to warn of potentially dangerous situations. In the longer term, applications may also actively intervene – for example, by conducting automatic collision avoidance maneuvers (Ferrara and Paderno 2006) or through vehicle crash preparation that can reduce injuries for vehicle occupants.

The key protocol challenges in enabling safety applications are:

Reliability and Timeliness in Sparse and Dense Networks

Protocols must deliver messages with high reliability to other nearby vehicles over a broad range of different network scenarios. It must provide reliable message delivery under very sparse traffic conditions – say, two vehicles on a rural highway, with shadowing effects from roadside structures, and in time for vehicles to allow avoiding accidents. Consider the case where a vehicle blocking a highway after a curve sends warning messages to approaching vehicles. Stopping distances at highway speed under wet conditions can exceed 250 m. Messages must also be reliably delivered under extremely dense conditions, when the network is primarily interference limited. In dense urban areas or around major highway intersections during rush hour, hundreds of vehicles may be within the nominal communication range and can potentially interfere with a transmission.

Authenticity and Anonymity

The communication system must be able to validate that messages were generated by a trusted agent. One step toward this goal is authentication of the transmitter. If vandals can spoof safety-critical messages such as a collision warning message, the warning system itself could create enormous psychological stress on drivers and occupants or even lead to rear-end collisions due to sudden breaking. At the same time, the communication system should protect driver's and vehicle's anonymity (except perhaps under well-defined circumstances for law enforcement). Periodically emitting a radio signal with a unique identifier would enable more efficient surveillance technologies that can monitor which vehicles arrive at certain sensitive locations (e.g., hospitals, political meetings, etc.).

To address these challenges, protocol research and development is carried out both in industry standard bodies and in academic venues. This section will first review physical and MAC layer standards, then discuss protocol design for safety applications, and finally review emerging geographic protocols.

8.3.1 Regulations and Standards

There exist a large number of standard activities that cover different aspects of vehicular network communications. This section will focus primarily on the more established spectrum regulations and physical and MAC layer standards.

Spectrum for automotive dedicated short-range communications (DSRC) has been allocated in several countries around the world. In the United States, for example, the Federal Communications Commission (FCC) has reserved spectrum in the 5.9 GHz band and regulates permissible transmission powers. Even though the FCC allows an Equivalent Isotropically Radiated Power (EIRP) as high as 44.8 dBm for public safety applications, regular vehicles are limited to an EIRP of 2 W using omni-directional antennas. Still EIRPs and the 800 mW maximum antenna input power is much higher than the maximum allowed by 802.11a to enable communications ranges up to about 1,000 m under line-of-sight conditions. Considering that the frequency band of 5.9 GHz is significantly affected by shadow fading, the effective range is often much less. Figure 8.7 shows the delivery probability for different distances simulated using ns-2 Rayleigh fading channel with parameters tuned from outdoor vehicle experiments. This

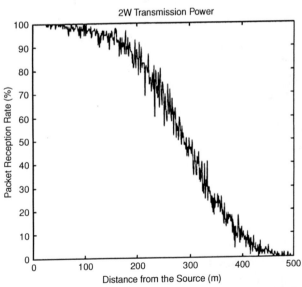

Figure 8.7. Simulated delivery probability versus distance in a Rayleigh fading channel using a 5.9 GHz 802.11p setup at 2W transmission power.

simulation shows that the effective range even at highest transmit powers can be expected to be a few hundred meters, with the range further reduced if more severe obstructions are present.

MAC and PHY Layer. Vehicular ad hoc communications are generally assumed to operate over an OFDM physical layer and CSMA/CA MAC. Such protocols are defined in the IEEE 802.11p working group (IEE 2006), which has adapted IEEE 802.11 protocols for vehicular characteristics. The physical layer remains very similar to an 802.11a OFDM PHY, except for the following changes. The 802.11p works in the band allocated for ITS applications, 5.850-5.925 GHz, allowing a total of eight channels of 10 MHz bandwidth. This differs from the 20 MHz channels in 802.11a, but some channels (i.e., channels 174,176 and 180,182) may be optionally combined to yield 20 MHz channels. The reduced channel bandwidth reduces bitrates to a maximum of 27 Mbps, but using the same number of subcarriers as in 802.11a makes 802.11p more robust to frequency selectivity of wideband channels. A higher OFDM Guard Interval of 1.6 μsec also makes 802.11p more robust to intersymbol interference caused by the high Root-Mean-Square (RMS) delay spreads that are encountered in vehicular environments 10 ns to 40 ns for vehicle separations of 10 m to 30 m and LOS, up to 400 ns in NLOS scenarios (Zang et al. 2005) as compared to 50 ns indoors. In addition, 802.11p has a longer preamble than 802.11a, allowing for better channel estimation.

The multiple-access mechanisms remain largely unchanged from 802.11a. The association mechanism has, however, been redesigned to account for the more dynamic nature of vehicular networks. To reduce the need for active scanning, 802.11p designates channel 178 as a control channel. The exact protocols are still under consideration, but it is expected that each station must periodically listen to this channel so that stations can negotiate the use of the other service channel. Frames on the control channel are always transmitted at a rate of 6 Mbps. Any prospective WAVE BSS user starts listening to the control channel for "WAVE Announcement action frames" that contain all the information required to join the BSS.

Wireless Access in Vehicular Environments (WAVE) Standards. The IEEE P1609 standard family defines an architecture and key services for vehicular networks. These include resource management, security, and multichannel operations. Finally, standards such as SAE J2735 contain a message dictionary that defines message formats for the exchange of vehicle and road information. It is typically assumed that vehicles know their own location, for example, through Global Positioning System (GPS) receivers. Using a message defined in this standard, vehicles can then transmit their current position and past trajectory to other nearby vehicles. Messages for advertising road topology and infrastructure also exist. Whereas the standards cover many aspects of the communication protocols, other aspects of the system, particularly antennas and applications, are left to car manufacturers. On new vehicles, the communication system may be connected to the vehicle bus to in-vehicle sensors including brake, traction control sensors, and the radar and lidar sensors deployed in some new vehicle models for adaptive cruise control. New vehicles with built-in systems carry the antenna on the center rear part of the roof. Other vehicles may mount the system and antenna near the rear-view mirror on the inside windshield, similar to current electronic toll tags. This position would enable quick deployment on legacy vehicles. Some systems may interpret the message simply to provide driver warnings, whereas other vehicles may use the information to configure vehicle systems such as the braking system.

8.3.2 Broadcast Protocols for Safety Applications

To support safety applications, each vehicle must have knowledge of the surrounding vehicle constellation, which is the position, speed, acceleration, and yaw rate of other nearby vehicles (typically vehicles within a 300 m radius). Thus, a key communication primitive for vehicular safety applications is a periodic broadcast from each vehicle to disseminate vehicle movement information. By receiving these position announcements, each vehicle can then combine all received reports to create a view of the surrounding vehicle constellation. The exact use of this information then depends on the specific safety application; it may be used to issue warnings to drivers or to take precautionary actions. Current U.S. standard deliberations are considering a messaging rate of 10 Hz for each vehicle.

8.3.2.1 Scalability and Density

Eventually, the vehicular network must scale to include all motor vehicles in the country. Although it may take many years of deployment efforts to reach this goal, it is worthwhile to consider how the technology could scale to such large and dense network scenarios early on to avoid costly recalls at a later time. According to the 2004 Highway Statistics (Federal Highway Administration 2004), there are about 240 million registered motor vehicles in the United States. The number of vehicles is similar to the number of phones supported by cell

Figure 8.8. Example regions with potentially high vehicle densities: (a) Junction of Freeway 110 and 105 (b) Intersection in New York City.

phone systems, but the network challenges are fundamentally different in that vehicles must dynamically organize themselves into local networks and allocate spectrum resources rather than relying on a carefully planned base station setup for coordination.

When deployed to a large number of vehicles, the system must meet its reliability requirements even in very-high-node density environments, which can be expected in rush-hour traffic on highways or in urban centers. Example regions that may encounter high densities are depicted in Figure 8.8.

The exact specifications are still under deliberation, but the Wireless Access in Vehicular Environment (WAVE) (IEE) and SAE standard groups are currently defining wherein each vehicle disseminates its position and vehicular dynamics information via periodic broadcasts. The broadcast rate is application-specific, but it is generally assumed to be one message every 100 ms per vehicle. The message size is typically less than 100 bytes but can reach larger sizes due to a large authentication header. In some cases, total message sizes can reach 500 bytes if optional payload information like path histories is included. Without the security overhead and optional information but considering MAC protocol overhead, this yields a typical data rate requirement of about 5 kbit/s/vehicle. With security overhead, it becomes approximately 10 kbit/s/vehicle. At first glance, this appears to be a very modest data rate requirement, but vehicular networks are interference-limited because the node densities that can be expected in vehicular scenarios make meeting this requirement challenging. For example, consider a congested, slow-moving, two-way highway with four lanes each and one car every 10 m. This results in 480 cars within a 600 m interference range of a 300 m transmission, thus requiring approximately 5 Mbps capacity if transmissions can be perfectly scheduled but more than 10 Mbps with the currently envisioned less efficient CSMA-based protocols. Unfortunately, current state-of-the-art DSRC technology cannot provide this capacity for the required communication range because radios have to operate at lower channel bandwidths of 10 MHz to reduce multipath delay spreads and Doppler effects in the vehicular environment.

These interference limitations motivate data aggregation approaches that can reduce bandwidth requirements while still achieving similar communication ranges.

8.3.3 Emerging Geo-Protocols

Vehicular networks are intricately linked to the physical world, and their applications require that each vehicle be able to monitor its position. These characteristics have also lead to a number of proposals that use geographic position information to improve network performance, rather than just dissemination vehicle positions over a location-agnostic network stack. One class of such protocols is geocasting – the delivery of messages to all nodes within a defined area. There exists a natural match to the typical requirement to disseminate vehicular movement information to all vehicles within a radius of say 300 m. Particularly when data aggregation and multi-hop forwarding are used, message propagation is no longer limited by a single vehicle's transmission range and requires other mechanisms to prevent flooding of the network. Establishing a geographic boundary for message propagation can fill this need.

In a multi-hop message forwarding scenario, geocast protocols could further increase network efficiency if safety applications can define smaller message delivery zones based on map information or recent vehicle trajectories. For example, consider the extended electronic brake light scenario illustrated in Figure 8.9. Here a brake message should be reliably delivered to all following

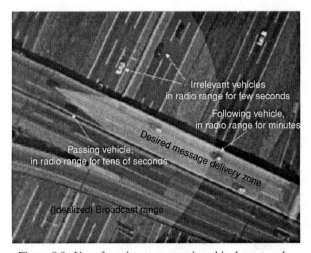

Figure 8.9. Use of persistent geocast in vehicular networks.

vehicles, where some vehicles might be out of communication range of the origin vehicle. Thus, some of the following vehicles must forward the message. The message is, however, only relevant to vehicles following on the same road; vehicles on the intersecting highway do not need to be notified. In this scenario, the intended message recipients are frequently changing, thus enumerating destination identifiers for each vehicle is cumbersome. A naive flooding approach with a time-to-live parameter might propagate in unintended directions, causing unnecessary network congestion. A geocast services provides a natural way to identify the destination vehicles through a geographic perimeter and can optimize message forwarding to only include nodes in the intended destination zone.

The concept of a geocast has first been proposed by Navas and Imielinsky of WINLAB in 1997 (Navas and Imielinsky 1997). For such highly mobile vehicular applications, this concept requires adaptation for ad hoc communication and persistence to notify new cars in the area. Whereas work in the MANET and sensor network field such as Mobicast (Mobile Just-in-time Multicasting) (Huang et al. 2002) has addressed multicasting in mobile ad hoc networks, protocols tailored specifically to vehicular networks have been proposed. For example, Maihöfer (2005) proposed *abiding geocast* that extends the earlier geocast models by including the notion of message validity duration. The abiding geocast protocol will not only deliver the message to all nodes present in the geocast region during the initial message transmission, but also continue to deliver the message to all vehicles that enter the geocast zone during the message validity duration. It also considers infrastructure-assisted geocast.

Another form of geographic protocols, *georouting*, can also find application in vehicular networks, for example, to transmit a message to a roadside unit that cannot be reached in a single hop. Rather than using the knowledge of logical link level associations to find a path from the source to the destination, as in typical topology-based routing schemes (Perkins and Royer 1999; Johnson and Maltz 1996), geographic routing finds the path by using location information of the destination and potential forwarding nodes. A node chooses as next-hop forwarder the neighboring node that is closest in geographic space to the destination node. The main advantage of location-aware routing is that it does not require route establishment and maintenance, which can be costly in highly dynamic vehicular networks. It does require, however, that the location of the destination node is known, which is easier to achieve for stationary roadside infrastructure nodes compared to mobile nodes.

Resiliency to mobility and channel variations in vehicular networks can be further improved through opportunistic protocol techniques. Standard routing protocols (including georouting) select a next-hop neighbor based on their routing metric and instruct the MAC layer to unicast a packet to this selected node.

This appears wasteful in dense wireless networks, because this particular destination may be unreachable due to fast or slow fading, whereas there are likely neighboring nodes that correctly receive the frame immediately but discard it due to the incorrect MAC address. A realization of an opportunistic protocol could use a soft destination, where any node close to these coordinates can forward the packet. This approach takes advantage of the additional resources in a dense network, and conceptually it realizes a form of cooperative diversity gain that recent works in the information theory community (e.g., Laneman and Wornell 2003; Laneman et al. 2004; Nostratinia et al. 2004) have shown to provide large gains in networks with idle nodes, or in a slow, fading environment. Ignoring all protocol overheads and assuming a Rayleigh channel, best-case gain estimates follow those for selection combining, which predicts a 12 dB diversity Signal-to-Noise Ratio (SNR) gain using two receivers (over one receiver) for an outage probability of 0.01 (Goldsmith 2005). Adding further receivers yields diminishing returns, but a third and fourth receiver still provides an additional 7 dB and 4 dB gain, respectively. By realizing similar diversity gains, cooperative protocols can operate at lower transmission power and thus increase spatial reuse. The key challenge in such cooperative protocols lies in low-overhead distributed forwarder selection algorithms. One approach is to allow multiple forwarders to contend through a backoff mechanism that skews the probability of channel access to forwarders in closer proximity of the destination (Kaul et al. 2008).

8.3.4 Security

Another key challenge in vehicular network protocol design is providing authenticated communication while maintaining anonymity or pseudonymity of the vehicles. Let us first consider the authentication mechanisms. To authenticate messages, the current standard for security services in DSRC/WAVE considers digital signature-based authentication primitives. Keys and certificates for vehicles could be issued during the vehicle registration process. Using elliptic curve cryptography, the overhead of these signatures on a packet amounts to a manageable few tens of bytes. The public keys for verifying messages could be distributed through certificates appended to the messages, which would increase message length noticeably, or could be periodically broadcast by each node.

> **Security of Aggregated Messages.** Aggregated messages, however, require additional protection because a spoofed or faulty message can misrepresent sensor information from a large number of vehicles. Consider again the position-monitoring application that collects the

location of nearby vehicles. Let us assume that a regular record contains the current location and speed of the vehicle, a timestamp, and a signature with certificate to authenticate the message. An aggregator could then combine multiple records syntacticly: By listing the vehicles' positions in a single record authenticated with a single signature and certificate. It could also aggregate information semantically, for example, by describing a bounding box that contains all vehicle positions. Again, to reduce overhead, the aggregated message ideally would contain only a single signature and certificate. Because authentication only establishes the source but not the correctness of the content, a regular message may contain incorrect or spoofed positions information for a vehicle. Simultaneously inserting a large number of spoofed vehicles would require having the same number of valid keys available. An aggregated message, however, could contain an arbitrary number (subject only to packet size constraints) of position claims using only a single key.

One approach to address this issue is probabilistic validation. A receiver can probabilistically verify the correct aggregation by requesting the original record for a randomly chosen identifier contained in the aggregated message. Full validation means obtaining the complete set of original records, checking their signatures, and confirming that applying the aggregation function to this yields the aggregated message. To reduce the bandwidth and resource consumption of this validation process, the receiver can use the probabilistic method. The randomly chosen identifier acts as a random challenge, so that the sender does not know beforehand which original message will be verified. If the same vehicle repeatedly does not supply a valid original message for the requested record, the receiver can assume that this record is spoofed. Note that this validation method can only catch spoofing of additional vehicles, not omission of existing vehicles. Omission of vehicles can, however, be more easily addressed by the nodes surrounding the aggregator. If they overhear an aggregated message that undercounts vehicles in the area, they can send a corrected aggregated message to add the additional vehicles (assuming the majority of vehicles are trustworthy). Removing spoofed vehicles would require collaboration between multiple nearby nodes, because no single node can be sure that the additional vehicles do not exist outside its radio range. Note also that probabilistic validation is most effective if a method for recourse exists, which penalizes the sender of the message. For example, the receiver could notify authorities of the suspected tampering with the vehicle communication system, who can track down repeated offenders.

Temporary Keys. Signatures and certificates are essentially pseudo-identifiers because a receiver knows that two messages originate from the same sender if both signatures can be verified with the same public key. Since DSRC communications are broadcast over the wireless medium other vehicles and any unauthorized parties can record such pseudo-identifiers from vehicles in the vicinity. By monitoring the identifiers at multiple locations, third parties could calculate a vehicles' average speed, or they could monitor the identifiers of vehicles visiting a sensitive location (e.g., medical clinic). Over time, these bits of information can create a profile that identifies the driver. To address these privacy concerns, inter-vehicle communication protocols should ideally be free of such static pseudo-identifiers but still provide basic authentication functions.

One approach to provide privacy while authenticating is switching among a large number of temporary keys. Because storage is relatively affordable, each vehicle could store a large number, say ten thousand, of certificates that can be used. Used keys could be replenished during vehicle maintenance or be remotely updated over a wide-area connection. It is critical that only one of these certificates is valid at any given time to prevent spoofing of other vehicles. The degree of privacy can be increased by using a higher switching frequency. More frequent key changes, however, make it more difficult to implement the safety applications that rely on tracking paths of nearby vehicles. It also creates a tension with the secure aggregation approach, because messages from a suspicious node cannot be filtered when it switches to a new key. A good solution must balance all these requirements.

8.4 The Role of the Infrastructure: MobiMESH and GLS

One of the unique features of the VANET is the omnipresence of the infrastructure. In fact, the wired infrastructure is accessed through a thin *wireless mesh* layer. It is thus important to understand the interaction and interdependence between vehicular networks, wireless mesh, and Internet. To start, VANET applications benefit from the support of Internet and wireless mesh network in many services:

Mobility: The infrastructure manages mobility. Mobility management (e.g., knowing where vehicle X is at time T) requires the registration with a location server (centralized or distributed) that is built in the infrastructure. The location server accepts registrations of participating vehicles as they roam the city and maintains the equivalent of a DNS mapping vehicle IDs to current estimated geolocation and AP to reach the mobile. Moreover, the infrastructure must facilitate AP to AP soft-session handoff of roaming mobiles. This is provided for both TCP sessions and stream and is critical for real-time applications (voice, videoconference, and interactive games).

Security and Authentication: An important overarching concern in VANETs is security and privacy. The infrastructure helps "authenticate" the alarms received from other vehicles – filtering bogus attacks, for example. It also helps preserve "location" privacy considering that vehicles potentially jeopardize such privacy by exchanging beacons, advertisements, and warnings with other drivers. Security and privacy guarantees require a certifying authority residing in the infrastructure.

Routing: The shortest path between two vehicles may go through the infrastructure. Using in part information supplied by the infrastructure (e.g., city map, vehicles density in various sectors, urban WiFi channel load, etc.), each vehicle can determine whether it is better to route a packet totally within the VANET or partly through the wired infrastructure (Gerla et al. 2006). Namely, the "Data Routing Advisory" is analogous of the Navigator Advisory for data packets.

Urban measurement repository: The infrastructure can serve as storage of various vehicle measurements ranging from traffic, pollution, and mobility pattern all the way to individual vehicle traces and bogus alarm attack reports. In particular, it keeps records of the data collected during VANET experiments.

Emergency operations: As drivers become progressively dependent on VANET services, such services should be maintained even when the infrastructure partially or totally fails. Critical VANET protocols (routing, capacity estimation, location service, resource allocation, and security management) must be carefully designed so as to allow a reliable transition from full Internet support to completely autonomous ad hoc operations in case of infrastructure facility failure or destruction. This is particularly important because in such situations, the VANET will be the only "infrastructure" available for emergency services such as vehicle evacuation and search-and-rescue team networking. It will offer an important backup to Public Emergency Networks like TETRA. The roadside wireless mesh will play an important role. APs powered by solar generators will use cognitive radio capabilities to reestablish a fixed, wireless emergency backbone throughout the affected urban area.

The services described here are best illustrated by describing the functionalities of the vehicular mesh, called MobiMESH (Capone et al. 2006), that is being installed at UCLA as part of the C-VeT testbed. The following properties make MobiMESH particularly suited for C-VeT support:

- *Broadband Backhauling* – the MobiMESH networks are able to build up and dynamically maintain a broadband wireless backbone that can be used to support/complement vehicle-to-vehicle communications;

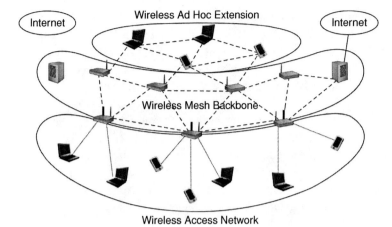

Figure 8.10. MobiMESH network architecture.

- *Mobility Support* – wireless devices are allowed to seamlessly roam within MobiMESH networks without losing active connections;
- *Flexibility* – the MobiMESH networks are self-configuring and self-managing.

A concise technical description of MobiMESH follows.

8.4.1 The MobiMESH Architecture

MobiMESH features a hybrid mesh network architecture. Indeed, the network consists of three main architectural building blocks shown in Figure 8.10:

- a mesh backbone composed of MobiMESH wireless mesh routers that provide the routing and mobility management infrastructure, and is further connected to gateways;
- an ad hoc extension responsible for extending MobiMESH functionalities to mobile nodes;
- an access network that can be used by standard WiFi clients to get connectivity.

The mesh backbone and the ad hoc extension are based on the ad hoc network paradigm, where all nodes and mesh routers collaborate to route traffic. Routing on the mesh is provided through a proactive ad hoc routing protocol based on OLSR (Clausen and Jacquet 2003) and properly modified to account for multiple radios at the mesh nodes, and for varying link quality metrics. The backbone network is also responsible for the integration with the wired network, through gateways equipped with a wired interface that can route traffic to the Internet. The access network is rather flexible and operates in the infrastructure mode, so that standard clients perceive the network as a standard WLAN and behave accordingly; in this way, MobiMESH can also be accessed by standard WLAN clients (e.g., pedestrians) with no specific software installed.

The MobiMESH Mesh Routers represent the main building block of the MobiMESH network because they are responsible for creating the broadband backhaul, further offering access to wireless mobile clients. The MobiMESH Mesh Routers can be equipped with two to four radio interfaces that can be flexibly used either as backbone or access interfaces. Moreover, any interface can be tuned to any available channel in the two frequency bands 2.4 GHz, 5.7 GHz, and 5.9 GHz (via DSRC). Mesh Routers with an interface dedicated to WiFi access are called Access Routers.

An important overarching concern in C-VeT environment is security and privacy. The MobiMESH network provides security functions, so that it can be safely employed to deliver any kind of traffic and to extend preexisting secure networks. In a Wireless Mesh Network (WMN), it is very important that only authorized devices can join the network; MobiMESH Mesh Routers are in fact authenticated through the use of X.509 certificates, and the backbone traffic is encrypted through a time-changing key encryption algorithm. Moreover, centralized MAC filtering and captive portal functionalities are supported.

MobiMESH architecture implements a proprietary mobility support daemon that dynamically handles the MAC-IP address association as clients roam throughout the network. Experiments carried out on real deployments have shown that the handover latency for a wireless client changing Access Router is upper bounded by 20 ms in most of the cases. Consequently, the handover is not perceived during VoIP calls.

In the following section, as an example of Infrastructure Service, we describe the Geo-Location Service (GLS) targeted for implementation in C-VeT.

8.4.2 The Geo-Location Service (GLS)

The Geo-Location Service (GLS) is a distributed service that maps any car ID to its most recent geo location. Exploiting MobiMESH, we propose an Overlay Location Service (OLS) implementation. As shown in Figure 8.11, an overlay structure is established in MobiMESH. Periodically (say, every minute) each car registers to the nearest MobiMESH APs with its ID (license#, IP address(es), time, owner name, owner IP address billing address, etc.) and the current geo-location. In normal operating conditions, OLS spans both the MobiMESH and the wired Infrastructure. In case of infrastructure failure, OLS can be completely supported (with some loss in performance) by MobiMESH, assuming the latter

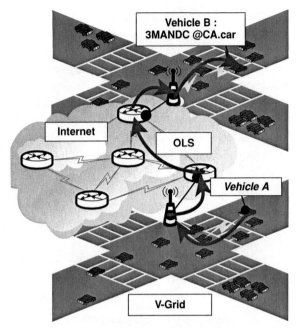

Figure 8.11. Location service and routing in the MobiMESH.

is fully connected by virtue of long-range Cognitive Radio links. OLS maintains an index of vehicle IDs. Each ID is mapped to the most recent geo coordinates (thus allowing motion prediction). The index is distributed across the overlay. It may be managed via DHT (Distributed Hash Table).

To illustrate the OLS operation, suppose that mobile host A wants to establish a TCP connection to mobile host B (see Figure 8.11). Host A injects in the nearest MobiMESH AP the query: 3MANDC@CA. It gets back the "most recent" set of time-tagged geo-locations of host B. From these, it can estimate vehicle speed and direction and thus infer the current location of B. A then selects the best AP to reach the destination. Host A encapsulates the message in an IPv6 network envelope with destination geo address in the extended header. The destination AP geo-routes the packet into the vehicular network to B using geo address, car ID, and MAC in the header. Upon successful delivery, car B responds with its own IP address and geo address. It directs its response (encapsulated in the overlay envelope) to the sender IP address.

8.5 Vehicular Testbeds

The primary goal of the vehicle testbed is to enable V2V and V2I experiments aimed at the evaluation of VANET protocols and applications in a realistic setting. It must allow external users to define, execute, and monitor various experiments. It must allocate resources so that users can efficiently share the testbed. It must assist the experimenters with software tools such as traffic generators, measurement collection, and preprocessing facilities, and possible interface to emulators and simulators.

In addition, the vehicle testbed must interact and interwork with the infrastructure so that the applications being tested can benefit from the various services of the latter. In particular, it manages coexistence of car-to-car 802.11p channel with WiFi-based mesh infrastructure; it interfaces with the infrastructure for support in mobility management, routing, traffic control, and congestion control; it facilitates transparent interconnection of vehicles across the city via the wired Internet; it enables the VANET to operate with and without infrastructure support with smooth transition between the two modes and phasing out of noncritical applications.

In this section, we present two vehicular testbed implementations: the UCLA C-VeT testbed and the ORBIT based Rutgers testbed.

8.5.1 C-VeT Architecture

C-VeT is an open platform that supports vehicular network and urban sensing research and related applications. It is inspired to the pioneering work done by Larry Peterson and Tom Anderson with Planet Lab (Peterson et al. 2002). It features an always-on, fully virtualized, Internet-accessible, sensor-equipped testbed infrastructure. The UCLA campus, with its 10 acres of urban development, reproduces many of the scenarios, propagation, and communication challenges typical of a city, in a realistic manner but yet relatively small-scale. In particular, the C-VeT architecture provides:

- A fully virtualized platform that runs both Linux-based and Windows-based operating system with full insulation among the guest virtual machines, and enables the users to redesign low-level protocols such as, for instance, MAC protocols. This feature will be key for network centric experiments.
- A Campus Wide Mesh network developed using OPEN WRT and optimized for the integration and support of the vehicular network. It will help cope with network disruptions (quite common in small-scale testbeds) and enable opportunistic, interactive, and delay-tolerant experiments that exploit the infrastructure.
- 30 facility management vehicles equipped with the C-VeT hardware/software, providing an always-on platform to run experiments and collect traces and measurements. The facility management vehicles perform both routine maintenance trips and on-demand interventions in response to

emergencies resulting in a varied mobility pattern that well approximates real city traffic.

- 30 commuting vans, equipped with the C-VeT-Census platform that will survey the environment gathering traffic and air quality, and stereoscopic images. The aim is to build a large micropollution database that enables new models and also facilitates visual environment surveys (see Figure 8.1).
- A number of downloadable, preconfigured virtual appliances to allow users to develop the protocols to be tested at home with a compatible software configuration.
- A large-scale emulator that will allow users to debug their algorithms and protocols on the same hardware as the actual C-VeT nodes but with an emulated network component developed with the Qualnet hybrid simulation.
- A robust Internet interface that will manage the users and deploy the experiments in a streamlined fashion. The Web server will provide the front-end for a number of user-friendly services and tools enabling users to focus on research rather than testbed implementation. For example, services to set up the experiments and gather the data; APIs to low-lever interfaces for hardware component virtualization; virtual MadWiFi layer for the support of virtual machines.
- The ability to develop algorithms, applications, and protocols that directly operate at Layer 2 using a TUN/TAP mechanism for both Windows and Linux OS. Recent research showed that the TCP/IP suite may not be the most appropriate choice for vehicular networks and a ground-up protocol stack redesign is needed.
- An organized live database of mobility traces, sensed environmental data, road traffic information, Vehicle CanBus statistics, MAC layer statistics (through MAD WiFi) and physical layer statistics taken using a variety of radios (Cognitive Radios, MIMO, etc.). This data collection will be made available to the research community in collaboration with existing trace collection programs and archives such as CRAWDAD (Kotz and Henderson 2005).

The testbed was designed using a top-down approach; the whole system can be described through a number of relatively simple building blocks: the C-VeT mobile node, the C-VeT mesh node, the C-VeT-Census platform, the Web-based control center, and the emulation platform.

The C-VeT infrastructure is designed to provide an always-on facility for research in wireless vehicular network. To achieve this goal, we chose to install our equipment in the UCLA campus facility management and van pool vehicles.

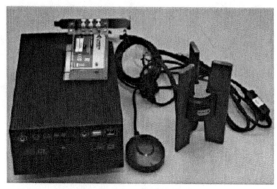

Figure 8.12. C-VeT mobile node.

Those cars and vans are driven everyday to fulfill the campus needs and perform both routine and nonroutine tasks. In addition to the permanent facility vehicles, there is a small pool of private vehicles equipped with C-VeT nodes that can be driven by the researchers themselves for customized, controlled experiments.

The C-VeT mobile node (Figure 8.12) is an industrial-strength Cappucino PC powered by an Intel Dual Core Duo processor at 2.5GhZ, 2GB of RAM, and 320GB of disk. Hard drive and internal parts are rugged to sustain physical stress (i.e., large temperature fluctuations, vibrations, etc.). The PC has three wireless interfaces: IEEE802.11a/b/g/n based on the Atheros AR9160 chipset; IEEE802.11p interface based on a Daimler-Benz customized chipset; and a standard Bluetooth interface mostly for internal communications.

Other radios can also be retrofitted in the mobile node platform. In particular, a few vehicles may be equipped with programmable Silvus SC2000 MIMO platforms (4x4 configuration) that provide full access to the physical layer and enable a new generation of experimental MAC layer research.

On Board Sensors: The C-VeT nodes are instrumented with a customized sensor platform designed to provide a flexible data collection. This includes Infrared-based CO_2 sensors; electrochemical CO sensors; SIRF III or Ublox-based GPS sensors; temperature, and humidity sensors; and a megapixel camera. Using the C-VeT cars as mobile air quality sensors will enable a new wave of atmospheric research aimed at the use of mobile sensing agents to study the air quality at the neighborhood level. Part of the fleet will feature high-performance exhaust particulate sensors DC2000CE by Echocem [ECO], thus being the first testbed able to support the currently leading research in microclimate air quality.

The C-VeT mesh node is based on MobiMESH hardware. C-VeT mesh nodes feature Open WRT OS and Atheros Chipset with MadWiFi support, thus easing up the integration with mobile nodes. The fixed infrastructure will be installed on

Figure 8.13. C-VeT infrastructure.

the roof tops of UCLA buildings aiming at full campus coverage and integration with the existing campus WiFi infrastructure. The mesh allows opportunistic Internet access from vehicles and also provides a control channel to the vehicles. The mesh network can be configured via the Web; e.g., customized routes can be set up by the network operator to perform particular experiments. This C-VeT integrated approach with infrastructure and vehicles broadens the experimental scenarios. In the initial phase, we will cover the south campus, and creating an initial backbone of six mesh points. The initial campus coverage map is shown in Figure 8.13.

To achieve seamless integration between the CVET-Mesh and the Vehicular network components, we will develop Layer 3 and Layer 2 routing and VLAN support. Level 3 network layer routing between moving vehicle and the fixed nodes will enable communications across campus and to the Internet. The Layer 2 routing will enable the experimenter to force mobiles to be in the same broadcast domain, ignoring the fact that there are several fixed nodes in between.

8.5.1.1 Testbed Deployment and Preliminary Results

Infrastructure Nodes Coverage

To find the best placement of the infrastructure nodes, we ran a campaign of coverage tests around the UCLA campus. The main focus is on the coverage of the roads. This represents a hard challenge because we experienced that the WiFi radio signal basically propagates only in Line of Sight (LOS). To assess the coverage of a single infrastructure node, we equipped a car with a laptop, a GPS receiver, and a IEEE802.11b/g wireless card. The car node would log every second its position and if it is in reach of the infrastructure node or not. Using this

Figure 8.14. Coverage experiment from the Ashe Center Building at UCLA.

information, we were able to plot the coverage map of each single infrastructure node. Figure 8.14 shows the coverage map for the infrastructure node placed on the top of the Ashe Center Building at UCLA. White dots represent the covered locations and red dots the unreachable ones. The results show that we were able to cover the whole area called Westwood Plaza that extends up to 700.

Video Streaming

As a preliminary experiment, we wanted to test the feasibility of a video transfer from a mobile node to an infrastructure node via the wireless mesh. The mesh consists of four nodes on the four corners of Engineer IV building at UCLA. In this configuration, each node could reach only the two nodes that are next to it. This means that to reach the farther node, two hops are required, as shown in Figure 8.15.We placed a webcam in the moving car and used VLC to stream

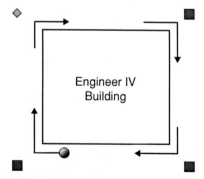

Figure 8.15. Video Streaming experiment: 1 moving video source (CSircle), 3 fixed nodes providing connectivity (Squares), and 1 fixed receiving node (Diamond) meters away from the infrastructure node. On the other hand, as soon as we lose the LOS, the connection breaks, as evidenced by traces on one of the crossing roads.

the video to one of the fixed nodes. With this setup, the car is always connected to the mesh and at most two hops away from the receiving node. To maintain connectivity and fresh routes, we used the OLSR (Clausen 2003) implementation provided by INRIA. The webcam was generating a video stream at resolution of 176 × 144 pixels at 15 frames per second. Thus the stream was generating an average of 128 Kbps (since the codec used was DIV3 the bitrate was not constant due to dynamic compression). The video was streamed using UDP, so the lost frames were not retransmitted. The VLC server was set with a cache of 200 ms.

On the receiving node, we were both saving and displaying the video. In the real-time video transfer, the missing frames were much more than 10 percent, but because we were saving the raw data received from the source, we were able to reconstruct and re-encode the video received. In Figure 8.16, we show the loss rate for the video after the reconstruction. As shown in Figure 8.16, the percentage of loss for both frames and blocks is approximately 10 percent. Such a loss still grants the possibility of actually displaying the video. For real-time delivery, the reconstruction buffer cannot be used. Forward error correction schemes and adaptive coding rate may be used in this case. Another important result of this experiment was the time when the frame losses occurred. In fact they occurred when the mobile node was swapping from one relay to another. This means that the refresh of the route is not fast enough to be transparent for the video stream. These experiments were useful to determine the impact of wireless mesh multihopping on real time traffic. Clearly, buffers and coding strategies must be properly matched to the topology and user requirements.

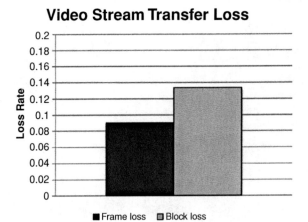

Figure 8.16. Loss rate for the video stream.

8.5.2 ORBIT Vehicular Testbed

ORBIT Indoor Testbed. The ORBIT laboratory testbed (Figure 8.17) comprises 800 IEEE 802.11a/b/g devices attached to 400 nodes in a 20-by-20 meter space that provides a controlled environment to generate reproducible results.

Mobile Outdoor Testbed. The ORBIT testbed also includes a vehicular outdoor field trial component. It comprises several building-mounted 802.11 base stations, vehicular nodes, programmable smart phones, and 3G data accounts for experimental purposes provided by a campus cellular network operator. Stationary nodes are deployed at five different locations, with ten nodes close to the ORBIT facility at the NJ Tech Center and three locations in Rutgers University Busch Campus. All outdoor nodes are connected through back-end Internet links using Ethernet tunnels to each other and the ORBIT control facility. The back-end interface can be used for experiment control, remote data collection, and to allow configuration of different network topologies.

As shown in Figure 8.17, the vehicular nodes use the same base node platform as used in the indoor ORBIT testbed to enable seamless moving of software by copying disk images between the testbeds. Every node is a custom-designed small form factor PC with 1GHz Via C3 CPU, 512 MB RAM, and 20 GB hard disk with remote management interface. The nodes include two IEEE 802.11 a/b/g interfaces whose PHY and MAC layers are similar to the ones defined in the DSRC/WAVE standards. For positioning, Garmin 18 5 Hz Global Positioning System receivers are used to obtain position updates at higher frequency (standard receivers provide only 1Hz samples, during which a vehicle can move

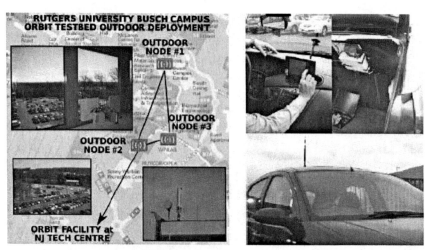

Figure 8.17. The ORBIT wireless research testbed: (a) Campus outdoor setup (b) Vehicular setup.

up to 30 m). The cars use magnetic mount omnidirectional external antennas for 2.4/5GHz. A 12-to-120V power inverter that serves as the power supply (via the car battery) and the setup includes optional keyboard and 7in LCD for experiment control.

Also available in the outdoor testbed are 10 Nokia N95 smart phones powered by an ARM11-based Texas Instruments OMAP2420 running at 330MHz. It is equipped with 64MB RAM, 160MB internal memory, and a flash memory that can be expanded up to 8GB. Short-range communication options include wireless LAN (802.11 b/g) and Bluetooth 2.0 EDR. The N95 also includes a built-in GPS receiver based on TI's GPS5300 NaviLinkô 4.0 single-chip solution for GPS and A-GPS. The Nokia N95 runs Symbian OS v9.2 and is programmable using C++, Java J2ME (MIDP 2.0, CLDC 1.1), and various scripting languages.

8.6 Conclusions

In this chapter, we have surveyed the emerging VANET applications, ranging from vehicular sensors to entertainment. We have contrasted VANET to traditional MANET design, identifying the unique VANET features and requirements. Given these unique features, we have proceeded to classify a representative set of VANET applications based on the vehicle's role in managing data: as source, consumer, source/consumer, or intermediary. We have then reported a vehicular sensing application – MobEyes – and a content distribution application – CarTorrent.

We have then introduced the protocol suite that makes such applications possible. The main focus was on routing and on emerging geolocation-based protocol architectures; on delay-tolerant routing; and on security and privacy.

We then identified the critical role of the infrastructure in the deployment of VANET applications; we introduced the notion a wireless mesh network and its role in support of mobility management.

Finally, we introduced the VANET testbeds that are being deployed at UCLA (C-VeT) and Rutgers (ORBIT-based Vehicular Testbed). We also reported preliminary experiments with live video uploads to an Internet client via a four-node mesh network.

The future of VANET research is bright. There are a number of compelling applications ready to be deployed, and users are eager to try them out. The protocols and the standards are nicely coming into place. The remaining roadblocks in VANET deployment and broad adoption are liability, privacy, and penetration. However, even these roadblocks will soon be removed. The liability is restricted to only a small class of applications (such as intersection crash prevention); moreover, rapid progress is being made in that area. Privacy issues have been practically resolved in two ways: by virtue of technology advances, and by the fact that users are getting accustomed to give up privacy for other benefits. Full penetration (say, of DSRC radios) is no longer critical for the deployment of many applications (such as navigation and Intelligent Transport) that are increasingly relying on 3G, WiFi, and WiMAX.

References

Burgess, J., Gallagher, B., Jensen, D., and Levine, B. N. 2006. MaxProp: Routing for Vehicle-Based Disruption-Tolerant Networks. *IEEE INFOCOM.*

Burke, J., Estrin, D., Hansen, M., A. Parker, A., Ramanathan, N., Reddy, S., and Srivastava, M. B. 2006. Participatory Sensing. *ACM WSW.*

Caesar, M., Castro1, M., Nightingale, E. B., O'Shea, G., and Rowstron, A. 2006. Virtual Ring Routing: Network Routing Inspired by DHTs. *SIGCOMM'06.*

Caliskan, M., Graupner, D., and Mauve, M. 2006. Decentralized Discovery of Free Parking Places. *ACM VANET, Los Angeles.*

Capone, A., Napoli, S., and Pollastro, A. 2006. Mobimesh: An Experimental Platform for Wireless Mesh Networks with Mobility Support. *Proc. of ACM QShine 2006 Workshop on Wireless Mesh: Moving Towards Applications.*

Chiu, D. M., Yeung, R. W., Huang, J., and Fan, B. 2006. Can Network Coding Help in P2P Networks? *NetCod'06.*

Clausen, T., and Jacquet, P. 2003. Optimized Link State Routing Protocol (OLSR). *RFC3626.*

Dikaiakos, M. D., Iqbal, S., Nadeem, T., and Iftode, L. 2005. VITP: An Information Transfer Protocol for Vehicular Computing. *ACM VANET.*

Eisenman, S. B., Ahn, G.-S., Lane, N. D., Miluzzo, E., Peterson, R. A., and Campbell, A. T. 2006. MetroSense Project: People-Centric Sensing at Scale. *ACM WSW.*

ElBatt, T., Goel, S. K., Holland, G., Krishnan, H., and Parikh, J. 2006. Cooperative Collision Warning Using Dedicated Short Range Wireless Communications. *Proceedings of the 3rd International Workshop on Vehicular Ad hoc Networks,* pages 1–9.

Eriksson, J., Balakrishnan, H., and Madden, S. 2008a. Cabernet: A Content Delivery Network for Moving Vehicles. *Technical Report TR-2008-003, MIT-CSAIL.*

Eriksson, J., Girod, L., Hull, B., Newton, R., Balakrishnan, H., and Madden, S. 2008b. The Pothole Patrol: Using a Mobile Sensor Network for Road Surface Monitoring. *MobiSys'08.*

Fan, L., Cao, P., and Almeida, J. 1998. Summary Cache: A Scalable Wide-Area Web Cache Sharing Protocol. *ACM SIGCOMM.*

Federal Highway Administration, U.S. Department of Transportation. 2004. *Highway statistics 2004.* http://www.fhwa.dot.gov/policy/ohim/hs04/htm/mv1.htm

Ferrara, A., and Paderno, J. 2006. Application of Switching Control for Automatic Pre-Crash Collision Avoidance in Cars. *Nonlinear Dynamics,* 46, 307–321.

Gerla, M., Zhou, B., Lee, Y.-Z., Soldo, F., Lee, U., and Marfia, G. 2006. Vehicular Grid Communications: The Role of the Internet Infrastructure. *WICON'06.*

Gibbons, P. B., Karp, B., Ke, Y., Nath, S., and Seshan, S. 2003. IrisNet: An Architecture for a Worldwide Sensor Web. *IEEE Pervasive Computing,* 2(4): 22–33.

Gkantsidis, C., and Rodriguez, P. 2005. Network Coding for Large Scale Content Distribution. *INFOCOM'05.*

Goldsmith, A. 2005. *Wireless Communications.* Cambridge University Press.

Guo, M., Ammar, M. H., and Zegura, E. W. 2005. V3: A Vehicle-to-Vehicle Live Video Streaming Architecture. *PerCom'05.*

Huang, Q., Lu, C., and Roman, G. 2002. Mobicast: Just-in-Time Multicast for Sensor Networks under Spatiotemporal Constraints. *Proc. of the 2nd International Workshop on Information Processing in Sensor Networks,* pages 442–457.

Hull, B., Bychkovsky, V., Chen, K., Goraczko, M., Miu, A., Shih, E., Zhang, Y, Balakrish-
   nan, H., and Madden, S. 2006. CarTel: A Distributed Mobile Sensor Computing System.
   ACM SenSys.
IEEE 1609 – Family of Standards for Wireless Access in Vehicular Environments
   (WAVE).
IEEE. 2006. "Draft Amendment to Standard for Information technology – Telecommuni-
   cations and information exchange between systems – Local and metropolitan networks –
   specific requirements – Part 11: Wireless LAN Medium Access Control (MAC) and
   Physical Layer (PHY) specifications: Amendment 3: Wireless Access in Vehicular Envi-
   ronments (WAVE).
Intanagonwiwat, C., Govindan, R., and Estrin, D. 2000. Directed Diffusion: A Scalable and
   Robust Communication Paradigm for Sensor Networks. *ACM MOBICOM.*
Johnson, D. B., and Maltz, D. A. 1996. Dynamic Source Routing in Ad Hoc Wireless
   Networks. *Mobile Computing,* 153–181.
Juang, P., Oki, H., Wang, Y., Martonosi, M., Peh, L.-S., and Rubenstein, D. 2002. Energy-
   Efficient Computing for Wildlife Tracking: Design Tradeoffs and Early Experiences with
   ZebraNet. *ACM ASPLOS-X.*
Kahn, R. 1977. The Organization of Computer Resources into a Packet Radio Network.
   IEEE Transactions on Communications, 25(1): 169–178.
Kaul, S., Gruteser, M., Onishi, R., Vuyyuru, R., and T.I.T. Center. 2008. GeoMAC: Geo-
   Backoff Based Co-operative MAC for V2V networks. *IEEE International Conference
   on Vehicular Electronics and Safety,* pages 334–339.
Kotz, D., and T. Henderson T. 2005. "Crawdad: A Community Resource for Archiving
   Wireless Data at Dartmouth. *IEEE Pervasive Computing,* 4(4), 12–14.
Laneman, J., Tse, D., and Wornell, G. 2004. Cooperative Diversity in Wireless Networks:
   Efficient Protocols and Outage Behavior. *IEEE Transactions on Information Theory,* 50,
   3062–3080.
Laneman, J., and Wornell, G. 2003. Distributed Space-Time-Coded Protocols for Exploiting
   Cooperative Diversity in Wireless Networks. *IEEE Transactions on Information Theory,*
   49, 2415–2425.
Lee, K. C., Lee, S.-H., Cheung, R., Lee, U., and Gerla, M. 2007. First Experience with
   CarTorrent in a Real Vehicular Ad Hoc Network Testbed. *MOVE'07.*
Lee, U., Magistretti, E., Gerla, M., Bellavista, P., Lio, P., and K.-W. Lee, K.-W. 2008a.
   Bio-inspired Multi-Agent Data Harvesting in a Proactive Urban Monitoring Environ-
   men. *Elsevier Ad Hoc Networks Journal, Special Issue on Bio-Inspired Computing and
   Communication in Wireless Ad Hoc and Sensor Networks,* 7(4), 725–741.
Lee, U., Magistretti, E., Zhou, B., Gerla, M., Bellavista, P., and Corradi, A. 2006a. MobEyes:
   Smart Mobs for Urban Monitoring with Vehicular Sensor Networks. *IEEE Wireless
   Communications,* 13(5): 51–57.
Lee, U., Magistretti, E., Zhou B., Gerla, M., Bellavista, P., and Corradi, A. 2008b. Dissemi-
   nation and Harvesting of Urban Data using Vehicular Sensor Platforms. *IEEE Transaction
   on Vehicular Technology.*
Lee, U., Park, J.-S., Amir, E., and Gerla, M. 2006b. FleaNet: A Virtual Market Place on
   Vehicular Networks. *V2VCOM'06.*
Lee, U., Park, J.-S., Yeh, J., Pau, G., and Gerla, M. 2006c. CodeTorrent: Content Distribution
   Using Network Coding in VANETs. *MobiShare'06..*
Maihöfer, C., Leinmüller, T., and Schoch, E. 2005. Abiding Geocast: Time-Stable Geocast
   for Ad hoc Networks. *Proceedings of the 2nd ACM International Workshop on Vehicular
   Ad hoc Networks,* pages 20–29.
Nadeem, T., Dashtinezhad, S., Liao, C., and Iftode, L. 2003. TrafficView: Traffic Data
   Dissemination Using Car-to-Car Communication. *ACM Mobile Computing and Com-
   munications Review (MC2R),* 8(3): 6–19.
Nandan, A., Das, S., Pau, G., Gerla, M., and M. Y. Sanadidi, M. Y. 2005. Co-operative
   Downloading in Vehicular Ad-Hoc Wireless Networks. *IEEE/IFIP WONS.*

Nandan, A., Tewari S., Das, S., Pau, G., Gerla, M., and L. Kleinrock. 2006. AdTorrent: Delivering Location Cognizant Advertisements to Car Networks. *IEEE/IFIP WONS, Les Menuires.*

Nath, S., Liu, J., and Zhao, F. 2006. Challenges in Building a Portal for Sensors World-Wide. *ACM WSW.*

Navas, J. C., and T. Imielinski, T. 1997. GeoCast: Geographic Addressing and Routing. *Proceedings of the 3rd Annual ACM/IEEE International Conference on Mobile Computing and Networking*, pages 66–76.

Nosratinia, A., Hunter, T., and Hedayat, A. 2004. Cooperative Communication in Wireless Networks. *Communications Magazine, IEEE*, 42, 74–80.

Ott, M., Seskar, I., Siracusa, R., and Singh, M. 2005. Orbit Testbed Software Architecture: Supporting Experiments as a Service. *Proceedings of IEEE Tridentcom*, pages 136–145.

Park, J.-S., Lee, U., Oh, S. Y., Gerla, M., and Lun, D. 2006. Emergency Related Video Streaming in VANETs Using Network Coding. *ACM VANET'06.*

Peterson, L., Anderson, T., Culler, D., and Roscoe, T. 2002. A Blueprint for Introducing Disruptive Technology into the Internet. *Proceedings of HotNets–I.*

Perkins, C., and Royer, E. 1999. Ad-hoc On-Demand Distance Vector Routing. *Proceedings of the 2nd IEEE Workshop on Mobile Computing Systems and Applications*, pages 90–100.

Ratnasamy, S., Karp, B., Yin, L., Yu, F., Estrin, F., Govindan, R., and Shenker, S. 2002. GHT: A Geographic Hash Table for Data-Centric Storage. *WSNA'02.*

Rheingold, H. 2003. *Smart Mobs: The Next Social Revolution.* Basic Books.

Riva, O., and Borcea, C. 2007. The Urbanet Revolution: Sensor Power to the People! *IEEE Pervasive Computing*, 6(2), 41–49.

Robinson, C., Caveney, D., Caminiti, L., Baliga, G., Laberteaux, K., and Kumar, P. 2007. Efficient Message Composition and Coding for Cooperative Vehicular Safety Applications. *IEEE Transactions on Vehicular Technology*, 56, 3244–3255.

Shah, R. C., Roy, S., Jain, S., and Brunette, W. 2003. Data MULEs: Modeling a Threetier Architecture for Sparse Sensor Networks. *Elsevier Ad Hoc Networks Journal*, 1(2–3): 215–233.

Small, T., and Haas, Z. J. 2003. The Shared Wireless Infostation Model – A New Ad Hoc Networking Paradigm (or Where There Is a Whale, There Is a Way). *ACM MOBIHOC.*

Sormani, D., Turconi, G., Costa, P., Frey, D., Migliavacca, M., and Mottola, L. 2006. Towards Lightweight Information Dissemination in Inter-vehicular Networks. *ACM VANET'06.*

Soroush, H., Banerjee, N., Balasubramanian, A., Corner, M., Levine, B., and Lynn, B. 2009. DOME: A Diverse Outdoor Mobile Testbed. *UMass Technical Report UM-CS-2009-23.*

Standard Specification for Telecommunications and Information Exchange Between Roadside and Vehicle Systems – 5 GHz Band Dedicated Short Range Communications (DSRC) Medium Access Control (MAC) and Physical Layer (PHY) Specifications, September 2003.

Vahdat, A., and Becker, D. 2000. Epidemic Routing for Partially-Connected Ad Hoc Networks. *Technical Report CS-200006, Duke University.*

Wang, Y., and Wu, H. 2006. DFT-MSN: The Delay/Fault-Tolerant Mobile Sensor Network for Pervasive Information Gathering. *INFOCOM'06.*

Wu, H., Fujimoto, R., Guensler, R., and Hunter, M. 2004. MDDV: A Mobility-Entric Data Dissemination Algorithm for Vehicular Networks. *ACM VANET.*

Yin, J., ElBatt, T., Yeung, G., Ryu, B., and Habermas, S. 2004. Performance Evaluation of Safety Applications over DSRC Vehicular Ad Hoc Networks. *ACM VANET'04.*

Yoon, H., Kim, J., Tan, F., and Hsieh, R. 2008. On-demand Video Streaming in Mobile Opportunistic Networks. *PerCom'08.*

Yoon, J., Noble, B., and Liu, M. 2007. Surface Street Traffic Estimation. *MobiSys'07.*

Zang, Y., Stibor, L., Orfanos, G., Guo, S., and Reumerman, H. 2005. An Error Model for Inter-vehicle Communications in Highway Scenarios at 5.9GHz. *Proceedings of the 2nd*

ACM International Workshop on Performance Evaluation of Wireless Ad hoc, Sensor, and Ubiquitous Networks, pages 49–56.

Zang, Y., Stibor, L., Reumerman, H., and Chen, H. 2008. Wireless Local Danger Warning Using Inter-vehicle Communications in Highway Scenarios. *14th European Wireless Conference*, pages 1–7.

Zhou, P., Nadeem, T., Kang, P., Borcea, C., and Iftod, L. 2005. EZCab: A Cab Booking Application Using Short-Range Wireless Communication. *IEEE PerCom'05*.

9

Opening Up the Last Frontiers for Securing the Future Wireless Internet

Wade Trappe, Arati Baliga, and Radha Poovendran

Abstract

Due to the low cost and ease of deployment associated with wireless devices, wireless networks will continue to be the dominant choice for connecting to the future Internet. Beyond serving as an edge-connecting medium, the rapid improvement in communication rates for emerging wireless technologies suggests that wireless networks will also play an increasingly important role in building the backbone of the future Internet. As wireless components become integrated into the design of future network architectures, one significant concern that will arise is whether their pervasiveness, affordability, and ease of programmability might also serve as a means to undermine the benefits they might bring to the future Internet.

Just as the future Internet initiative has brought new perspectives on how protocols should be designed to take advantage of improvements in technology, the future Internet initiative also allows us to reexamine how we approach securing our network infrastructures. Traditional approaches to building and securing networks are tied tightly to the concept of protocol layer separation. For network protocol design, routing functions are typically considered separately from link layer functions, which are considered independently of transport layer phenomena or even the very applications that utilize such functions. Similarly, in the security arena, MAC-layer security solutions (e.g., WPA2 for 802.11 devices) are typically considered as point-solutions to address threats facing the link layer, while routing and transport layer security issues are dealt with in distinct, nonintegrated protocols like IPSEC, TLS, or even in the abundance of recent secure routing protocols.

Although traditional security solutions, that is, cryptographic protocols that work in isolation, are an essential step to understanding how to secure networks, they do not represent a holistic approach. Just as there are significant performance gains to be achieved when combining information from multiple layers to build improved MAC and routing functions,[1] so too is there the potential to significantly improve the security of the future Internet by considering cross-layer approaches to security. The network modality that promises the most opportunities for cross-layer design is wireless. Physical properties outside the normal purview of the network, such as the device itself, or the location of communicating entities, or even the physical properties of the signals being transmitted, can serve as cross-layer information for enhanced security. In this chapter, we will examine the use of cross-layer mechanisms that pull information from the device, from location, and from the physical layer itself to open up the *last frontier* of security design.

9.1 Security Challenges Facing the Future Wireless Internet

Before commencing with this discussion, we briefly describe the potential threats and security opportunities that we envision are possible in the future wireless Internet. As a starting point, we must recognize that wireless devices are inherently commodity items – they are generally low-cost, highly portable, very heterogeneous in their forms, and are becoming increasingly more programmable. One of the great success stories behind wireless networking is that wireless technologies have made networking and communication connectivity available to the broader society. Even the most technically unsavvy person can purchase a wireless router from their local department store for a very accessible price, and deploy their own network, while it requires a far more technically astute person (or team) to deploy and administer a wired router. Not only does this imply that wireless devices are readily available for legitimate purposes, but it also implies that wireless devices could become an ideal platform for illegitimate purposes. This fact, when combined with the fact that wireless devices are small, often hand-held, and allow their users to connect to the broader network anywhere at anytime, means that wireless will be an ideal modality to launch a variety of threats against the broader network and its users. As if this were not a harsh enough scenario, there is a movement to make wireless devices increasingly programmable. Already there are a handful of programmable smart phones, such as the Google Android[2] and Apple iPhone[3] and supporting SDKs[4,5] that promise to make it easier to develop new software for good and bad purposes. At the same time, new radio platforms, like software-defined radios and cognitive radio (CR) platforms, are being developed to open up the lower layers of the protocol stack for general development. Consequently, many threats that might have been prevented are now easily possible because firmware restrictions are no longer in place.

We may decompose the threats facing the future wireless Internet in terms of a classical CIA (confidentiality, integrity, and availability) framework. In the discussion that follows, we list several CIA threats made possible by the commodity nature of wireless technologies.

- Confidentiality: Wireless communications between entities are especially susceptible to confidentiality threats from attackers interested in snooping over the message contents. As messages are broadcast over the air, malicious adversaries can easily intercept and interject packets. For unencrypted communication, an adversary can easily decipher packet contents by listening to broadcast packets, consequently violating confidentiality. Alternatively, man-in-the-middle attacks are possible by injecting false packets, thereby allowing an adversary to decipher traffic crossing the network.
- Integrity: Due to the commercial nature of wireless devices, integrity needs to be established at various levels. This involves integrity of the wireless device itself, integrity of the software running on the device, and integrity of message communication between the sender and the receiver. Attacks can be carried out at different levels. For example, an attacker can manipulate the device hardware to alter its behavior. He can alter the device software to install malicious versions of system and application software. In case of cognitive radios, a malicious attacker may tamper with the installed policies while maintaining unaltered version of the software. Finally, in systems that do not verify message integrity, malicious adversaries can inject false messages or carry out man-in-the-middle attacks.
- Availability: More malicious attacks could involve a user programming a CR device to give him/her advantage relative to neighboring devices. For example, a greedy user might seek to decrease the back-off window size in an 802.11 implementation, and as a result obtain a larger fraction of the channel utilization. Generally, such greedy attacks can take a variety of forms, ranging from bypassing agreed-on MAC-layer behavior to ignoring implementations of fairness in spectrum-etiquette policies. A deleterious adversary might seek to turn the CR platform into a jamming platform by listening to channel utilization and emitting short blocker packets to prevent the reception of packets.

9.2 The Final Frontier: Introducing the Physical into Security

The traditional approach to security involves layer-specific protocols that are unaware of the platform or the physical medium on which their associated messages rely. Throughout this chapter, we take the view that the *physical* world represents an important aspect of communication that must be addressed in order to properly have a holistic approach to securing devices on the future Internet. By physical world, we mean the physical platform associated with the communication, the physical medium over which communications are carried, and also the physical context of the communication in terms of the locations of the communicating entities.

We now highlight several types of new security services that may be built by using the physical aspects of communication into the security framework.

- Physical Device Integrity Services: The programmability allowed in cognitive radios necessitates an architecture that does not allow the devices to violate high-level spectrum etiquettes. The physical devices integrity services should be able to verify the integrity of the physical device and the policy enforcement code that runs on top of it.
- Authentication/Identification Services: The uniqueness of the channel between two locations provides a means for uniquely identifying wireless entities. Devices may authenticate themselves based on their ability to produce an appropriate received signal at the recipient.
- Confidentiality Services: The fact that pairwise radio propagation laws between two entities are unique and decorrelate quickly with distance can serve as the basis for establishing shared secrets. These shared secrets may be used as encryption keys for higher-layer applications or wireless system services that need confidentiality.
- Availability Services: RF-specific denial-of-service attacks targeting the ability of radio devices to transmit or receive messages may be launched against wireless networks. Detecting RF interference, or jamming, attacks must be performed at lower layers. Spectral evasion strategies may be integrated into the devices so as to assure the availability of the wireless network in the presence of interference attacks.
- Verifiable Location Services: Radio communications do not exhibit brick-wall propagation, and consequently wireless networks may be accessed from locations other than their intended coverage region. This phenomena can facilitate threats to the security of both wireless networks and the broader Internet. However, we may also use lower-layer functionalities, such as power control, to provide mechanisms to verify the location of mobile entities.
- Non-repudiation Services: RF energy naturally radiates, and wireless entities within the radio coverage pattern may serve as witnesses for the actions of the transmitter. This makes it harder for radio entities to deny receiving a message or having performed an action. We may introduce communication auditors into the wireless infrastructure to assist in quantifying the trust of wireless entities.
- Forensic Services: The wireless medium is perturbed by the introduction of new entities, whether physical objects or other radio transmitters. Lower-layer

information can serve as forensic evidence for detecting an environmental change and possibly even identifying the cause of such a change. Wireless forensic services identify unauthorized intrusions in the radio environment and serve to actuate responses, such as adjusting system level security policies.

In the rest of this chapter, we shall examine potential physical security mechanisms that operate at the device level, take advantage of location contexts, and use information from the physical layer.

9.3 Platform and Device-Level Assurance

A starting point for protecting against attacks is to realize that if all wireless devices were following their proper hardware and software instructions, then no attacks would be present. Consequently, all attacks originating from wireless (or other) devices originate from devices that are executing their supposed functions improperly. For example, the software associated with networking functions may have been altered, and such malicious code may then be used to subvert the network. Malicious code may corrupt routing updates, selectively drop messages, mount slander attacks, and allow nodes to collude to hurt other nodes.

As a first line of defense, it is natural to attempt to verify that the code running on a network node is approved. Checking the integrity of hardware and software thus can act as a first filter in preventing attacks. Typically, the research in this arena has pursued two different directions: hardware- and software-based attestation techniques. On the hardware front, trusted platform modules (TPMs) have been used to establish a dynamic root of trust, and hardware protection can be used to prevent unauthorized access to the secure loader block, identify whether code execution occurs after a reboot, and allow for code to be executed in an isolated environment.

Recent research[6] has shown that hardware-based mechanisms can provide a powerful abstraction to implement dramatically improved secure network protocols. The basic premise is that if one can trust the code that has generated an output, and further that this code includes input verification, then the output can be trusted. This new approach for designing secure networking protocols promises to greatly enhance the security and efficiency of distributed systems, and will be an important component to securing the future wireless Internet. Unfortunately, advancements in hardware-based attestation cannot address security threats being conducted from devices that do not employ TPMs. In such a case, software-based code attestation can be used. Recently, the SWATT and Pioneer systems have shown that it is possible to provide software attestation on legacy platforms.[7,8]

Building on such work, we now explore how trusted platform technology may be used to ensure that future programmable wireless platforms, like the cognitive radio, exhibit trustworthy behavior.

9.3.1 Security and Cognitive Radios

There has been considerable effort directed at developing "cognitive radio" (CR) platforms, which will expose the lower layers of the protocol stack to researchers and developers.[9] This initiative is supported by two separate technical efforts: First is a wealth of research devoted to uncovering the gains that are possible by letting the lower protocol layers become programmable and adaptable; second are the recent advances in programmable integrated circuits that have significantly increased the amount of computation that can be done without requiring specialized hardware/firmware components. By being able to scan the available spectrum, select from a wide range of operating frequencies, adjust modulation waveforms, and perform adaptive resource allocation – all of these in real time – these new "cognitive" radios will be able to adapt to a wide variety of operational settings, supporting a true "anywhere-anytime" vision of the future Internet.

Although there is great potential for such a radio platform, some caution regarding their ubiquitous use in wireless systems is warranted, because their deployment will not be limited to the laboratory. Already, the GnuRadio platform[10] is available for general use, and supporting this platform is an open-source software effort to develop GnuRadio "blocks"[11] – software modules capable of conducting a broad range of functions associated with the reception/transmission of radio signals. Other CR platforms, such as the Xilinx-based Rice WARP cognitive radio platform[12] and the WINLAB WiNC2R platform,[13] will also reach a large consumer base with similar open-source efforts supporting lower-layer protocols.

The open-source nature of cognitive radio software is empowering but also dangerous. It is easily conceivable that inexpensive and widely available cognitive radios could become an ideal platform for abuse since the lowest layers of the wireless protocol stack are accessible to programmers. Thus, the gains promised by adaptive resource-allocation schemes and good spectrum-etiquette policies can be negated if cognitive radio devices can be reprogrammed to violate or bypass locally fair-spectrum policies either maliciously or inadvertently. If fail-safe mechanisms are not employed, individual devices could use the wireless medium to their advantage at the expense of the greater good.

To regulate this future radio platform, we present a framework, known as TRIESTE (Trusted Radio Infrastructures for Enforcing SpecTrum Etiquettes), which can guarantee that a cognitive radio behaves according to acceptable communal policies.[14]

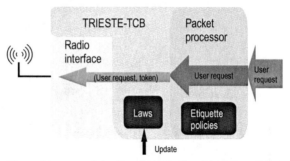

Figure 9.1. The architecture of the Cognitive Radio with on-board TRIESTE-TCB.

We begin by assuming the presence of a third party, known as the Spectrum Law Makers, which give general guidelines on how the cognitive radios should operate. For example, the Federal Communications Commission (FCC) might be such an entity, and would provide rules describing how spectrum should be accessed, which bands are not allowed to be transmitted on, and requirements on interference between cognitive radios and other *primary* wireless modalities. To enforce these rules, we believe it is necessary to have an on-board Trusted Computing base/module (TRIESTE-TCB) in each cognitive radio that enforces the spectrum laws and etiquettes.

The TRIESTE-TCB, as depicted in Figure 9.1, includes all the hardware and software in the cognitive radio that enforces universal laws and etiquette policies passed down by the Spectrum Law Makers. The TRIESTE-TCB can be thought of as a control gate that user processes have to go through to access the radio. In TRIESTE, typically, before the user can transmit information over a certain radio spectrum band, the user/process would send a spectrum access request, which includes information about the target radio frequency band, the spectrum etiquette the user will follow, the transmission power, transmission duration, and so forth, to the packet processor. Here, we note that we shall abuse terminology and, for simplicity, collectively refer to the packet processor as an entity consisting of multiple processors handling packets, such as the Network Processor, the CR Policy Processor, and so on. The packet processor shapes the user's radio access request according to the spectrum-etiquette policies programmed by the user or spectrum owner, then passes the modified user request to TRIESTE-TCB. The TRIESTE-TCB in turn will validate the request against the laws available to it and will allow the request to go through only if it does not violate any of those laws.

The TRIESTE-TCB would evaluate the access request along with the user's credentials and checks it against the spectrum laws. If the request and credential combination is valid in the context of spectrum laws, then the TCB would issue a privilege token for that request. The privilege token is a tuple consisting of the ⟨spectrum-access-details, timestamp and a signed hash of [spectrum-access-details‖timestamp]). The spectrum-access-details might specify, for example, the radio frequency, duration, and spectrum access limitation granted. If the user's credentials do not permit the privilege level of the request or if the combination somehow violates some spectrum law, then the TRIESTE-TCB could either try to find a permissible modification of the request that is in compliance with the spectrum laws or reject the request if such a modification is not feasible. We note that the user's credentials may change over time, and each request would be evaluated in the context of the credentials presented with it. For example, a user with emergency-responder credentials would have higher privilege spectrum access during an emergency situation as opposed to during a nonemergency one.

TRIESTE-TCB would compare the access request with the spectrum laws, and only if the request does not violate any spectrum laws would the request be validated and update-privilege tokens issued to the user, otherwise the request will be rejected or modified. An access token, which specifies the radio frequency, duration, and spectrum access limitation granted, together with the user request, will be passed to the radio interface processor via a tamper-proof path. Typically, to further prevent a user from bypassing the TRIESTE-TCB and forging the token itself, authentication mechanisms are necessary to assure the token is granted by the TRIESTE-TCB.

Inside TRIESTE-TCB would be a monitoring component, known as the monitor verifier, which will monitor the on-board radio activity and observe the radio environment, and check any potential violation by comparing "spectrum laws." If the user does not follow the rules it claims to obey, the TRIESTE-TCB will stop the radio operation and revoke the user's token/privilege.

We now discuss a few challenging issues on which the TRIESTE-TCB could depend. First of all, the law should be stored in a secure storage to ensure protection against tampering. Additionally, after the token has been issued to the user, the association relationship of user request and token should not be altered as the "(user request, token)" pair passes between cognitive radio components. To achieve integrity, encryptions can be used, though at the cost of additional computational overhead. Alternatively, we can design the cognitive radio in such a way that after the creation of "(user request, token)," the pair travels among components via trusted paths. Thus, the data pair cannot be intercepted on the way, nor can the content of the user request be changed. As the spectrum laws will evolve over time, it is thus desirable to make the law extendable.

Since the cognitive radio is a programmable wireless platform that will support a wide range of radio network scenarios, from autonomous agile radios to those that use higher-layer protocols to share spectrum, it is wise to consider a generic high-level architecture, such as shown in Figure 9.2. Here, a CR consists of Flexible RF units, a baseband processor, a network processor, and a cognitive radio policy processor (which also functions as the host). Besides those components, we have added a logical component, the TRIESTE-TCB, to enforce

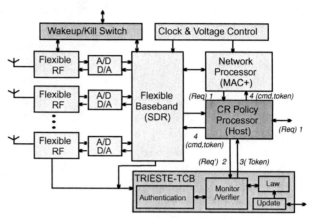

Figure 9.2. A generic SDR/CR platform involving RF processors, baseband processor, network processor, and the cognitive radio processor. Note that TRIESTE regulates via a TCB component and an externally accessible authenticated kill-switch.

the spectrum laws. Here we want to point out that the law/policy enforcement activities are likely to be performed at several functional places within the CR, because law/policy enforcement is potentially related to every network protocol that will access the spectrum. Although we show the TRIESTE-TCB in one monolithic block, in implementation, the functions of the TRIESTE-TCB will be located in firmware in different processors.

As noted earlier, the TRIESTE-TCB can be thought of as the controlled gate that users have to go through to access radio. The basic structure of TRIESTE-TCB consists of a generic *Controller* that can interpret and enforce any well-formed *Law*. As we pointed out earlier, the TRIESTE-TCB is a virtual block, and the real functions of the TRIESTE-TCB will be located in hardware or software on different components of the CR. In particular, many of the proposed functionalities of the TRIESTE-TCB might require a secure, tamper-proof chip on board the CR platform that is dedicated to providing a hardware-based root of trust. Recent efforts by the Trusted Computing Group have mapped out specifications for the Trusted Platform Module, to enable trusted computing functions such as platform attestation/integrity, hardware-based cryptographic functionality, and secure storage. [15] Manufacturers, such as Atmel, have already produced TPM chips that find use in digital rights management services, and such technologies warrant application to securing CRs.

In the TRIESTE framework, cognitive radios must adhere to the Spectrum Laws published by agencies, such as the FCC. Future cognitive radios should be able to adapt to new laws/policies dynamically, as laws/policies tend to change over time. A starting point for defining such laws would be to use XGPL (XG Policy Language) [16] to express spectrum policies formally. XGPL is part of the XG (neXt Generation Communications) research program that aims to let radios utilize available spectrum intelligently and dynamically based on the knowledge of actual conditions and spectrum policies. In particular, the XG project chose OWL (Web Ontology Language) as its XG Policy Language for several reasons. First of all, OWL provides the structure and richness needed to express policies. Secondly, general theorem proving/reasoning engines for deductive inference are already available. Finally, OWL is an efficient language for describing data and passing it around different systems.

OWL is originally designed for processing information on the Web and is designed to be interpreted by computers. It is written in XML (Extensible Markup Language). We note that OWL is not another programming language, but is a structured way to build representations for information and policies for machine understanding. For example, the OWL expression of magnitude 10 is as follows:

```
<xgparam:magnitude>
    <xsd:integer rdf:value="10" />
</xgparam:magnitude>
```

The paragraph above defines a property "magnitude" in the name space "xgparam." The value of the property magnitude is 10, the type of the value is integer, which is defined in namespace rdf. More detailed and precise exposition on OWL can be found in OWL Web Ontology Language Guide.[17]

For the remainder of this chapter, we use the shorthand notion described in XG Working Group Document.[16] The shorthand notation yields representations equivalent to OWL representations. For example, we describe the previous "magnitude is 10" in the following way:

```
(magnitude 10)
```

Detailed mapping from OWL to shorthand notion can be found in XG Working Group Document.[16]

A spectrum policy rule is composed of three facts: a selector description, an opportunity description, and a usage constraint description, as shown below:

```
(PolicyRule (id Policy_name)
    (SelDesc S)
    (OppDesc SomeOpp)
    (UseDesc SomeUseDesc)
```

The first part in a spectrum policy rule is a selector description, which is used to filter policy rules to the subset of rules that may apply to a given situation. The selector description contains one or more facts that describe the frequency, time, and region the policy covers, the authority that defines the policy, and the radio device to which the policy rule applies. For example, a selector description may include filters such as "applies to operation in U.S.A" or "applies to operations in the 3.6 *GHz* to 3.7 *GHz* bands."

The second part in a policy rule is an opportunity description, which is used to evaluate whether the transmission request is valid or not based on whether or not a given environment and device state match the opportunity description in the filtered subset rules. For example, the opportunity description can be "if a beacon is heard at 823 *MHz*," or "peak received power is less than −80 *dBm*."

A valid opportunity indicates transmission that conforms to the usage constraint description is permitted. Usage constraint description constrains the radio behavior, such as "transmit with a maximum power of −10 *dBm*" or "maximum continuous on-time must be 1 second and the minimum off-time must be 100 *msec*."

We envision that usually, a spectrum policy rule is first defined in XGPL, then each element (a selector description, an opportunity description, and a usage constraint description) is defined in a format similar to the format used to specify policy rules.

The spectrum law includes both spectrum access laws and punishment laws. We have discussed how to express spectrum access rules using XGPL. In the original XG project, XGPL is designed to describe spectrum access control. XGPL was not used to specify any form of punishment for spectrum abuse. In particular, the underlying idea of the XG project is that the regulatory policy does not tell the radio what to do; it only defines what constitutes authorized use of the spectrum. Punishment, however, tells the radio what should be done once violation has occured.

We believe that it is necessary to define punishment rules as part of the spectrum laws, because punishment can serve to prevent potential spectrum violation as well. Although it might be a challenge, XGPL can be extended to define punishment rules. One way to define punishment is to add one more description, the punishment description, into the policy rules as shown:

```
(PolicyRule (id Policy_name)
    (SelDesc S)
    (OppDesc SomeOpp)
    (UseDesc SomeUseDesc)
    (PunDesc SomeAction)
```

One possible way to perform the punishment is as follows. If the punishment rule is selected and activated, then new punishing rules with certain expiration period will be generated based on the level and type of punishment, and inserted into the existing spectrum polices for specified amount of time. For example, the newly generated punishing spectrum access rules could be that the radio device cannot access to band 3.6−3.7 *GHz* for two hours. Of course, precedence mechanisms are needed to resolve conflict. Detailed techniques for defining punishment and precedence require further investigation.

One concern regarding a TPM-based approach to building the TRIESTE-TCB is that TPMs are generally focused on software rather than hardware attacks, and simple hardware-based man-in-the-middle attacks can compromise the boot sequence. Because this attack does not use TPM die probes, the vulnerability is not overcome with stronger chip-level tamper resistance. As a consequence, hardware-related security challenges for CR include: (1) deductions made in the software layer may no longer hold when the hardware layer is accessible, and (2) hardware-protected information may not necessarily be localized to a single TPM chip. Absolute physical protection of integrated circuits is difficult because testing is required after packaging. The state-of-the-art in tamper and probing resistance involves proprietary commercial techniques. Regardless of the physical protection methods employed, combining as many functions as possible on one chip is desirable because it increases the cost of a physical attack (since external pin probes may be insufficient) and decreases the cost of protection (since fewer chips need to be tamper-resistant). Current trends in CR design suggest that the most suitable platform design involves FPGAs, which have the needed adaptability and logic capacity for CR functions. Security aspects of the interfaces and functionality assigned to various CR components, and the FPGA in particular, are new system-level partitioning constraints that need to be developed. In the hardware domain, the design principles used to improve performance, namely nonsharing of computation, communication, and memory resources, also promote system security by restricting access to private information. Hardware-based access restrictions are generally simpler to assure than software or software-managed hardware (such as memory management units). These guarantees are diminished when the hardware is shared between different processes, because cached private information often exists prior to a context switch. A further area of investigation is the enforcement of basic operational policies using hardware-layer "interlocks" that cannot be overridden by software layers. This would require analyzing the interfaces and dependencies between hardware and software layers, selecting the policies to be enforced with hardware, formal state analysis of the hardware blocks responsible for policy enforcement, and a mechanism for securely updating policy enforcement circuits.

9.4 Location as an Enabler for Security Services

Radio signals in wireless networks may be accessed from locations other than their intended coverage region. This fact poses several security threats to the deployed wireless networks because it can be accessed by malicious users from outside the perimeters. Therefore, location information and position verification methods are crucial to the deployment of a security framework, which can provide different types of access control policies depending on the physical location of the device. However, we may also use lower-layer functionalities, such as power control, to provide different kinds of security services in wireless networks. In this section, we focus on two different ways in which location can be used to enhance security: First, we examine the use of location as a means to detect the spoofing of a wireless entity; and, second, we examine a key management scheme that uses location information and power control to allow for secure multicast in wireless ad hoc networks.

9.4.1 Location-Based Recognition of Spoofing Attacks

Spoofing attacks are serious threats because they can facilitate a variety of traffic injection attacks against networks. These attacks are particularly easy to conduct at the edge of the Internet, where wireless devices, such as sensor nodes and wireless LANs, cannot employ appropriate authentication mechanisms to detect the injection of false messages. It is thus desirable to detect the presence of spoofing and eliminate them from the network. The traditional approach to address spoofing attacks is to apply cryptographic authentication. However, authentication requires additional infrastructural overhead and computational power associated with distributing and maintaining cryptographic keys. Due to the limited power and resources available to wireless devices on the edge, it is not always possible to deploy authentication. In addition, key management often incurs significant human management costs on the network.

We will now examine how the physical properties associated with where wireless transmissions are being sent from can be used to detect spoofing. Specifically, we present a scheme for both detecting spoofing attacks and localizing the positions of the adversaries performing the attacks. The approach that we summarize utilizes the Received Signal Strength (RSS) measured across a set of monitoring nodes (e.g., access points) to perform spoofing detection and localization.

9.4.1.1 Formulation of Spoofing Attack Detection

In a spoofing attack on a wireless (edge) network, an adversarial node will claim the identity of another, legitimate node (e.g., by altering its MAC address).[18] Unless the adversary is located at precisely the same location as the legitimate node, it should be possible to distinguish between the two communication streams by localizing each transmission and noticing that packets coming from the claimed address appear to come from multiple, simultaneous locations.* There can be multiple nodes spoofing the same MAC address.

---

* We note that the methods that we described are most suited for scenarios where the legitimate entity is present at the same time as the adversarial entity.

RSS is a physical parametes, widely available in deployed wireless communication networks, and is intimately tied to the location of a device in physical
space. For this reason, RSS is a common physical property used in localization
algorithms, [19-21] and can be used to detect communication spoofing.

Spoofing attack detection can be formulated as a statistical significance test
where the null hypothesis is:

$$\mathcal{H}_0 : \text{normal (no attack)}.$$

In significance testing, a test statistic $\mathbf{T}$ is used to evaluate whether observed data
belongs to the null hypothesis or not. If the observed test statistic $\mathbf{T}^{obs}$ differs
significantly from the hypothesized values, the null hypothesis is rejected and
we claim the presence of a spoofing attack.

Although affected by random noise, environmental bias, and multipath
effects, the RSS value vector, $\mathbf{s} = \{s_1, s_2, \ldots s_n\}$ (n is the number of landmarks/
access points [APs]), is closely related to the transmitter's physical location and
is determined by the distance to the landmarks. [21] We will describe the collection
of vectors $\mathbf{s}$ as constituting a signal space. When there is no spoofing, for each
MAC address, the sequence of RSS sample vectors will be close to each other
and will fluctuate around a mean vector. However, under a spoofing attack, there
is more than one node at different physical locations claiming the same MAC
address. As a result, the RSS sample readings from the attacked MAC address
will be mixed with RSS readings from at least one different location. Based on
the properties of the signal strength, the RSS readings from the same physical
location will belong to the same cluster points in the n-dimensional signal space,
whereas the RSS readings from different locations in the physical space should
form different clusters in signal space.

This observation suggests that we may conduct cluster analysis on the RSS
readings from each MAC address to detect spoofing. For example, the K-means
algorithm is an easy-to-use and efficient candidate algorithm for clustering. If
there are M RSS sample readings for a MAC address, the K-means clustering
algorithm partitions M sample points into K disjoint subsets S_j containing M_j
sample points so as to minimize the sum-of-squares criterion:

$$J_{min} = \sum_{j=1}^{K} \sum_{\mathbf{s_m} \in S_j} \|\mathbf{s_m} - \mu_j\|^2 \qquad (9.1)$$

where $\mathbf{s_m}$ is a RSS vector representing the mth sample point and μ_j is the
geometric centroid of the sample points for S_j in signal space. Under normal
conditions, the distance between the centroids should be close to each other
because there is basically only one cluster. Under a spoofing attack, however,
the distance between the centroids is larger because the centroids are derived

from the different RSS clusters associated with different locations in physical space. We thus choose the distance between two centroids as the test statistic $\mathbf{T}$ for spoofing detection,

$$D_c = ||\mu_i - \mu_j|| \qquad (9.2)$$

with $i, j \in \{1, 2..K\}$.

The thresholds used in defining the critical regions of the significance test can either be set empirically or via an analytical model. To illustrate, we use the following definitions: *an original node* P_{org} is referred to as the wireless device with the legitimate MAC address, while *a spoofing node* P_{spoof} is referred to as the wireless device that is forging its identity and masquerading as another device. We now present results from an experimental validation that shows that position information can be a valuable tool for detecting spoofing.

We will evaluate the performance of a cluster-based spoofing detector by analyzing the resulting detection rate and false-positive rate. The detection rate is defined as the percentage of actual spoofing attack attempts that are correctly classified as being an attack. Note that when the spoofing attack is present, the detection rate corresponds to the probability of detection P_d. Under normal (non-attack) conditions, a detection corresponds to a false alarm, and hence we are also interested in the false-positive P_{fa} rate.

Table 9.1 presents the detection rate and false-positive rate for an 802.11 network and a 802.15.4 network under different threshold settings (details describing the experimental set up can be found in Yang et al. 2009).[22] The results show that for false-positive rates less than 10 percent, the detection rates are above 95 percent. Even when the false positive rate goes to zero, the detection rate is still more than 95 percent for both 802.11 and 802.15.4 networks.

Table 9.1. *Detection Rate and False-Positive Rate of the Spoofing Attack Detector. Two Different Types of Wireless Networks Were Used (802.11 and 802.15.4) to Show the Feasibility of Using Location to Detect an Identity Attack Against a Wireless Network, Without Resorting to Cryptographic Mechanisms*

Network, Threshold	Detection Rate	False Positive Rate
802.11, $\tau = 5.5$dB	0.9937	0.0819
802.11, $\tau = 5.7$dB	0.9920	0.0351
802.11, $\tau = 6$dB	0.9884	0
802.15.4, $\tau = 8.2$dB	0.9806	0.0957
802.15.4, $\tau = 10$dB	0.9664	0.0426
802.15.4, $\tau = 11$dB	0.9577	0

9.4.2 Location-Oriented Multicast Key Management

When sending an identical message to multiple receivers, adopting the multicast communication model reduces the network traffic and allows the sender to conserve energy consumed for data processing. Several critical network operations such as routing, neighbor discovery, key distribution, and topology control can benefit from multicast by efficiently distributing protocol status updates or any other required data. Furthermore, in a wireless environment, due to the broadcast nature of the wireless medium, multicasting has the potential to not only reduce the network traffic in number of messages, but also reduce the network energy expenditure. A single broadcast transmission will reach any receiver within the communication range* of the source. However, anyone in range can listen to an information broadcast over the wireless medium. Hence, it is important to ensure that only the intended receivers have access to the group communication at any given time.

Encrypting the information transmitted over the open wireless channel is the most common technique for securing the multicast communication.† The use of cryptography requires all the valid receivers to hold the decryption key in order to decrypt a common message. The shared decryption key is called Session Encryption Key (SEK). To preserve the secrecy of the multicast data, the SEK needs to be updated each time a membership change occurs. For updating the SEK, multicast members share additional keys called Key Encryption Keys (KEK) that allow the secure update of the SEK to valid members. The *key management* problem is to ensure that only the legitimate members of the multicast group hold valid keys at any time during the session. In the presence of group members that may join or leave the multicast group, the key management problem is equivalent to the problem of finding efficient mechanisms to generate, assign, and distribute cryptographic keys. Hence, the key management problem can be reduced to the *key distribution* problem, which addresses the secure and efficient distribution of the cryptographic keys to valid members.

We will show that it is possible to provide energy-efficient key distribution scheme for implementing group access control for multicast communication in wireless ad hoc networks. To reduce the energy expenditure (physical layer parameter) of the key distribution (application layer operation), we propose a cross-layer design approach that incorporates (a) the network topology (location of the nodes) and (b) the propagation medium characteristics (physical layer). We note that the use of network topology for efficient multicast key management

---

* The communication range is defined as the maximum distance from the transmitter to a receiver, so that the signal-to-noise ratio (SNR) is above the required threshold for communication.
† Additionally, cryptography can also support group access control for dynamic multicast groups through secure management of the cryptographic keys.[23]

of cellular networks has been examined in Sun et al. (2002) and Sun et al. (2003).[24,25]

9.4.2.1 Network Model Assumptions

Network generation. The network consists of N multicast members plus the Group Controller (GC) randomly distributed in a specific area. The GC is also randomly placed within the network region. We consider a single-sender multiple-receiver communication model. We assume that all users can act as relay nodes and therefore relay information to any user within the communication range. We also assume that the network nodes have the ability to generate and manage cryptographic keys.

Node location acquisition. For our main analysis, once generated, the nodes of the network are assumed to be in a fixed location. We also assume that nodes have a mechanism to acquire their location information. Such information is often obtainable through the Global Positioning System (GPS).[26]

However, in many cases, GPS may not be available due to the expensive hardware required (e.g., sensor networks) or the lack of obstacle-free communication (indoor setting). Several approaches have been proposed for acquiring location information without a GPS receiver.[19,27,28,29,30] After a node correctly acquires its location, it can report it to other nodes through a location service algorithm.[31,32] However, to prevent denial-of-service attacks, the use of a secure location verification algorithm is important.[33]

Network initialization. We assume that the network has been successfully initialized and initial cryptographic quantities (pairwise trust establishment) have been distributed through secure channels.[34,35] We further assume that the underlying routing is optimized in order to minimize the total energy required for broadcast. Although it is known that finding the optimal solution for power-optimal broadcast is NP-complete,[36,37] several heuristics resulting in routing trees with satisfactory performance have been proposed in the recent literature.[38,36,39]

Wireless medium and signal transmission. We consider the cases of a homogeneous and heterogeneous medium separately, because the complexity and inputs of the algorithms that we propose differ depending on the type of the medium. In the case of the homogeneous medium, we assume that the transmission power $P(d_{i,j})$ required for establishing a communication link between nodes i and j is proportional to a constant exponent (attenuation factor γ) of the distance $d_{i,j}$, that is, $P(d_{i,j}) \propto d_{i,j}^{\gamma}$. For simplicity, we set the proportionality constant to be equal to 1. An example of a homogeneous path loss medium is an obstacle-free, open space terrain with line-of-sight (LOS) transmission.

For a heterogeneous medium, no single path loss model may characterize the signal transmission in the network deployment region. Even when node locations are relatively static, path loss attenuation can vary significantly when the network is deployed in mountains, dense foliage, urban region, or inside different floors of a building. We consider the following two models of varying path loss for calculating the power attenuation at a distance d from the transmitter: [40] (a) suburban area – a slowly varying environment where the attenuation loss factor changes slowly across space; (b) office building – a highly heterogeneous environment where the attenuation loss factor changes rapidly over space.

Antenna model. We assume that omnidirectional antennas are used for transmission and reception of the signal. [38] The omnidirectionality of the antennas results in a property unique in the wireless environment known as the *broadcast advantage*. [38] When the sender transmits a message to a node, any other nodes that lie within the transmission range can receive the broadcasted message for free. Hence, when an identical message needs to be sent to multiple receivers, the sender can significantly reduce the energy expenditure by directly transmitting the data to the farthest member. However, omnidirectional antennas require more power to transmit a signal at distance d than directional ones. [40] We further assume that signal transmission is the major component of energy expenditure and ignore any energy cost due to computation and information processing. [34,41]

Our aim is to develop energy-efficient key management scheme for wireless ad hoc networks, which accounts for the following factors: (a) A_1 : The network topology, that is, node location and relative position to minimize the energy consumption; (b) A_2 : The characteristics of the medium where the network is deployed (path loss parameter, homogeneous as well as heterogeneous, or power measurements); and (c) A_3 : Scalability of communication overhead in both bandwidth and required key storage space with respect to group size N.

9.4.2.2 A New Evaluation Metric for Measuring the Communication Overhead of Key Management

In order to incorporate the features A_1 to A_3 into the key management scheme, we first define a suitable performance evaluation metric that reflects the energy expenditure associated with the key distribution overhead. We then show that if key graphs are evaluated with the new metric, their performance is dependent not only on key graph structure, but also on node location and medium type (homogeneous or heterogeneous).

In wired networks, the communication overhead associated with the key management is measured as the number of messages sent by the GC to the group members in order to complete a key update. In key trees, a higher-degree tree requires a larger number of messages to be transmitted by the GC in case of a member leave.[42] For logical key hierarchies as proposed in Wang et al. (2000),[42] the communication cost to update the SEK and compromised KEKs after a member deletion is equal to $(\alpha \log_\alpha N - 1)$, where α is the degree of the tree. In Canetti et al. (1999),[23] the authors propose the use of key trees in conjunction with one-way functions, to reduce the communication cost to $(\alpha - 1) \log_\alpha N$ messages per member deletion. None of the metrics that have been developed for the wired networks, with the exception of time delay, is calculated collectively on the whole network. We propose a metric that calculates the energy expenditure of the whole network, occurring due to member leave/deletion.

In wireless ad hoc networks, each message has an associated energy cost that depends on the location of the receiver relative to the source, the routing path that connects them, and the path-loss model assumed. Since in an ad hoc setup, messages with different recipients have different energy requirements, a small number of transmitted messages by the GC does not necessarily translate to low energy expenditure. Hence, a higher-degree tree may have a lower energy cost compared to a lower-degree tree, even if more messages need to be transmitted in the case of the tree of higher degree to update compromised keys after a member deletion. To capture the energy dimension of the key management, we propose a new performance metric called *Average Update Energy*, as defined below.

Definition: Average update energy: Let $\tilde{E}_{M_i}$ denote the energy expenditure for updating the compromised keys after the deletion of the ith member. Also, let $p(M_i)$ denote the distribution of the member leaves/deletions from the multicast group. Then, we define the average update energy required for key update after a member leave/deletion as:

$$E_{Ave} = \sum_{i=1}^{N} p(M_i)\tilde{E}_{M_i} \qquad (9.3)$$

We define the update energy in the average sense, since a member deletion triggers transmissions to different subgroups of the multicast group. Hence, $\tilde{E}_{M_i}$ depends on the member that is being deleted. For example, in Figure 9.3(a), if member M_1 were to be deleted, the messages that need to be transmitted are shown in Figure 9.3(b). To further reinforce the idea, if M_8 were to be deleted, the following message transmissions have to take place:

$$GC \to M_7 : \quad \{K'_{2.4}\}_{K_{3.7}}, \{K'_{1.2}\}_{K_{3.7}}$$
$$GC \to \{M_5, M_6\} : \quad \{K'_{1.2}\}_{K_{2.3}}$$
$$GC \to \{M_5 - M_7\} : \quad \{K'_0\}_{K'_{2.1}}$$
$$GC \to \{M_1 - M_4\} : \quad \{K'_0\}_{K_{1.1}}$$

Note that the member M_7 will receive keys $K'_{2.4}$, $K'_{1.2}$ both encrypted with key $K_{3.7}$. Although the GC can concatenate both keys into one message and

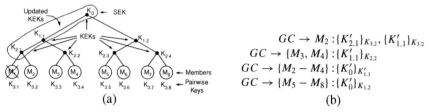

Figure 9.3. (a) A binary logical hierarchical key tree. Members are placed at the leaf nodes. Each members holds the keys traced along the path from the leaf to the root of the tree. If M_1 leaves the multicast group, all keys known to it (keys traced along the path from the leaf $[M_1]$ to the root of the tree) are updated (b) Update messages sent by the GC after M_1 leaves the multicast group.

update both keys to M_7 with one transmission, we show that two messages are transmitted from the GC to M_2. We intentionally note the messages separately for counting purposes. Assuming that all the keys have the same length and the concatenated message is twice as long when compared to a single message, the two representations are equivalent in both bits transmitted and energy consumed by the network. A key concatenation has the advantage of guaranteeing that both keys are received with same delay by M_7. Sending the keys through separate messages is suitable in a highly lossy medium where frequent retransmissions occur.[43] Because our scheme is concerned with the energy consumption, we use a representation that allows us to count the amount of energy spent, using the parameter of one new key encrypted per message.

From these two examples, for an ad hoc network with random node distribution, $\tilde{E}_{M_1} \neq \tilde{E}_{M_8}$, that is, the deletion of M_1 and M_8 result in different energy expenditures. As mentioned earlier, we consider the case of a member leave, because significantly higher communication cost occurs during a member leave than a member join.[44] We now examine the properties of E_{Ave}.

9.4.2.3 Dependency of the Average Update Energy on the Group Size N, the Tree Degree α and the Deployment Region

In this section, we examine the dependency of the E_{Ave} on the group size N, the degree of the key distribution tree α, and the network deployment region. We do so by extracting an upper bound on E_{Ave} that does not depend on the distribution $p(M_i)$ of the member leaves/deletions, or the network topology.

E_{Ave} depends on the energy $\tilde{E}_{M_i}$ required for the deletion of each member M_i from the multicast group MG. Regardless of which member is deleted, the number of messages sent by the GC for updating keys after a member leave, is equal to $(\alpha \log_\alpha N - 1)$. These messages are routed to different subgroups SG of the multicast group MG (see Figure 9.3). However, the energy for sending a message from the GC to any subgroup SG of the multicast group MG, cannot exceed the energy for sending a message to the whole group MG if the same routing tree is used in both cases. Though true for any routing tree, assuming that the optimal routing tree in total transmission power is used for message delivery,

$$E_{SG} \leq E_{MG}^*, \quad \forall \, SG \subseteq MG \tag{9.4}$$

where E_{MG}^* denotes the minimum energy consumed for sending a message from the GC to the whole multicast group, calculated according to the optimal routing tree, and E_{SG} denotes the energy consumed for sending a message from the GC to all members of the subgroup SG, calculated with the same routing tree. Note that E_{SG} need not necessarily be optimal. Using the inequality in (9.4), we can bound the energy $\tilde{E}_{M_i}$, for routing $(\alpha \log_\alpha N - 1)$ update messages to different subgroups of the multicast group MG by:

$$\tilde{E}_{M_i} \leq E_{MG}^*(\alpha \log_\alpha N - 1) \tag{9.5}$$

By combining (9.3) and (9.5), we can bound the average update energy E_{Ave} for a multicast group MG of size N and a key distribution tree of degree α:

$$
\begin{aligned}
E_{Ave} &= \sum_{i=1}^{N} p(M_i)\tilde{E}_{M_i} \\
&\leq \sum_{i=1}^{N} p(M_i)E_{MG}^*(\alpha \log_\alpha N - 1) \\
&\leq E_{MG}^*(\alpha \log_\alpha N - 1)\sum_{i=1}^{N} p(M_i) \\
&\leq E_{MG}^* \left(\alpha \log_\alpha N - 1\right) \tag{9.6}
\end{aligned}
$$

The bound in (9.6) has two different components. The first component is the minimum energy E_{MG}^* required for sending a message to the whole multicast group MG. E_{MG}^* depends on the wireless medium characteristics and the network topology/routing protocol that defines the routing tree. However, we can relax the network topology dependency by bounding E_{MG}^* using only the wireless medium characteristics and size of the deployment region.

Let γ_{max} be the maximum value of the attenuation factor for the heterogeneous medium where the network is deployed, and let d_{max} be the size of the deployment region, defined by the physical distance between the GC and the farthest member.[*] Assuming that omnidirectional antennas are used, the GC can broadcast a message to all members of the multicast group, just by transmitting to the farthest member located at d_{max}.[38] Under this routing strategy, the

---

[*] The size of the deployment region may also be defined as the maximum physical distance between any two nodes of the network. However, such a definition leads to a looser upper bound and is not considered.

transmission power of the GC for sending one message to all members of MG cannot exceed:

$$P_{max} \leq (d_{max})^{\gamma_{max}} \qquad (9.7)$$

Hence, the energy expenditure, $E_{MG}^{broadcast}$ for broadcasting a message from the GC to all members of MG can be bounded as:

$$E_{MG}^{broadcast} = P_{max} T_{trans}$$

$$\leq (d_{max})^{\gamma_{max}} T_{trans} \qquad (9.8)$$

where T_{trans} is the duration of the transmission of one message, fixed by the size of the message and the transmission bit rate. However, E_{MG}^{*} is optimal for sending a message from the GC to *all* members of MG. Hence, the optimal energy E_{MG}^{*} corresponding to the minimum total power strategy, should not be higher than $E_{MG}^{broadcast}$. Therefore, E_{MG}^{*} in (9.6) can be bounded by:

$$E_{MG}^{*} \leq E_{MG}^{broadcast}$$

$$\leq (d_{max})^{\gamma_{max}} T_{trans} \qquad (9.9)$$

The second component of the bound in (9.6) is the number of update messages sent by the GC for deleting a member from the multicast group. Whereas the number of messages grows logarithmically with the group size N, and N is not a design parameter, we can calculate the tree degree α^{*} that minimizes the number of update messages:

$$\frac{d}{d\alpha}(\alpha \log_{\alpha} N - 1) = 0$$

$$\frac{\ln \alpha - 1}{\ln \alpha^2} = 0$$

$$\alpha^{*} = e \qquad (9.10)$$

The degree of the tree has to be an integer number, and hence the lowest upper bound for E_{Ave} is achieved when $\alpha = 3$. The lowest upper bound for the average update energy, independent of the network topology and distribution of member leaves/deletions, is:

$$E_{Ave} \leq (3 \log_3 N)(d_{max})^{\gamma_{max}} T_{trans} \qquad (9.11)$$

Note that if we optimize the tree degree α, to minimize the number of rekey messages when both joins and leaves are taken into account, and assuming that they occur equally likely, it can be shown that $\alpha^{*} = 4$.[42] Also, if we consider key trees using one-way functions as in Canetti et al.(1999)[23], the optimal tree degree α that minimizes the number of rekey messages is equal to $\alpha^{*} = 2$. The analysis presented in this section holds for both one-way function trees as in Canetti et al. (1999),[23] and joint consideration of joins and leaves as in

Wong et al. (2000),[42] with the upper bound in (9.11) adjusted according to the optimal tree degree in each case. In general, the optimal tree degree in (9.11) can be adjusted to correspond to any assumed model of joins and leaves, or any other key tree structure.

9.4.2.4 Impact of "Power Proximity" on the Key Management Overhead under the New Metric

In this section, we investigate the impact of the power proximity on the energy efficiency of the key distribution. By observing that in the homogeneous medium case (constant attenuation factor γ), power-proximity between two nodes is a monotonically increasing function of the physical distance between them, we show that we can perform energy-efficient key distribution by taking into account only the physical proximity of the nodes. We then show that when the medium is heterogeneous, location information alone is not sufficient for constructing an energy-efficient key-distribution scheme. In the case of a heterogeneous medium, we show that location has to be combined with path loss model information or power measurements in order to extract the power proximity between pairs of nodes, which allows us to construct energy-efficient key trees.

Network Deployed in a Homogeneous Medium (Constant Attenuation Factor)
In a homogeneous medium, the transmission power for communication between nodes i, j is a monotonically increasing function of the distance $d_{i,j}$. Under the assumption that routing is optimally selected to minimize the total transmission power, spatially correlated nodes are connected in the routing tree or receive information through similar routing paths.[38] Intuitively, given the node location, members that are physically close should be grouped together and receive similar key updates to reduce the energy expenditure.

To illustrate the need for designing a location-aware key distribution, we consider the ad hoc network in Figure 9.4(a), which is deployed in a homogeneous medium. The routing tree shown in Figure 9.4(a) is optimal in total transmit power. In the key tree of Figure 9.4(c), denoted as Tree A, we randomly place the four members of the multicast group in the leaves of the key tree, independent of the network topology as in wired networks. Assume that key K_0 needs to be updated. On the first row of Table 9.2, for Tree A, we indicate the messages sent by the GC for the update of K_0 to the appropriate subgroups, and the corresponding energy expenditure. The energy is computed according to the optimal routing tree structure of Figure 9.4(a).

Assume now that the members are grouped according to their physical proximity. Then, M_1 is grouped with M_4, and M_2 with M_3, resulting in the location-aware key tree of Figure 9.4(d), denoted as Tree B. On the second row of Table 9.2, we indicate the messages sent by the GC to update K_0 to the appropriate subgroups, and the corresponding energy expenditure for Tree B. The energy

Table 9.2. *Messages Sent by the GC for the Update of K_0 and Associated Energy Expenditure for the Key-Distribution Trees of Figure 9.4(c), (d), (e). E^X_{rekey} Denotes the Energy Required for Updating K_0 in Key Tree X. $E_{A \rightarrow B}$ Denotes the Energy Required for Transmission of a Key from A to B*

Key Tree	Messages Sent by the GC	Energy Expenditure
Tree A	$GC \rightarrow \{M_1, M_3\} : \{K_0'\}_{K_{1.1}}$ $GC \rightarrow \{M_2, M_4\} : \{K_0'\}_{K_{1.2}}$	$E^A_{rekey} = E_{M_1 \rightarrow M_4} + 2E_{GC \rightarrow M_2} + E_{M_2 \rightarrow M_3}$
Tree B	$GC \rightarrow \{M_1, M_4\} : \{K_0'\}_{K_{1.1}}$ $GC \rightarrow \{M_2, M_3\} : \{K_0'\}_{K_{1.2}}$	$E^B_{rekey} = E_{GC \rightarrow M_1} + E_{M_1 \rightarrow M_4} + E_{GC \rightarrow M_2} + E_{M_2 \rightarrow M_3}$
Tree C	$GC \rightarrow \{M_1, M_2\} : \{K_0'\}_{K_{1.1}}$ $GC \rightarrow \{M_3, M_4\} : \{K_0'\}_{K_{1.2}}$	$E^C_{rekey} = E_{GC \rightarrow M_1} + E_{M_2 \rightarrow M_3} + E_{GC \rightarrow M_2} + E_{M_3 \rightarrow M_4}$

saved by performing a rekey operation with the location-aware key Tree B over the random key Tree A for the network of figure 9.4(a) is computed as:

$$E^A_{rekey} - E^B_{rekey} = E_{GC \rightarrow M_2} - E_{GC \rightarrow M_1} > 0 \qquad (9.12)$$

where E^X_{rekey} denotes the energy required for updating K_0 in key Tree X and $E_{A \rightarrow B}$ denotes the energy required for transmission of a key from node A to

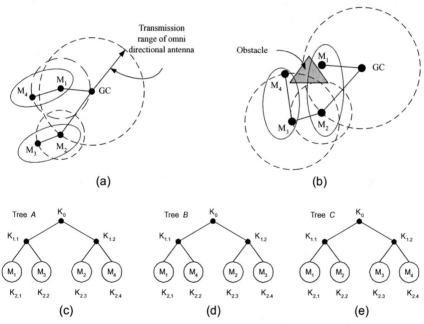

Figure 9.4. An ad hoc network and the corresponding routing tree with the minimum total transmission power, deployed in (a) a homogeneous medium and (b) a heterogeneous medium (c) A random key-distribution tree, Tree A. (d) A key-distribution tree based in physical proximity, Tree B, (e) A key-distribution tree based on "power proximity," Tree C.

node B. Non-negativity follows from the fact that $d_{GC,M_2} > d_{GC,M_1}$ and from the homogeneity of the medium (γ is constant). Hence, $P(d_{GC,M_2}) > P(d_{GC,M_1})$.

Network Deployed in a Heterogeneous Medium (Variable Attenuation Factor)
We now consider the case of an ad hoc network deployed in a heterogeneous medium, where the attenuation factor γ varies significantly over different regions of the network. Under heterogeneous path loss, physical proximity of two nodes does not necessarily imply that the power needed for establishing a communication link is lower than the power needed for two nodes located farther apart. Thus, closely located nodes do not necessarily receive messages through similar routing paths. Hence, node location information alone is not sufficient for constructing an energy-efficient key tree.

To illustrate the preceding observation, we consider the ad hoc network shown in Figure 9.4(b), in which nodes have the same locations as in Figure 9.4(a). However, there exists a physical obstacle between nodes M_1 and M_4. Thus, the attenuation factor for signal transmission between M_1 and M_4 is significantly higher than the rest of the obstacle-free network regions. Therefore, the optimal routing tree in total transmission power connects M_4 to the network through M_3.

We now show that in an environment with variable path loss, we are able to construct an energy-efficient key tree by correlating nodes according to their power proximity rather than physical proximity. We may acquire such information either by using path loss information in addition to the node location, or by measuring the required transmission power for communication between pairs of nodes. Members that are closely located in terms of power are grouped together (placed adjacently to the key tree).

For the network in Figure 9.4(b), we construct the key distribution tree in Figure 9.4(e), denoted as Tree C. We place members adjacently to the key tree according to their power proximity. M_1 is grouped with M_2, and M_3 with M_4 in order to minimize the total communication power variance of clusters of two members. In the third row of Table 9.2, we indicate the messages sent by the GC for the update of K_0 to the appropriate subgroups and the corresponding energy expenditure for Tree C. The energy saved for performing a rekey operation by incorporating location as well as the path loss information instead of location alone is computed as the energy gain due to use of Tree C over Tree B:

$$E^B_{rekey} - E^C_{rekey} = E_{M_1 \to M_4} - E_{M_3 \to M_4} > 0. \tag{9.13}$$

Non-negativity follows from the observation that due to the obstacle between M_1 and M_4, $E_{M_1 \to M_4} > E_{M_3 \to M_4}$. Based on our analysis in Sections 9.4.2 and 9.4.2, we make the following conclusions:

Conclusion 1: When the medium is homogeneous, the transmission power $P(d_{i,j})$ is a monotonically increasing function of the distance $d_{i,j}$ between nodes

i and j, $(P(d_{i,j}) \propto d_{i,j}^{\gamma}$, γ constant). Hence, closely located nodes require less power for communication and therefore are connected in the routing tree that minimizes the total transmission power. By exploiting the physical proximity information, we can develop an energy-efficient key tree hierarchy.

Conclusion 2: In the case of a heterogeneous medium, no single function can map the distance to transmission power. Different functions with variable attenuation factor γ hold for different network regions. Hence, the use of physical proximity does not necessarily result in the minimum total power-routing tree. Instead, we use power proximity to create an energy-efficient key tree hierarchy.

Based on conclusions 1 and 2, we develop our key distribution algorithms for the homogeneous and heterogeneous cases.

9.4.3 Location-Aware Key Distribution for a Homogeneous Medium

In this section, we develop an energy-efficient key-distribution algorithm for the homogeneous medium, based on node location information. Note that updating the keys after a member deletion requires multicast transmissions to subgroups of various sizes (see Figure 9.3). For energy-efficient key distribution, we need to fully utilize the broadcast advantage when we distribute keys to subgroups.

In Section 9.4.2.4, we showed that placing closely located nodes adjacently on the key distribution tree results in significant savings in energy resources when the medium is homogeneous. In order to systematically construct a key tree hierarchy, we need to be able to cluster nodes based on the location information. The clustering of the nodes should allow us to form a hierarchy. Then we can translate the physical clustering of the nodes into a key tree hierarchy, thus obtaining an energy-efficient key distribution tree. Hence, the task of developing a location-aware key distribution scheme is reduced to the task of identifying (a) a location-based clustering mechanism, and (b) building a cluster hierarchy that utilizes the location-based clustering. We discuss both tasks in the following sections.

9.4.3.1 Location-Based Clustering for Energy-Efficient Key Distribution

For the homogeneous medium, we have set the constraint that the only information available to us is node location, without any explicit parametric model assumptions for our clustering. Hence, our clustering technique should be model-free while taking the location information into account. We also note that for the homogeneous case, physical proximity is a suitable metric because the attenuation factor γ is a constant. Hence, the Euclidean distance between the nodes is a natural metric for identifying and grouping neighbor nodes. Certainly some other distance metric, such as the Minkowsky metric, [45] can be used as well, but the monotonicity of the power to the distance in the case of constant γ makes the Euclidean a very attractive one, since it leads to low-complexity algorithms.

Our effort is focused on finding a clustering technique that (a) requires only location information as an input, (b) identifies the physical network clusters with high success and, (c) generates clusters of equal size.

Problem Formulation for Location-Based Clustering
Let the coordinates of a node i be $x_i = (x_{i_1}, x_{i_2})$. The squared Euclidean distance between two nodes i and i' is equal to:

$$d_{i,i'}^2 = \sum_{j=1}^{2}(x_{ij} - x_{i'j})^2 = \|x_i - x_{i'}\|^2 \tag{9.14}$$

If C denotes an assignment of the nodes of the network into α clusters, the dissimilarity function expressing the total intercluster dissimilarity $W(C)$ is:

$$W(C) = \sum_{k=1}^{\alpha} \sum_{C(i)=k} \|x_i - m_k\|^2 \tag{9.15}$$

where $C(i) = k$ denotes the assignment of the ith point to the kth cluster, and m_k is the mean (centroid) of cluster k. Intercluster dissimilarity refers to the dissimilarity between the nodes of the same cluster. We wish to find the optimal cluster configuration C^* that minimizes (9.15), subject to the constraint that the sizes of the resulting clusters are equal:

$$C^* = \arg \min_{C} \sum_{k=1}^{K} \sum_{C(i)=k} \|x_i - m_k\|^2, \quad \ni \quad |C(i)| = |C(j)|, \quad \forall i, j \tag{9.16}$$

Note that this formulation provides an optimal way to create α subclusters from one cluster. This location-based clustering has to be iteratively applied to generate the desired cluster hierarchy.

Solution Approach
If we relax the constraint $|C(i)| = |C(j)|$, $\forall i, j$, in (9.16), and allow clusters of different sizes, the solution to the optimization problem in (9.16), can be efficiently approximated by K-means algorithm.[45] K-means uses squared Euclidean distance as a dissimilarity measure to cluster different objects. It also generates clusters by minimizing the total cluster variance (minimum square error approach). Note that K-means may result in a suboptimal local minimum solution depending on the initial selection of clusters, and hence, the best solution out of several random initial cluster assignments should be adopted.[45] However, K-means is easily implemented and hence, is an ideal solution for computationally limited devices. Algorithmic details on solving (9.16) without any constraint on the cluster size are given in Hastie et al. (2001).[45]

To satisfy the constraint posed in (9.16), we need a refinement algorithm (RA) that balances the cluster sizes while taking advantage of the low-complexity of K-means algorithm. According to (9.16), the RA should result in balanced clusters with the lowest total intercluster dissimilarity. In the binary tree case, given two clusters A, B with $|A| > |B|$, the refinement algorithm moves objects $i_1, i_2, \ldots, i_k \in A$, with $k = \lfloor \frac{|A|-|B|}{2} \rfloor$, from cluster A to cluster B, such that the intercluster dissimilarity after the refinement is minimally increased. We choose the objects $i_1, i_2, \ldots, i_k \in A$ such that:

$$i_j = \arg\min_{i \in A} \left[d^2_{i,m_B} - d^2_{i,m_A} \right], \quad j = 1 : \left\lfloor \frac{|A| - |B|}{2} \right\rfloor \qquad (9.17)$$

where m_A and m_B refer to the centroids of clusters A and B, respectively. Note that in K-means, objects are assigned to the closest centroid, and hence, $[d^2_{i,m_B} > d^2_{i,m_A}]$, $\forall i \in A$. By moving objects from A to B that increase $W(C)$ by the minimum possible amount, we achieve the optimal solution for the constrained optimization problem in (9.16) in the case of binary trees. *However, optimality is not guaranteed if more than two subclusters need to be balanced (d-ary tree).*

9.4.3.2 An Energy-Efficient Key-Distribution Scheme Based on Physical Proximity

We now develop an algorithm that maps the location-based clustering into a hierarchical key tree structure. Assume that we wish to construct a key tree of fixed degree α. Initially, the global cluster is divided into α subclusters using K-means. Considering that we want to construct a fixed-degree tree, every cluster *must* have equal number of members. Hence, we employ the RA algorithm to balance the cluster sizes. The RA leads to the construction of a balanced key tree when $N = \alpha^n$, $n \in \mathbb{Z}$, and allows us to construct a structure as close to the balanced as possible when $N \neq \alpha^n$. Each cluster is subsequently divided into α new ones, until clusters of at most α members are created (after $\log_\alpha N$ splits). Figure 9.5 presents the pseudo-code for our *Location-Aware Key Distribution Algorithm* (LocKeD) using K-means. We now describe the notational and algorithmic details of Figure 9.5.

Let $\mathcal{P}$ denote the set containing all the two-dimensional points (objects) corresponding to the location of the nodes. Let $C = \{C(1), C(2), \ldots, C(n)\}$ denote a partition of $\mathcal{P}$ into n subsets (clusters), that is, $\bigcup_i C(i) = \mathcal{P}$. Initially, all objects belong to the global cluster $\mathcal{P}$. The function *AssignKey()* assigns a key to every subset (cluster) of its argument set. For example, *AssignKey(*$\mathcal{P}$*)* will assign the SEK to every member of the global cluster $\mathcal{P}$.

Location-Aware Key Distribution – LocKeD

$C = \{\mathcal{P}\}$
$AssignKey(C)$
$index=1$
$while\ index < \lceil \log_\alpha(N) \rceil$
$\quad C\_temp = \{\emptyset\}$
$\quad thres = \lceil \frac{N}{\alpha^{index}} \rceil$
$\quad for\ i = 1 : |C|$
$\quad\quad R=Kmeans(C(i), \alpha)$
$\quad\quad R=Refine(R, thres)$
$\quad\quad AssignKey(R)$
$\quad\quad C\_temp = C\_temp \bigcup R$
$\quad\quad index++$
$\quad end\ for$
$C = C\_temp$
$end\ while$

(a)

Refinement Algorithm – RA

$C_{Low} = \{C(i) \in C : |C(i)| < thres\}$
$C_{High} = \{C(i) \in C : |C(i)| > thres\}$
$repeat\ until\ C_{High} = \emptyset$
$\quad find\ x^* \in A,\ A \in C_{High}$

$$x^* = \arg\min_{x \in A}[diss(x, m_B) - diss(x, m_A)],$$

$$\forall x \in A,\ \forall A \in C_{High},\ \forall B \in C_{Low}$$

move x^* to cluster B

$C_{Low} = \{C(i) \in C : |C(i)| < thres\}$
$C_{High} = \{C(i) \in C : |C(i)| > thres\}$

end repeat

(b)

Figure 9.5. Pseudo-code for (a) the location-aware key-distribution algorithm (LocKeD) and (b) the Refinement Algorithm (RA). Repeated application of *Kmeans()* function followed by the Refinement Algorithm *Refine()* for balancing the clustering sizes, generates the cluster hierarchy. Function *AssignKey()* maps the cluster hierarchy into a tree hierarchy by assigning appropriate keys to cluster members.

The *index* variable counts the number of steps required until the termination of the algorithm. The *thres* variable holds the number of members each cluster ought to contain at level $l = index$ of the key tree construction. The root of the tree is at level $l = 0$. The $Kmeans(C(i), \alpha)$ function divides the set $C(i)$ into α clusters and returns the cluster configuration to variable R. The $Refine(R, thres)$ function balances the sizes of clusters in R according to the *thres* variable. Then, $AssignKey()$ is applied to assign different keys to every cluster in R. The process is repeated until $\lceil \log_\alpha N \rceil$ steps have been completed.

In terms of algorithmic complexity, the LocKeD algorithm iteratively applies K-means up to N times in the worst case (generation of a binary tree). K-means has algorithmic complexity of $O(N)$.[45] Hence, the complexity of the LocKeD is $O(N^2)$. We note that LocKeD requires only location information as input, assuming that the tree degree is fixed a priori.

Application of LocKeD on a Sample Network Deployed
in a Homogeneous Medium
Consider the network in Figure 9.6(a), deployed in a homogeneous medium with an attenuation factor $\gamma = 2$. Assume that we wish to construct a location-aware key distribution tree of degree $\alpha = 2$ with nodes $\{2, 3, \ldots, 9\}$ being the members $\{M_2, M_3, \ldots, M_9\}$ of the multicast group, respectively. Initially, all members belong to the global cluster $\mathcal{P}$.

Note that the GC does not participate in the clustering. The hierarchical key tree is constructed by executing the following steps:

Step 1: Assign the SEK K_0 to every member of the global cluster $\mathcal{P}$.

Step 2: Create two clusters by splitting the global cluster. The two clusters that yield minimal total cluster dissimilarity are:

$$C_1 = \{M_2, M_3, M_4, M_6, M_8, M_9\}, C_2 = \{M_5, M_7\}.$$

Considering that we seek to construct a balanced key tree, apply the refinement algorithm to balance the clusters sizes. Move M_2 and M_6 to cluster C_2. Assign two different KEKs to members of clusters C_1 and C_2. Members of C_1 are assigned KEK $K_{1.1}$ and members of C_2 are assigned KEK $K_{1.2}$.

Step 3: Create clusters of two members by splitting the clusters of four members. The four created clusters are:

$$C_3 = \{M_2, M_6\}, \ C_4 = \{M_3, M_4\}, C_5 = \{M_8, M_9\}, \ C_6 = \{M_5, M_7\}.$$

Again, different KEKs are assigned to members of clusters C_3-C_6. Members of C_3 are assigned KEK $K_{2.1}$, members of C_4 are assigned KEK $K_{2.2}$, members of C_5 are assigned KEK $K_{2.3}$, and members of C_6 are assigned KEK $K_{2.4}$. At this point, we have completed the $\lceil \log_\alpha N \rceil$ steps required by LocKeD and the algorithm terminates.

The hierarchical key tree constructed using LocKeD is shown in Figure 9.6(b).

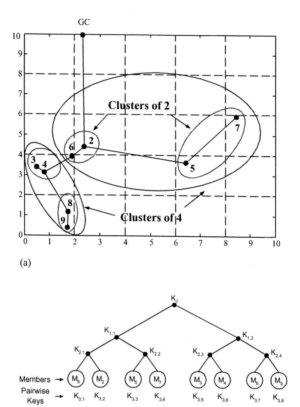

Figure 9.6. (a) An ad hoc network deployed in a homogeneous medium and the corresponding routing paths. Iterative application of the location-based clustering and the resulting cluster hierarchy. (b) The key-distribution tree resulting from the application of LocKeD.

When the medium is homogeneous, the key distribution algorithm discussed earlier makes use of the node location to securely and efficiently distribute keys to valid multicast group members. When the medium is heterogeneous, the node location information alone is not sufficient for energy-efficient clustering, and details are discussed in Lazos and Poovendran (2007) and Salido et al. (2007). [46,47] Location-based clustering algorithms were found to be power-efficient and secure when applied to obstacle-free open-space, suburban, and indoor environments. [46,47]

9.5 Using the Physical Layer to Enhance Security

The final component we will examine involves integrating the physical layer into the design of security protocols. The physical layer is responsible for the transmission and reception of signals between two or more entities. In the wireless context, the richness of the multipath environment associated with typical usage

scenarios (e.g., indoor or urban scenarios) implies that the physical characterization of the communication channel is a unique and hard-to-predict source
of information shared between two communicators. More specifically, channel
frequency responses decorrelate from one transmit-receive path to another if the
paths are separated by the order of an RF wavelength or more.[48] The fact that
pairwise radio propagation laws between two entities are unique and decorrelate
quickly with distance can serve as the basis for establishing shared secrets. These
shared secrets may be used as encryption keys for higher-layer applications or
wireless system services that need confidentiality or secret keys.[49,50] Similarly,
the wireless channel can enable wireless entities to authenticate other transmitters by tracking each other's ability to produce an appropriate received signal at
the recipient.[51,55–58,60]

In this section, we shall focus on how authentication can be achieved at
the lowest possible layer for a general wireless transmitter-and-receiver pair
involving multiple transmit and receive antennas. For those interested in how
the physical layer can be used for confidentiality, such as key establishment or
secrecy dissemination, we refer the reader to Mathur et al. (2008).[50]

Prior work[52] on physical layer authentication has focused on single-antenna
systems. However, with the ability to provide diversity gain and/or multiplexing gain, multiple-input multiple-output (MIMO) techniques will be widely
deployed in future wireless networks – for example, IEEE 802.11n and
WiMAX – to improve traffic capacity and link quality. The first caveat that must
be understood when employing physical layer authentication is that channel-
based authentication can only used to discriminate between different transmitters, and must be combined with a traditional handshake authentication process
to completely identify an entity. In other words, identity is inherently a higher-
layer function. Throughout this section, we assume that an entity's identity is
obtained at the beginning of a transmission using traditional higher-layer authentication mechanisms. Consequently, the role that channel-based authentication
plays is to ensure that all signals in both the handshake process and data transmission are actually from the same transmitter. Thus, this may be viewed as a
cross-layer design approach to authentication.

9.5.1 System Model

As shown in Figure 9.7, Alice, Bob, and Eve are assumed to be located in
spatially separated positions. Alice is the legal client with N_T antennas, initiating
communication by sending signals to Bob. As the intended receiver, Bob is the
legal access point (AP) with N_R antennas. Their nefarious adversary, Eve, will
inject undesirable communications into the medium with N_E antennas, in the
hopes of impersonating Alice.

We assume that Alice sends pilots from N_T antennas, and Bob uses these
to estimate channel responses (we note that these pilots are typically used for

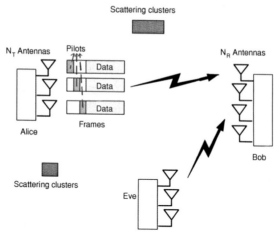

Figure 9.7. The adversarial multipath environment involving multiple scattering surfaces. The transmission from Alice with N_T antennas to Bob with N_R antennas experiences different multipath effects than the transmission by the adversary, Eve. Bob uses pilot symbols to estimate channel responses from the transmitters, and thus discriminate between Alice and Eve.

equalization purposes, and hence the security functions we describe can dual-use such physical layer information). In the authentication process, Bob tracks the channel responses to discriminate between legitimate signals from Alice and illegitimate signals from Eve.

A legal transmission from Alice to Bob in Figure 9.7 will involve a MIMO system with N_T transmit (Tx) antennas and N_R receive (Rx) antennas. Bob measures and stores channel frequency response samples at M tones, across an overall system bandwidth of W, where each subband has bandwidth $b (\leq W/M)$, and the center frequency of the system is f_0.

We consider channel frequency responses for two frames, which may or may not come from the same transmitter, and denote them by:

$$\mathbf{H}_i = \left[\underline{H}_i(1, 1), \underline{H}_i(1, 2), \dots, \underline{H}_i(N_T, N_R)\right]^T, \quad i = 1, 2, \qquad (9.18)$$

where $\underline{H}_i(j_t, j_r) = [H_{i,1}(j_t, j_r), \dots, H_{i,M}(j_t, j_r)]^T$, $1 \leq j_t \leq N_T$, $1 \leq j_r \leq N_R$, and $H_{i,m}(j_t, j_r) = H_i(j_t, j_r, f_o + W(m/M - 0.5))$ is the channel response at the mth tone in the ith frame, connecting the j_tth Tx antenna and j_rth Rx antenna. The $N_T N_R M$ elements in $\mathbf{H}_i$ are independent and identically distributed. Considering the phase rotation and receiver thermal noise, one may model the estimated channel frequency response as $\hat{\mathbf{H}}_i = \mathbf{H}_i e^{j\phi_i} + \mathbf{N}_i$, where $\phi_i \in [0, 2\pi)$ denotes the unknown phase measurement rotation, and $\mathbf{N}_i$ is the receiver thermal noise vector with $N_T N_R M$ elements, which are independent and identically distributed complex Gaussian random variables, $CN(0, \sigma^2)$, where σ^2 is the receiver noise power per tone.

MIMO-assisted channel-based authentication compares channel frequency responses at consecutive frames, and we should report spoofing if the channel responses from the same *claimed* user are significantly different in two frames. We note that we are assuming that the terminals are essentially stationary and that the channel does not vary significantly with time. Recent work[51] has examined the more general cases of time-variant channels.

We note that Eve can conduct an authentication attack only if she knows N_T and uses N_T transmit antennas. Following Kirkhoff's Principle for security analysis, we assume that Eve knows N_T. Assuming Bob obtains channel responses of $\hat{\mathbf{H}}_1$ and $\hat{\mathbf{H}}_2$, respectively, for two frames with the same identity, we build a simple hypothesis test for the purpose of transmitter discrimination. In the null hypothesis, $\mathcal{H}_0$, two estimates are from the same terminal, and thus the claimant is the legal user. Otherwise, Bob accepts the alternative hypothesis, $\mathcal{H}_1$, and claims that a spoofing attack has occurred.

The following test statistic is used to cope with unknown phase quantities ϕ_1 and ϕ_2:

$$L = \frac{1}{\sigma^2}||\hat{\mathbf{H}}_1 - \hat{\mathbf{H}}_2 e^{j\phi}||^2, \tag{9.19}$$

where

$$\phi = \arg\min_x ||\hat{\mathbf{H}}_1 - \hat{\mathbf{H}}_2 e^{jx}|| = Arg(\hat{\mathbf{H}}_1 \hat{\mathbf{H}}_2^H) \tag{9.20}$$

It can be shown under $\mathcal{H}_0$ that:

$$L_{\mathcal{H}_0} \approx \frac{1}{\sigma^2}||\mathbf{N}_1 - \mathbf{N}_2||^2 \sim \chi_S^2 \tag{9.21}$$

when the SNR is high, where $S = 2N_T N_R M$ degrees of freedom. Otherwise, when $\mathcal{H}_1$ is true, L is a noncentral Chi-square variable, given by:

$$L_{\mathcal{H}_1} \approx \frac{1}{\sigma^2}||\mathbf{H}_1 - \mathbf{H}_2 e^{j\phi} + \mathbf{N}_1 - \mathbf{N}_2||^2 \sim \chi_{S,\mu}^2 \tag{9.22}$$

where the noncentrality parameter, μ, is written as:

$$\mu = \frac{P_T}{P_N N_T}||\mathbf{H}_1 - \mathbf{H}_2 e^{j\,Arg(\mathbf{H}_1 \mathbf{H}_2^H)}||^2 \tag{9.23}$$

For fixed P_T, the dimension of $\mathbf{H}_i$ is proportional to MN_R, and thus μ rises with both N_R and M. On the other hand, the impact of N_T is more complex, depending on the specific value of $\mathbf{H}_1$, $\mathbf{H}_2$, and P_T.

The rejection region of $\mathcal{H}_0$ is defined as $L \leq k$, where k is the test threshold, which is selected according to an appropriate performance target.

The performance of a physical layer authentication scheme should examine the "false alarm rate" for a given k as:

$$\alpha = Pr(L > k|\mathcal{H}_0) = 1 - F_{\chi_S^2}(k) \tag{9.24}$$

where $F_X(\cdot)$ is the CDF of the random variable X, as well as the "miss detection rate" for given k, which is given by:

$$\beta = Pr(L \leq k|\mathcal{H}_1) = F_{\chi_{S,\mu}^2}(k) \tag{9.25}$$

It can be seen that α rises with k, while β decreases with k, and further that the miss rate decreases with P_T.

The use of multiple antennas has a twofold impact: It improves security performance by increasing the frequency sample size from $2M$ to $2MN_TN_R$, and the use of multiple transmit antennas reduces the transmit power per antenna, leading to performance loss of some degree.

Note that the frequency sample size, $M \in [1, M_s]$, is selected for security purposes, where M_s ($\geq M$), the total number of subbands is determined by non-security issues such as data-decoding accuracy. The average transmit power per tone is determined by M_s, with $P_T = P_{total}/M_s$, where P_{total} is the total system transmit power. Hence, P_T is independent of any other parameters mentioned, and we assume constant P_T in the comparison of system configurations.

In wideband systems, b is fixed and the detection performance improves with W, since channel responses decorrelate more rapidly in space with higher system bandwidth. It can be seen that β increases with b, since the power of measurement noise is proportional to b.

To illustrate the performance of channel-based authentication, we present simulation results that were obtained using the WiSE ray-tracing tool.[59] WiSE was used to generate typical channel responses for different locations in a typical office building. A brief description of the scenario is now provided, and we refer the reader to Xiao et al. (2008a)[53] for more detailed descriptions. In the office building, we deployed Bob as an access point roughly in the middle of the building, and varied the locations for Alice and Eve throughout a region of the building. For every Alice and Eve location, we calculated β for a given α. We then aggregated the results over all Alice-Eve location pairs to understand the overall feasibility for an adversary to conduct spoofing in this environment.

In the simulations, we consider MIMO, single-input multiple-output (SIMO), multiple-input single-output (MISO), and single-input single-output (SISO) systems, with separation of two neighboring antennas of 3 cm (i.e., half wavelength), $\alpha = 0.01$, $f_0 = 5$ GHz, $N_F = 10$, $b = 0.25$ MHz, and $P_T \in \{0.1, 1, 10\}$ mW, if not specified otherwise. The per tone SNR ranges from -16.5 dB to 53.6 dB, with a median value of 16 dB, using transmit power per tone $P_T = 0.1$

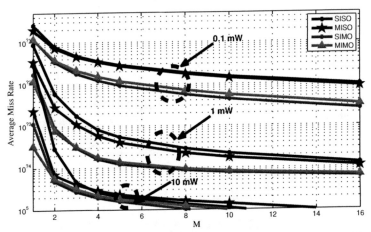

Figure 9.8. Average miss rate of spoofing detection in wideband systems, in SISO, 2×1 MISO, 1×2 SIMO, and 2×2 MIMO systems, respectively, with $\alpha = 0.01$, $M = 5$, $b = 0.25$ MHz, $W = 20$ MHz, and $P_T \in \{0.1, 1, 10\}$ mW.

mW, $b = 0.25$ MHz, and $N_T = N_R = 1$. Figure 9.8 shows that the average miss rate decreases with the frequency sample size, M, with a system bandwidth of $W = 20$ MHz, indicating that we should use all of the channel estimation data and set $M = M_s$. In addition, it can be seen that the security gain of MIMO decreases with M, when $P_T > 0.1$ mW. If using high power and small M (e.g., $M = 1$), the SISO system has accurate but insufficient channel response samples. Thus the additional dimensions of channel samples in MIMO systems allow for much better performance. On the contrary, if using high P_T and large M, the performance of SISO systems is too good to be significantly improved on.

We can also see that the security gain for a MIMO system over a SISO system slightly rises with M, when P_T is as low as 0.1 mW. This is because, when the channel estimation is not accurate due to low SNR, the system needs much more data to make a correct decision. Similarly, the impact of P_T on the MIMO security gain also depends on the value of M. The gain rises with P_T, under small M, whereas under large M, the security gain decreases with P_T.

We now summarize by examining how the above results on physical layer authentication should be interpreted in the context of future wireless networks. First, MIMO communication is becoming a dominant modality for wireless communication. Already technologies like 802.11n, which employ MIMO, are becoming prolific because of the communication rate and reliability improvements MIMO offers. It is possible to dual-use information associated with channel characterization, which is naturally obtained for the purposes of channel estimation for normal coding and decoding of signals, to provide a method for distinguishing between transmitters. The typical false alarm and miss rates that one can expect from physical layer authentication, for example $\alpha \approx 0.1$ and $\beta \approx 10^{-4}$, suggests that physical layer authentication will not replace the strong authentication guarantees of classical cryptographic methods (such as message authentication codes and digital signatures). However, these values for α and β do imply that it is possible to use the physical layer as a lightweight authentication service, which can serve as an initial filter lightening the load on higher-layer authentication services,[54] or even as an anomaly-detection scheme that flags a network administrator of potential intrusions on an open network.

9.6 Concluding Remarks

In this chapter, we have explored several new approaches to integrating security into the design of a future Internet. The methods discussed all share the common feature that physical properties must be used in order to enhance security and, for the most part, do not employ traditional cryptographic mechanisms/protocols. These physical properties can vary from assuring the integrity of network devices themselves to making use of position information to check that transmissions are occurring where they should, or as a means to improve the efficiency of security functions (e.g., key management), or even to using the raw properties of the signals being transmitted and received so as to derive signatures that can discriminate between transmitters. The discussion in this chapter has primarily focused on wireless networks, and hence is targeted at the *edge* of the future Internet. To a large part, our discussion has focused on wireless networks because they represent the primary access technology that users will employ in the future (and hence security mechanisms at the edge are paramount to establishing a first line of defense for the broader network). Beyond this, though, wireless technologies are also the most rapidly evolving of communication technologies, where interfaces to all layers of the protocol stack are being made available to programmers for development, and thus such platforms also allow for an easy path to experiment with such nontraditional approaches to security. We would note, though, that there is no reason why the methods described in this chapter could not be employed on other networks, such as optical networks, with appropriate modifications.

Lastly, we would remark that the objective of this chapter is to highlight a complementary set of tools that can be used to provide additional security, and we emphasize that a starting point for securing any network should be a collection of properly designed security protocols at the various layers of the protocol stack, and which utilize cryptographic primitives that correctly interlock with each other across the various network layers. Unfortunately, this is generally a daunting task, and the methods outlined in this chapter can be viewed as tools that can assist when there are weaknesses inherent in the underlying security protocols.

References

[1] Wu, S. G. Z., and Raychaudhuri, D. 2006. Irma: Integrated Routing and MAC Scheduling in Multihop Wireless Mesh Networks. *Proceedings of IEEE WiMesh Workshop.*

[2] Google android smart phone. http://www.android.com/

[3] Apple iphone. http://www.apple.com/iphone/

[4] Apple iphone sdk. http://developer.apple.com/iphone/

[5] Google android sdk. http://code.google.com/android/

[6] Shi, E., Perrig, A., and Doorn, L. V. 2005. BIND: A Time-of-Use Attestation Service for Secure Distributed Systems. *Proceedings of IEEE Symposium on Security and Privacy.*

[7] Seshadri, A., Perrig, A., van Doorn, L., and Khosla, P. 2004. Swatt: Software-Based Attestation for Embedded Devices. *Proceedings of the IEEE Symposium on Security and Privacy.*

[8] Seshadri, A., Luk, M., Shi, E., Perrig, A., van Doorn, L., and Khosla, P. 2005. Pioneer: Verifying Integrity and Guaranteeing Execution of Code on Legacy Platforms. *Proceedings of ACM Symposium on Operating Systems Principles*, pages 1–16.

[9] Mitola, J., and Maguire, G. Q. 1999. Cognitive Radio: Making Software Radios More Personal. *Personal Communications, IEEE*, 6(4): 13–18.

[10] The GNU software radio. http://www.gnu.org/software/gnuradio/

[11] How to write a signal processing block for gnu radio. http://www.gnu.org/software/gnuradio/doc/howto-write-a-block.html

[12] Rice University WARP – wireless open-access research platform. http://warp.rice.edu

[13] WINLAB WIN2CR platform. http://www.winlab.rutgers.edu/docs/focus/WiNC2R.html

[14] Xu, W., Kamat, P., and Trappe, W. 2006. TRIESTE: A Trusted Radio Infrastructure for Enforcing Spectrum Etiquettes. *IEEE Workshop on Networking Technologies for Software Defined Radio (SDR) Networks.*

[15] Trusted computing group. http://www.trustedcomputinggroup.org/

[16] BBN Technologies. 2004. XG Policy Language Framework, Version 1.0. *XG Working Group Document.*

[17] Owl web ontology language guide. http://www.w3.org/TR/owl-guide/

[18] Bellardo, J., and Savage, S. 2003. 802.11 Denial-of-Service Attacks: Real Vulnerabilities and Practical Solutions. *Proceedings of the USENIX Security Symposium*, pages 15–28.

[19] Bahl, P., and Padmanabhan, V. N. 2000. Radar: An In-building Rf-based User Location and Tracking System. *Proceedings of the IEEE International Conference on Computer Communications (INFOCOM)*, pages 775–784.

[20] Youssef, M., Agrawal, A., and Shankar, A. U. 2003. WLAN Location Determination via Clustering and Probability Distributions. *Proceedings of the First IEEE International Conference on Pervasive Computing and Communications (PerCom)*, pages 143–150.

[21] Elnahrawy, E., Li, X., and Martin, R. P. 2004. The Limits of Localization Using Signal Strength: A Comparative Study. *Proceedings of the First IEEE International Conference on Sensor and Ad hoc Communcations and Networks (SECON 2004)*, pages 406–414.

[22] Yang, J., Chen, Y., and Trappe, W. 2009. Detecting Spoofing Attacks in Mobile Wireless Environments. *Proceedings of the Sixth Annual IEEE Communications Society Conference on Sensor, Mesh and Ad Hoc Communications and Networks.*

[23] Canetti, R., Garay, J., Itkis, G., Micciancio, D., Naor, M., and Pinkas, B. 1999. Multicast Security: A Taxonomy and Some Efficient Constructions. *Proc. of IEEE INFOCOM 99.*

[24] Sun, Y., Trappe, W., and Liu, R. 2002. An Efficient Key Management Scheme for Secure Wireless Multicast. *Proc. of IEEE Int. Conf. on Communication (ICC'02),* vol. 2, pages 1236–1240.

[25] Sun, Y., Trappe, W., and Liu, R. 2003. Topology-Aware Key Management Schemes for Wireless Multicast. *Proc. of IEEE GLOBECOM.*

[26] Dommety, G., and Jain, R. 1996. Potential Networking Applications of Global Positioning Systems (GPS). *Technical report TR-24, the Ohio State University.*

[27] He, T., Huang, C., Blum, B. M., Stankovic, J. A., and Abdelzaher, T. 2003. Range-Free Localization Schemes in Large Scale Sensor Networks. *Proc. of the Ninth Annual ACM International Conference on Mobile Computing and Networking (MOBICOM).*

[28] Niculescu, D., and Nath, B. 2001. DV Based Positioning in Ad Hoc Networks. *Journal of Telecommunication Systems.*

[29] Priyantha, N. B., Chakraborty, A., and Balakrishnan, H. 2000. The Cricket Location-Support System. *Proc. of the Sixth Annual ACM International Conference on Mobile Computing and Networking (MOBICOM).*

[30] Savvides, A., Han, C., and Srivastava, M. 2001. Dynamic Fine-Grained Localization in Ad-Hoc Networks of Sensors. *Proc. of the ACM Annual International Conference on Mobile Computing and Networking (MOBICOM 2001).*

[31] Li, J., Jannotti, J., De Couto, D., Karger, D., and Morris, R. 2000. A Scalable Location Service for Geographic Ad-Hoc Routing. *Proc. of the Sixth Annual ACM International Conference on Mobile Computing and Networking (MOBICOM).*

[32] Xue, Y., Li, B., and Nahrstedt, K. 2001. A Scalable Location Management Scheme in Mobile Ad-Hoc Networks. *Proc. of the IEEE Conference on Local Computer Networks (LCN 2001).*

[33] Sastry, N., Shankar, U., and Wagner, D. 2003. Secure Location Verification. *Proc. of the ACM Workshop on Wireless Security (Wise 2003).*

[34] Carman, D. W., Cirincione, G. H., and Matt, B. J. 2002. Energy-Efficient and Low-Latency Key Management for Sensor Networks. *Proc. of the 23rd Army Science Conference.*

[35] Stajano, F., and Anderson, R. 1999. The Resurrecting Duckling: Security Issues for Ad-hoc Wireless Networks. *Security Protocols, 7th International Workshop.*

[36] Cagalj, M., Hubaux, J. P., and Enz, C. 2002. Minimum-Energy Broadcast In All Wireless Networks: NP-Completeness and Distribution Issues. *Proc. of the 8th ACM Annual International Conference on Mobile Computing and Networking (MobiCom 2002).*

[37] Liang, W. 2002. Constructing Minimum-Energy Broadcast Trees in Wireless Ad Hoc Networks. *Proc. of the 3rd ACM International Symposium on Mobile Ad Hoc Networking and Computing (MobiHoc 2002).*

[38] Wieselthier, J. E., Nguyen, G. D., and Ephremides, A. 2000. On the Construction of Energy Efficient Broadcast and Multicast Trees in Wireless Networks. *Proc. of IEEE INFOCOM 2000,* pages 586–594.

[39] Lee, S., Su, W., Hsu, J., Gerla, M., and Bagrodia, R. 2000. A Performance Comparison Study of Ad Hoc Wireless Multicast Protocols. *Proc. of the IEEE Conference on Computer Communications (INFOCOM 2000)*, pages 565–574.

[40] Rappaport, T. 1996. *Wireless Communications: Principles & Practice*, Prentice Hall.

[41] Raghunathan, V., Schurgers, C., Park, S., and Srivastava, M. B. 2002. Energy-Aware Wireless Microsensor Networks. *IEEE Signal Processing Magazine,* 19(2), 40–50.

[42] Wong, C. K., Gouda, M., and Lam, S. 2000. Secure Group Communications Using Key Graphs. *IEEE/ACM Trans. On Networking* 8(1), 16–31.

[43] Mittra, S. 1997. Iolus: A Framework for Scalable Secure Multicasting. *Proc. of SIGCOMM.*

[44] Wallner, D. M., Harder, E. C., and Agee, R. C. 1998. Key Management for Multicast: Issues and Architectures. *INTERNET DRAFT.*

[45] Hastie, T., Tisbshirani, R., and Friedman, J. 2001. *The Elements of Statistical Learning, Data Mining, Inference and Prediction*, Springer Series in Statistics, New York.

[46] Lazos, L., Poovendran, R. 2007. Power Proximity Based Key Management for Secure Multicast in Ad Hoc Networks. *Wireless Networks,* 13(1), 127–148.

[47] Salido, J., Lazos, L., and Poovendran, R. 2007. Energy and Bandwidth-Efficient Key Distribution in Wireless Ad Hoc Networks: A Cross-Layer Approach. *IEEE/ACM Transactions on Networking,* 15(6), 1527–1540.

[48] Jakes, W. C. 1994. *Microwave Mobile Communications.* Wiley-IEEE Press.

[49] Li, Z., Trappe, W., and Yates, R. 2007. Secret Communication Via Multi-Antenna Transmission. *Information Sciences and Systems, 2007. CISS '07. 41st Annual Conference on,* pages 905–910.

[50] Mathur, S., Trappe, W., Mandayam, N., Ye, C., and Reznik, A. 2008. Radio-Telepathy: Extracting a Secret Key from an Unauthenticated Wireless Channel. *MobiCom '08: Proceedings of the 14th ACM International Conference on Mobile Computing and Networking,* pages 128–139.

[51] Xiao, L., Greenstein, L., Mandayam, N., and Trappe, W. 2008. Using the Physical Layer for Wireless Authentication in Time-Variant Channels. *IEEE Trans. on Communications,* pages 2571–2579.

[52] Xiao, L., Greenstein, L., Mandayam, N., and Trappe, W. 2007. Fingerprints in the Ether: Using the Physical Layer for Wireless Authentication. *Proc. IEEE International Conference on Communications (ICC),* pages 4646–4651.

[53] Xiao, L., Greenstein, L., Mandayam, N., and Trappe, W. 2008. MIMO-Assisted Channel-Based Authentication in Wireless Networks. *Proc. Conference on Information Sciences and Systems (CISS),* pages 642–646.

[54] Xiao, L., Greenstein, L., Mandayam, N., and Trappe, W. 2008. A Physical-Layer Technique to Enhance Authentication for Mobile Terminals. *Proc. IEEE International Conference on Communications (ICC).*

[55] Xiao, L., Greenstein, L. J., Mandayam, N. B., and Trappe, W. 2008. A Physical-Layer Technique to Enhance Authentication for Mobile Terminals. *ICC '08. IEEE International Conference on Communications,* pages 1520–1524.

[56] Xiao, L., Greenstein, L. J., Mandayam, N. B., and Trappe, W. 2008. Mimo-Assisted Channel-Based Authentication in Wireless Networks. *Proceedings of the International Conference on Information Sciences and Systems (CISS).*

[57] Xiao, L., Greenstein, L. J., Mandayam, N. B., and Trappe, W. 2009. Fingerprints in the Ether: Using the Physical Layer for Wireless Authentication. *CoRR,* abs/0907.4877.

[58] Xiao, L., Greenstein, L. J., Mandayam, N. B., and Trappe, W. 2009. Using the Physical Layer for Wireless Authentication in Time-Variant Channels. *CoRR,* abs/0907.4919.

[59] Li, Z., Xu, W., Miller, R., and Trappe, W. 2006. Securing Wireless Systems Via Lower Layer Enforcements. *Proceedings of the 2006 ACM Workshop on Wireless Security*, pages 33–42.

[60] Yang, J., Chen, Y., and Trappe, W. 2008. Detecting Sybil Attacks in Wireless and Sensor Networks Using Cluster Analysis. *Proceedings of the Fourth IEEE International Workshop on Wireless and Sensor Networks Security*.

10

Experimental Systems for Next-Generation Wireless Networking

Sachin Ganu, Max Ott, and Ivan Seskar

10.1 Introduction

With the evolution of wireless technologies that continue to offer higher data rates using both licensed and unlicensed spectrum, the number of portable, handheld computing devices using wireless connectivity to the Internet has increased dramatically. Another major category for growth in wireless devices is that of embedded wireless devices or sensors that help monitor and control objects and events in the physical world via the Internet. Vehicular networking is an emerging application for wireless networking with a focus on increased road safety.

The broad architectural challenge facing the wireless and network research communities is that of evolving the Internet architecture to efficiently incorporate emerging wired and wireless network elements such as mobile terminals, ad hoc routers, and embedded sensors and to provide end-to-end service abstractions that facilitate application development. A top-down approach to the problem starts by identifying canonical wireless scenarios that cover a broad range of environments such as cellular data services, WiFi hot spots, mobile peer-to-peer (P2P), ad hoc mesh networks for broadband access, vehicular networks, sensor networks, and pervasive systems. These wireless application scenarios lead to a rich diversity of networking requirements for the future Internet that need to be analyzed and validated experimentally. One of the key challenges faced in characterization and evaluation of these complex wireless scenarios is the lack of generally available tools for modeling, emulation, or rapid prototyping of a complete wireless network. It has been observed that much of this work relies on a formal separation between the radio and networking layers. As a result, most of the contemporary research in wireless networks is primarily based on simulations or in-house small-scale emulators.

Pure simulation-based approaches rely on discrete event network simulation providing support for simulation of transport protocols, network layer protocols, and multicasting, as well as routing protocols over both wired and wireless links. The link layer is characterized based on various parameters such as loss, latency, and bit rate. The simulation tools allow the highest flexibility and programmability by abstracting these various physical layer effects, thus facilitating quick development and evaluation of novel layer-2 and networking protocols. Simulation-based techniques provide a cost-effective and repeatable method for large-scale system evaluation. Because all the processing is software-based, even large-scale topologies can be simulated in reasonable time using parallel computation, if necessary. The main drawback of simulation-based evaluation is simplistic modeling of certain key parameters that impact the accuracy of the evaluation. For example, as described in Kotz et al.[1] and Pawlikowski et al.,[2] most of the simulation models assume that the radio's transmission area is perfectly circular or that the wireless links between two communicating entities are symmetric in terms of link losses. Also, some simulation models incorrectly assume fixed-link bandwidths over fixed distances. In addition, some of the traffic models may be too simplistic to capture the characteristics of actual usage patterns, thereby resulting in simulation results not reflecting real-world constraints and conditions.

One of the common network simulators is ns,[3] which is popular because of variety of contributed models based on its open-source distribution. Wireless extensions to this simulator were developed[4] to enable simulation of mobile nodes connected by wireless links, including the ability to simulate multihop wireless ad hoc networks. To improve the simulation efficiency of discrete event simulation, GlomoSim[5] offers scalable simulation environment for wireless and wired network systems based on parallelizing execution of discrete-event simulations on different computing infrastructure. Commercially available network simulators such as Opnet Modeler[6] and Qualnet[7] are software tools for network modeling and simulation, and have been extended to support application performance management and network planning tools.

Emulation techniques combine software-based simulation and hardware in order to introduce realistic physical link layer under more controllable environments. This approach introduces some degree of realism by using devices for actual communication but still simulating some network behavior in software. This approach can be used to test protocols under certain realistic network conditions. For example, network topologies can be emulated by setting up packet filters in software to selectively drop packets or via artificial attenuation of the transmit power levels. Although this approach is more realistic than simulations, due to the use of actual hardware, it is limited to small-scale, controlled experiments that still lack the full degree of realism in terms of actual user mobility patterns and real physical layer link conditions such as attenuation, multipath, and fading effects.

Small-scale experimental setups such as the Ad-Hoc Protocol Evaluation Testbed (APE)[8] have attempted to address the deficiencies of the simulation-based approaches. The APE testbed provides software that can be installed on laptops with wireless NICs, as well as script-based mobility patterns that can be used to conduct repeatable mobility ad hoc wireless networking experiments. Software tools have been provided to create arbitrary topologies using MAC layer filtering of packets, allowing multihop links to be created in a controlled manner. Similarly, MobiEmu[9] emulates mobile ad hoc network environment with a fixed network of machines in a lab environment and supports mobility scenarios without actually moving nodes physically by dynamically installing or removing packet filters, thereby emulating network dynamics. Kiess and Mauve[10] provide a good summary of various testbeds.

Real-world experimental setups provide a completely realistic environment with complex interactions between different protocol components and enable evaluation of protocols in practical environments that consider software and hardware limitations. Examples include the Roofnet[11] network at MIT, which is an outdoor deployment of 802.11b-based wireless nodes on rooftops to create multihop wireless links to characterize wireless link behavior useful for study of novel multihop routing protocols. However, these setups are still limited in size and lack the ability to perform controlled repeatable experimentation.

Even though there are drawbacks to each approach, they are complementary to each other, and a general evaluation methodology can encompass all three approaches. To support this, there is a need for a framework that allows portability of protocol implementations from simulation, to emulation, to real-world testbeds. This way, some of the earlier studies can be done in more controlled simulation environments and the final protocol can be evaluated on emulators and testbeds. Results from actual experiments can help further improve the protocol design as well as simulation models, and is critical in supplementing simulation results with real-world experimental data that captures the complexities of various components, cross-layer interactions, and dependencies between various layers of the protocol.

This chapter describes the requirements, design considerations, and challenges for large-scale, open-access networking research testbeds. Key components including programmability at different layers of the networking stack and control-and-management software to enable multiple simultaneous experimenters to access the resources using virtualization are also discussed. Several existing testbeds including the ORBIT testbed at Rutgers University, CitySense project at University of Massachusetts, Amherst, Kansei sensor network testbed at Ohio State University, and others are described in terms of capability and deployment scale, as well as experiments that they support. This leads into a discussion of ongoing efforts to federate several such different substrates and existing wired experimental infrastructure to create a globally distributed experimental network resource as a part of NSF-sponsored Global Environment for Network Innovations (GENI).

10.2 Future Wireless Networking Testbeds: Requirements and Challenges

10.2.1 Design Requirements

The experimental infrastructure described earlier should be designed to incorporate a wide range of wireless networking capabilities in order to provide experimenters with access to emerging radio technologies that are becoming increasingly important at the Internet edge. Such an infrastructure should support various types of wireless technologies including 802.11 (WiFi), emerging WiMAX for metro area wireless access, sensor networking, and the more recent software radio platforms providing programmable means to enable a new generation of "adaptive wireless networks."

To conduct insightful experiments on these different kinds of networks, it is extremely important to provide the ability to program various radio nodes and network elements at all levels to run protocols and software supplied by the user. The user should also be able to collect real-time measurements at different granularities, dynamically change parameters at run-time, and configure the network topology in a flexible manner.

In addition, wireless communication is highly influenced by the environment in which it takes place. The same hardware configuration inside an office building, on a factory floor, or outdoors may result in very different behavior. This leads to a potential requirement for multiple similar testbeds in various representative environments.

Another requirement of the system is efficient resource utilization that can be achieved by simultaneously supporting more than one "virtual network" or protocol implementations. One approach to accomplish this has been demonstrated in the PlanetLab testbed[12,13] and involves the concept of partitioning of node and link resources ("slices") in a nonoverlapping manner across multiple virtual networks. Similar concepts can be applied to wireless networks taking care to account for shared-media interactions between wireless links and nodes. The degree of virtualization that can be practicably achieved depends on the hardware platforms used and the specific wireless networking scenario under consideration.

A final requirement of the experimental infrastructure is supporting repeatability, which is defined as the ability of the infrastructure to help predict within certain error boundaries the performance of the system under test under certain parameters.

10.2.2 Design Challenges

Supporting the previously described requirements in a multiuser wireless experimental facility presents some interesting challenges. These include routine ones related to management and user interface for experimentation (user account maintenance, access control, user portal for experimenters) as well as complex ones related to optimizing the usage of the testbed by accommodating as many users as possible in a given time duration.

In wired experimentation, resources for individual users can be segregated either at the MAC layer using VLANs, or IP layer using firewalls, or a combination of both. Wireless experimentation poses an interesting challenge due to the inherent broadcast nature of the medium, thereby affecting the other nodes in the vicinity. This results in complex interference and interactions between different "independent" allocated resources, making it far more difficult to set up a reproducible wireless networking experiment. Other factors unique to wireless environments include:

- Radio channel properties depend on specific wireless node locations and surroundings.
- Physical layer bit-rates and error-rates are time-varying.
- Shared medium layer-2 protocols on the radio link have a strong impact on network performance.
- There are complex interactions between different layers of the wireless protocol stack and currently their mutual interaction cannot be studied easily.
- User's exhibit random mobility and their location also plays a role.

In addition, visibility and programmability of different radio layer parameters are challenging due to limited software ability to change existing protocol behavior.

10.2.3 Key Components for Flexible Wireless Experimentation

We describe several key components that address the requirements and meet the design challenges outlined earlier to provide multiuser open-access wireless networking infrastructure.

10.2.3.1 Open WiFi/WiMAX Hardware and Software

Most of the commercially available wireless devices provide limited support to change the behavior of the radio as well as the link layer. However, it is important to provide programmable access at various layers of the network stack. With the advent of WiMAX as a wide-area wireless alternative to cellular networks, it is important to have support for outdoor deployments with dual mode devices that can switch from indoor WiFi to outdoor WiMAX mode based on their location.

10.2.3.2 Virtualization of Wireless MAC

Virtualization is about slicing a resource into smaller portions with minimal interactions or interference between them. This is one way to efficiently utilize the experimental system resources. Testbeds, such as PlanetLab, virtualize their existing resources to simultaneously support multiple applications or simulated networks. Unlike traditional wired networks, it is difficult to virtualize wireless environments because the performance and characteristics of wireless networks are greatly influenced by their MAC and PHY layers. Some level of virtualization can be achieved by allocating different frequency spectrum resources to different users. Virtualization of the medium access layer is even more challenging. Recent research[14] indicates that it is feasible to build inexpensive, virtualized wireless interface that can also be used for efficient overlay networks for experiments at the MAC layer using available software and off-the-shelf wireless network interface cards. SoftMAC[15] is a software library that provides software control over a specific wireless network hardware interface. The basic SoftMAC layer can be extended to create MultiMAC.[15] Packets are received using a common PHY layer implemented by the underlying radio and are then presented to the different MAC layers. The resulting MultiMAC implementation can be used to build both virtualized and overlay wireless networks at the MAC layer, and to do so using commodity hardware, although on a limited set of computing platforms.

10.2.3.3 Cognitive Radio Platforms

Software radios are quickly emerging as the enabling platform for future wireless communications systems, from commercial,[16] to open source,[17] to military.[18] In a software radio, many signal-processing functions such as modulation, coding, and spreading are performed in software. This enables agile radios – a software radio that uses its flexibility to dynamically change waveform characteristics and behavior in response to instruction. A cognitive radio extends these concepts to a radio that senses and reacts to its operating environment. In addition to the ability to virtualize the medium access layer and providing software control to tweak the different parameters in commercially available WiFi platforms, programmability of physical layer will be very important in future wireless networks. Cognitive radio platforms for research are currently being developed at several organizations including University of Utah,[17] University of Kansas,[19] Rutgers University,[20] and Rice University.[21]

10.2.3.4 Wireless Network Monitoring and Measurement

A common requirement for many routing algorithms, congestion avoidance, admission control, and other protocols or applications is the need to estimate certain performance metrics such as bottleneck bandwidth, available bandwidth, and bulk transfer capacity over wireless links. Many of these require measurements, but measurements are also instrumental in evaluating and benchmarking system performance. Thus, wireless measurement plays an important role in the design, development, and operation of wireless networking protocols and architecture.

For this reason, measurements should be viewed as an important core systems service. Often, obtaining the raw measurements is hardware and system dependent, requiring specific additional processing as well as initial configuration. A service approach will also avoid duplication when multiple components require the same measurements. From an experimentation point of view, it needs to be designed so that measurement collection and its delivery causes minimal disruption to the actual system being evaluated. One approach to address this is separating the experimental plane and control/management plane or temporarily storing the measurement data locally and collating at the conclusion of the experiment.

10.2.3.5 Mobility Support

Mobility of wireless devices, and the dynamic component it adds, is one of the most important aspects in many experiments. Mobility in the context of testbeds falls roughly into two categories: making devices move and measuring their movements. Accordingly, there are two types of testbeds supporting mobility: controlled and observed. In controlled mobility environments, the device is usually mounted on a mobile platform, such as a robot that can be directly controlled by the experimenter. In other environments, where devices are, for instance, carried by buses[34] or people, mobility can either be directly observed by measuring a device's location, or the statistical properties of the underlying mobility patterns are known, or the environment is categorized as being representative of a specific environment.

10.2.3.6 Wireless Network User Models and Measurements Repository

Many research publications report experimental evaluations of various wireless protocols under specific user mobility models and traffic modeling parameters. It is important to efficiently collect and preserve large amounts of user and systems

Table 10.1. *Summary of Sample Wireless Experimental Testbeds*

Testbed	Classification	Scale	Features
ORBIT	Indoor wireless grid	400 nodes with dual radio 802.11 a/b/g radios, Zigbee radios and GNU radio platforms (USRP and USRP2)	High-density indoor wireless experimentation Multihop topology emulation using noise injection Limited mobility emulation using software Dedicated wired network for experiment control, resource management, and data collection.
Kansei	Indoor sensor grid	210 sensor nodes with 802.11b radios and sensing radio	Support for sensor data generation, real-time event creation, data storage Limited mobility support using robot-mounted nodes
CitySense	Outdoor distributed sensor	100 pole-mounted nodes with modular sensing radio and 802.11b/g radio	802.11b/g radio and OLSR routing protocol for remote monitoring and management
DieselNet	Outdoor mobile networking	40 buses with wireless nodes	Supports wireless access on one radio and opportunistic peer to-peer-or dedicated relay-based communication on other radios
Emulab	Indoor office environment	40 nodes with 802.11 a/b/g and GNU Radio platform (USRP and USRP2) scattered around the two floors of a large building	Multiplexed virtual node implementation allowing emulation of larger networks
VT-CORNET	Indoor office environment	48 USRP2-based nodes with custom RF daughterboard	Use of Software Communications Architecture (SCA) framework for waveform generation.

observations in a consistent manner. This will enable the development of new inference and data-mining algorithms to extract new insights from previously conducted experiments.

The Crawdad project[22] at Dartmouth University is an example of an archive with the capacity to store wireless trace data from many contributing locations and tools for collecting, data anonymity, and analysis. Sample data sets include association and mobility trails of users at conferences. It is also important to take into account the privacy concerns for long-term experimentation involving real users.

10.3 Existing Wireless Testbeds

In this section, several experimental testbed facilities that have been built as part of the NSF-sponsored initiative for open wireless multiuser experimental facility (MXF) testbeds are described. Specifically, we include testbeds purpose-built for indoor wireless ad hoc and mesh networking research (ORBIT), sensor networking (Kansei), citywide outdoor sensor deployment for environment monitoring (CitySense), vehicular networking testbed (DieselNet), network emulation testbed (EMULAB), and cognitive radio testbed (VT-CORNET). Table 10.1 broadly summarizes the capabilities, scales of deployments, and other features of each testbed.

10.3.1 ORBIT: Indoor Wireless Testbed for Ad hoc and Mesh Networking

10.3.1.1 ORBIT System

The ORBIT large-scale radio grid emulator[23] consists of an array of 20×20 programmable nodes, each with multiple 802.11a,b,g,n or additionally other radios cards such as Bluetooth, Zigbee, USRP, and USRP2. The radio nodes are connected to an array of back-end servers over a switched Ethernet network, with separate physical interfaces for data, management, and control. Interference sources and spectrum monitoring equipment are also integrated into the radio grid, as shown in Figure 10.1. Users of the grid use an *Experiment Controller* to conduct experiments involving network topologies and protocol software specified using an ns-2 like scripting language. Experiments are described in a domain-specific language, called OEDL.[28] A *radio mapping algorithm*[27] that uses controllable noise sources spaced across the grid to emulate the effect of physical distance is used to map real-world wireless network scenarios to specific nodes in the grid.

10.3.1.2 Hardware Components

The ORBIT nodes in this setup are suspended from the ceiling with grid spacing of 1 m, as shown in Figure 10.2; each node is equipped with two 802.11a/b/g radio cards and optionally with Bluetooth, Zigbee, and GNU radios. The custom-built radio node is a programmable platform including a 1 GHz VIA C3 processor with 512MB of RAM and 20 GB of hard disk memory. This configuration is necessary to support heavy-duty protocol processing, computation, and storage

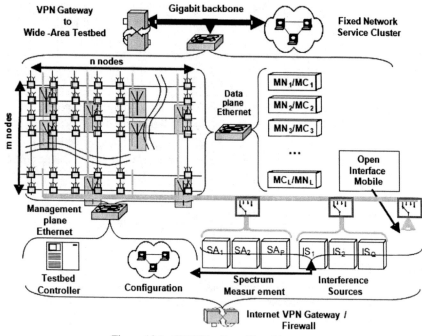

Figure 10.1. ORBIT radio grid architecture.

at high throughputs. Each platform also includes two wired 1,000 Base-T Ethernet interfaces for experimental data and control. In addition, each node contains a custom-designed chassis manager module with Ethernet connectivity to allow for remote monitoring and rebooting of nodes.

RF instrumentation: The ORBIT grid includes equipment for measurement of radio signal levels and to create various types of artificial RF interference. This can be used to create different topologies in the

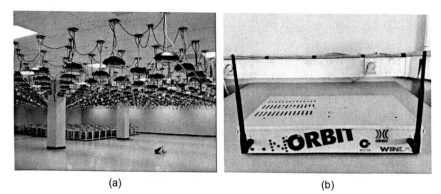

(a)   (b)

Figure 10.2. ORBIT radio grid, ORBIT nodes.

static grid (e.g., multihop mesh) with the same physical positioning of the nodes.

WLAN Monitoring: An independent WLAN monitoring system provides a MAC/network layer view of the radio grid's components using a number of WLAN "observers" spread across the system.

Support Servers: The testbed's back-end equipment includes several front-end servers for Web services, experiment support, and data storage. The database servers support multiterabyte storage capacity. There is also an Ethernet switching array with approximately 1,400 ports necessary to switch traffic from 3×400 grid node interfaces and the servers.

ORBIT Sandboxes: The ORBIT system includes several "sandboxes," each with two radio nodes connected by RF cable, to permit users to debug their experimental code before using the large grid for experiments at scale. This helps improve testbed utilization by off-loading most-early-stage experiments and software development to sandboxes rather than the main radio grid.

10.3.1.3 Software Architecture

The ORBIT system uses a time-slot-based reservation mechanism to schedule experiments (Figure 10.3). Users log into a portal to reserve future time slots

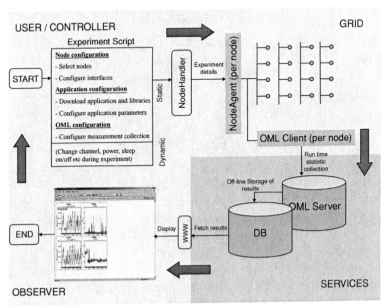

Figure 10.3. Experiment lifecycle on ORBIT testbed.

for experimentation and are granted access to their slots only at the allotted time. Users log into the system during their allocated slot and can launch the experiment via scripts. The entire experiment is usually captured in a script that is descriptive: It defines the wireless nodes, their roles, various network parameters such as traffic pattern, protocol, and packet sizes, as well as wireless settings such as channel, power levels, and the like. ORBIT uses open-source drivers MadWiFi and ath5k for Atheros and IPW2200 for Intel 802.11a/b/g interfaces. Additionally, users can specify the statistics (at different layers of the stack) to be collected at configurable intervals (time- or sample-based). The script is interpreted by the experiment controller software that powers up the wireless nodes, installs custom images if necessary, configures network parameters on these nodes, and starts the experiment work flow. It also initialized databases necessary for collecting the results of the experiment based on user-selected statistics and granularity of collection. The ORBIT Measurement Framework Library (OML)[26] is responsible for data collection in a nonintrusive manner over dedicated Ethernet interface and network. In addition to time- and sample-based filters, the OML framework allows run-time filters to be applied to either of these measurement techniques to report minimum, maximum, average, or sum of time-based or sample-based measurements.[29] The readers are referred to Raychaudhuri et al.,[23] Ganu et al.,[24] Ott et al.,[25] Singh et al.,[26] Lei et al.,[27] Rakotoarivelo et al.,[28] and White et al.[29] for further details on the ORBIT testbed and Management Framework (OMF).

10.3.2 Kansei Sensor Networking Testbed

The Kansei testbed[30] is hosted at the Ohio State University and supports a two-dimensional grid of experimental sensor hardware with support for sensor data generation, real-time event creation, data storage, and the related experiment management software.

10.3.2.1 Hardware Infrastructure

The Kansei testbed consists of three types of sensor arrays: stationary, portable, and mobile. The stationary array (as shown in Figure 10.4) has 210 sensor nodes placed in a 15 × 14 meter rectangular grid with 1 meter spacing. Each node consists of two hardware platforms: the Extreme Scale Motes (XSMs) has a 7.3 MHz CPU and 4KB of RAM and uses a 433 MHz single-channel radio supporting a data rate of up to 38.4 KBps. All this is supported by the TinyOS operating system. The other platform is a Stargate[31] node with a 400 MHz CPU, PCMCIA-based 802.11b wireless card, and direct interface with the XSM sensor platform, whereas Linux is used for the operating system. The stationary array infrastructure can be coupled with one or more portable arrays for recording sensor data

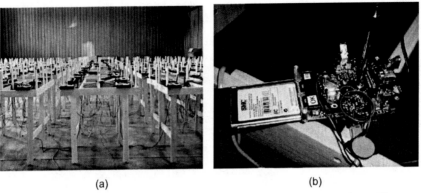

(a)   (b)

Figure 10.4. Kansei Sensor testbed and hardware. (Picture source: http://ceti.cse.ohio-state
.edu/kansei).

in situ and for field-testing sensor-network applications. The mobile platform
consists of robot-mounted mobile nodes that can move on a Plexiglas base.

10.3.2.2 Software Infrastructure

The testbed uses the Kansei Director software to manage experiments. This
provides support for scheduling of experiments, script-based execution, monitoring, data collection, and storage. For larger-scale experiments, the software
also supports a hybrid simulation model that runs large-scale simulations on
a host machine, interfaces with the actual testbed for physical events such as
sensing or traffic delivery (while pausing the simulation state) until the event
is completed on the real testbed, and reinserts the results from the physical
testbed into the simulation engine. The testbed has a Web-based interface on
which experiments can be scheduled and the results retrieved. Further details on
the experimental framework and usage model can be found on the Kansei Web
site.[30]

10.3.2.3 CitySense: Urban-Scale Wireless Network Testbed

CitySense testbed[32] is hosted in the city of Cambridge, Massachusetts, and
provides an urban-scale wireless network testbed for experimental purposes
(Figure 10.5). The deployment consists of about 100 radio nodes mounted at
different locations in the city on street lights. These nodes have the ability to
monitor physical events via sensors and are powered by street-light mountings.
In contrast to traditional sensor platforms that are battery-operated with severe
power constraints, this platform enables newer applications that can make use
of the increased power and processing capability of the hardware.

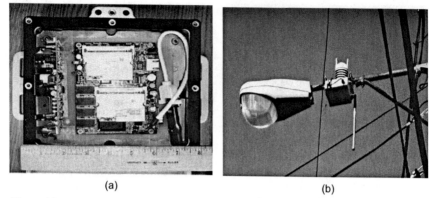

(a)   (b)

Figure 10.5. CitySense Sensor Node hardware; mounted on light pole. (Picture source: http://www.citysense.net/).

10.3.2.4 Hardware Infrastructure

The experimental platform used for the wireless nodes is based on Soekris net4826 motherboard running Linux OS with mini-PCI-based 802.11a/b/g radio cards. The nodes interface with a modular sensing hardware that can serve multiple modes (weather sensing using Vaisala WXT510 unit, or CO_2 sensing using Vaisala GMP343 units). These sensing units connect to the Soekris motherboard via its serial port.

10.3.2.5 Software Infrastructure

Unlike the previous testbeds that use wired network for the control plane activities including experiment deployment, management, monitoring, and data collection, CitySense architecture uses multihop wireless mesh techniques running optimized OLSR routing protocol on a dedicated 802.11 radio. Nodes periodically monitor their connectivity to the Web-based management software and reboot the connection with the system in case of missing network *heartbeats*. The other 802.11 radio is available for experimental purposes. A Web-based user interface provides real-time access to network statistics.

10.3.2.6 DieselNet: Outdoor Mobile Networking Testbed

The DieselNet testbed[33] in the city of Amherst, Massachusetts, is an outdoor mobile testbed with a focus on disruption tolerant networking (Figure 10.6). The testbed comprises wireless nodes installed in city buses that ply different routes. The nodes communicate to establish ad hoc connectivity with each other when in physical proximity, as well as through the use of dedicated relay nodes (stationary) deployed at strategic locations.

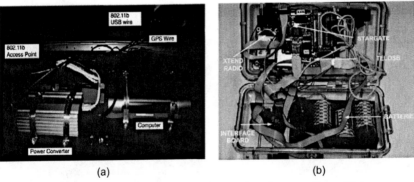

Figure 10.6. DieselNet Mobile Node in a bus and stationary relays. (Picture source: http://prisms.cs.umass.edu/dome/umassdieselnet).

10.3.2.7 Hardware Infrastructure

The wireless nodes installed in buses are HaCom Open Brick nodes with 577 MHz CPU, 256MB RAM, 40GB hard drive running Linux OS. The node has three radios: an 802.11b Access Point (AP) to wireless access to bus passengers, a second USB-based 802.11b interface that constantly scans the RF neighborhood for peer nodes in nearby buses, and a longer-range MaxStream XTend 900MHz radio to connect to the stationary relay nodes. The nodes also have GPS receivers to keep track of location. The relay nodes use a combination of COTS low-power platforms (Stargate PXA255 platform) and low-power microcontrollers. They are equipped with MaxStream long-distance radios to communicate with the wireless nodes in the buses.

10.3.2.8 Software Infrastructure

The developed testbed control software allows the deployment of applications and tracking of node mobility, connectivity, and throughput. Further details on the testbed and common usage models can be found in Banerjee et al.[33] and University of Massachusetts DieselNet Web site.[34]

10.3.2.9 Emulab: Network Emulation Testbed

Emulab is an open-access testbed[35] hosted at the University of Utah in Salt Lake City (Figure 10.7). Originally designed as a wired network emulation testbed that offered controlled and deterministic network topologies that were configurable via scripts, this has now been expanded to include wireless devices including 802.11 nodes, USRP (Universal Software Radio Peripheral)[36]-based software radios, as well as sensor networks. A common user interface has been created to control access to the resources.

(a)   (b)

Figure 10.7. Emulab and wireless node deployment. (Picture source: http://emulab.net)

10.3.2.10 Hardware Infrastructure

The wireless nodes scattered throughout the building have two basic configurations: a) 3 GHz Pentium 4 based machines with 1 GB of DDR-400 RAM, 120 GB hard drive, and 2 802.11 a/b/g cards; and b) 600 MHz Pentium III machines with 256MB of PC100 RAM, 13GB hard drive, and 1 802.11 Atheros chipset-based a/b/g wireless device; 35 nodes have Zigbee and USRP/USRP2 devices attached. The 25 USRP devices have 900 MHz daughterboard whereas 10 USRP2 devices have 2.4 and 5 GHz (ISM band) daughterboard.

10.3.2.11 Software Infrastructure

To be able to use the wireless resources in parallel with minimal impact on other concurrent experiments, the software enables partitioning of the wireless medium by using separate frequency bands that do not overlap. Configurability is restricted to the options permitted via the Atheros chipset such as link rate selection, mode of operation, transmit power levels, and so forth.

In addition, users have access to USRP-based radios that have been deployed as shown in Figure 10.7, and the infrastructure allows users to utilize the bare cognitive radio hardware and download custom software onto the USRP nodes or as a managed wireless interface. Further details can be found at EMULAB Web site.[35,37,38]

10.3.3 VT-CORNET

VT-CORNET is a wireless communication network testbed based on 48 radio nodes scattered on four floors in a building at Virginia Tech. The primary focus of this testbed is to provide a tool for developing methods that allow efficient

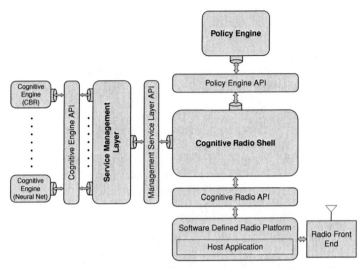

Figure 10.8. VT-CROSS architecture. (Picture source: http://wireless.vt.edu/coreareas/cognitive.html)

sharing of spectrum resources among heterogeneous devices and to maximize the speed and reliability of communications.

10.3.3.1 Hardware Infrastructure

The hardware infrastructure is based on USRP2 with custom daughterboard covering frequency range of 100 MHz to 4 GHz.

10.3.3.2 Software Infrastructure

The testbed uses modular open-source software framework[39] shown in Figure 10.8, which is based on a set of modules that use socket as intercomponent communication method. This allows for individual modules to be implemented in any programming language that supports TCP/IP socket abstraction. Currently, the system has the following components:

- Cognitive Radio Shell (CRS)
- Cognitive Engine (CE)
- Policy Engine (PE)
- Service Management Layer (MSL)
- Software-Defined Radio Host Platform.

The software design is based on Software Communications Architecture (SCA),[40] which is an architectural framework for standardizing the deployment, management, interconnection, and intercommunication of software application components in embedded, distributed-computing communication systems. This enables programmable radios to load waveforms, run applications, and be networked into an integrated system, and provides a standard operating environment across all hardware, leading to interoperability by enabling the same waveform to be ported to all radios.

10.4 Global Environment for Network Innovations (GENI)

The testbeds described in the earlier sections provide an important research tool to study indoor wireless ad hoc and mesh networks, sensor networks, cognitive radios, and outdoor mobile scenarios in isolation. However, many of the new Internet application concepts involve a mixture of wireless end-users, physical sensing, and user mobility, thereby requiring integrated research infrastructure.

A recent NSF-sponsored multiphase initiative known as The Global Environment for Network Innovations (GENI)[41] addresses the need for building such an experimental network research infrastructure. The objective is to understand complex interactions between several different network technologies and protocols based on realistic usage models and identify next steps toward a novel architecture to meet the requirements of the evolving networking trends. The first phase of the project was a "Planning phase" (some time between 2005 and 2007) that resulted in initial design documents, draft funding proposals, and establishment of the GPO (GENI Project Office) at BBN.

The GENI vision that came out of the planning phase is to support a wide range of experimental protocols over heterogeneous substrates including fiber optics, high-speed routers, citywide experimental urban radio networks, computing clouds, and sensor networks. This will be in a form of a shared global facility available to experimenters with support for experiment deployment, measurements, and data analysis and sharing of information. To achieve this goal, GENI proposes the following core components:

- **Programmability** to enable experimenters to deploy custom software onto the experimental infrastructure and provide ability to control and modify various parameters and behavior of underlying network elements
- **Resource Sharing** to allow the simultaneous use of the infrastructure by multiple users and experiments while ensuring maximum isolation between these experiments. Thus, network bandwidth and CPU resources will be proportionally shared among various instances of experiments. This can be achieved using virtualization techniques where every unique experiment can get a "slice" of distributed network and computing resources proportional to the requirements of the experiment.
- **Federation** will enable an integration of heterogeneous resources that may be owned and operated by various organizations to create a global distributed

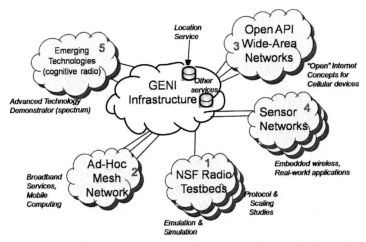

Figure 10.9. Experimental networks for future Internet research – Overview.

network of resources. This will require wireless testbeds to be integrated with the wide-area Internet infrastructure in order to facilitate end-to-end experimentation with new protocols (shown in Figure 10.9).

- **Integrated control and management framework** will allow discovery of resources for experimentation, creating a slice based on user request, initializing the experiment state, providing mechanisms for users to deploy and launch their experiments, define measurement points, and collect measurements at run-time from the experiment. Such an integrated control-and-management framework will require the ability to coordinate and control various heterogeneous network resources that may be governed by different usage policies and present them as a virtual slice to the experimenter upon request.

The strategy of the GENI GPO is based on spiral development, flexible funding model with large community involvement and organic network growth. This development plan is currently being executed with the goal of a large-scale network deployment over approximately two to three years. GENI phases (or spirals) last a year and are based on open solicitation process and peer review. In general, insights and experience from earlier phases are used to set specific goals for the later ones. To date, GENI phases consist of three sets of projects:

- **Planning** projects (approximately 2005–2007) focused on identifying research challenges and developing preliminary designs for programmable network deployments.
- **Spiral 1** projects[42] (approximately from October 2008 to September 2009) focused on proof-of-concept prototypes for technology risk reduction and to develop, integrate, and attempt to operate very rudimentary, end-to-end working prototypes.

- **Spiral 2** projects[43] (approximately from October 2009 to September 2010) with increased focus on deployment and experimental network operations.

We next describe ongoing efforts to meet the various requirements and design goals described earlier, as well as wireless-related projects from the two spirals.

10.4.1.1 Phase 1: Integration of Wireless and Wired Experimental Networks

As a first step toward achieving the global network of experimental resources, we describe ongoing efforts to integrate two different experimental network testbeds operated through two different control frameworks – PlanetLab[12] and ORBIT[23] – and support the execution of experiments that span both wired and wireless network elements. Each of the existing testbeds provides support for experimental flexibility and some level of network virtualization, and they need to be harmonized to work within the proposed future GENI model.

Key technical requirements for such an integrated testbed are as follows:

(1) virtualization of network resources (wired and wireless) to provide capabilities for support of multiple concurrent experiments ("slices") on the same set of nodes;
(2) integration of control and management across wired and wireless networks, providing research users with a single programming interface and experimental methodology.

10.4.1.2 Network Virtualization

Virtualization techniques are intended to share a set of computing and communication resources (CPUs, routers, links, networks) among multiple users with the appearance of dedicated, noninterfering allocation of resources. While VMWare[44] provides full virtualization where multiple OSs can be run on the same machine, its memory requirements restrict the number of simultaneous experiments per node. Paravirtualization techniques such as Xen[45] similarly use a "hypervisor" layer that can host multiple guest operating systems by controlling access to hardware and scheduling across physical CPU's access to controls, but suffers from similar memory usage constraints. A third approach, adopted by PlanetLab[12,13] and User Mode Linux,[46] is to virtualize at the level of system calls and provide reduced isolation in favor of supporting a much larger number of users per node. With regards to the virtualization of network resources, VLANs enable a network of computers to communicate even though they may be connected to different physical segments of a LAN. More recently, PlanetLab's virtual network access module isolates the traffic of multiple users from one another.

Virtualization of a wireless network is inherently different from the wired counterpart because the wireless medium is a broadcast medium, and isolating the wireless physical layer per user involves eliminating co-channel as well as adjacent channel interference from other collocated wireless devices. Several methods to extend the concept of virtualization into the wireless domain are described next, with the objective of supporting certain classes of service software and protocol experiments over a common set of experimental wireless networking resources.

10.4.1.3 Wireless Virtualization Approaches

Wireless network virtualization can be approached using a combination of methods that range from simple "virtual MAC" (VMAC) techniques (for restricted 802.11-based access point topologies) to more general space division multiple access (SDMA), frequency division multiple access (FDMA), and time division multiple access (TDMA) methods for separating "slices" in the network.

- **Virtual MAC (VMAC):** The VMAC technique is based on logical portioning of the radio channel based on individually assigned slice BSSIDs. Current 802.11 radio drivers provide support for up to 16 BSS per radio; however, topologies are restricted to fixed-star scenarios where an AP is configured in the normal infrastructure WLAN mode, and devices with WLAN cards connect to the AP with a single-hop wireless connection. Virtualized use of such a star topology wireless network may be appropriate for certain classes of long-term service or protocol experiments, but may not be particularly useful for wireless mesh or other experiments involving more complex or dynamically changing topologies.
- **Spatial diversity + VMAC:** To increase the number of slices that can be supported by an experimental wireless network, spatial separation can be combined with the previously described VMAC method. In this approach, multiple AP-based star networks coexist on the radio grid in which the AP-based clusters are separated by a distance greater than the radio interference range allowing them to coexist (on the same channel) without interference.
- **Frequency diversity + VMAC:** VMAC can also be combined with frequency diversity to support a larger number of simultaneous slices for a given set of radio nodes. One way of implementing this technique would be to use wireless nodes with multiple ($n > 1$) radio cards where the cards operate on orthogonal frequencies.
- **Time multiplexing:** The idea in TDMA-based approach is to share a given wireless node among multiple experiments using different time slots. For example, a wireless node can be logically partitioned into three virtual nodes by allocating three nonoverlapping time slots to three experiments. Initial

work from University of Wisconsin[49] presents a virtual TDMA-like multiplexing built on top of existing 802.11 MAC.

A combination of these techniques can also be used to increase the number of supported users per set of devices. The virtualization described earlier will allow multiple users to share the wireless grid resources with minimal interference. This model can be extended further so as to support wired networking experiments with large numbers of emulated wireless nodes at the edge.

10.4.1.4 Integrated Control and Management Plane

Currently operational wired testbeds such as PlanetLab employ "service oriented" network architecture and provide the users with the ability to run long-term experiments. This experimental testbed runs on top of the Internet as an overlay, thereby giving researchers access to (1) a large set of geographically distributed machines; (2) a realistic network substrate that experiences congestion, failures, and diverse link behaviors; and (3) the potential for a realistic client workload. Experimenters get access to a long-running slice of these distributed resources via a central slice creation and management utility.

In contrast, wireless experimental testbeds such as ORBIT employ an "exclusive access for limited time" model, via the ORBIT Radio Resource Management and Scheduling facility, where a user can get exclusive access to the entire resources for a limited duration of time. The ORBIT system also provides tools to define and describe experiments, including their required resources, deployment options, and measurement collection settings. This promotes a repeatable mode of running experiments and systematic data archiving.

To facilitate experimentation over these globally distributed wired and wireless resources, it is important to integrate these two different experimental models to provide a common and consistent abstraction for the experimenter. This requires the various control-and-management frameworks to coordinate access to the network resources, as well as a simple interface to enable users to run different experiments. To integrate these two different experimental models, the following approaches[48] are considered:

- wireless experiments requiring access to wired nodes;
- wired experimenters requiring access to wireless edge as a part of their experiment.

In viewing the wired network as an extension of the wireless testbed, the first approach (Figure 10.10) involves the use of a long-term "slice" in the wired infrastructure and interfacing the slice to the wireless testbed using a gateway. The ORBIT experimental infrastructure (described in Section 10.1) has been extended to communicate with nodes on the local subnet (wireless nodes) as well as remote PlanetLab nodes. Specifically, the naming/addressing scheme

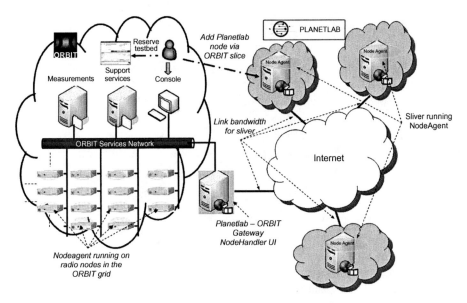

Figure 10.10. Wired extension to wireless experiment.

and communication protocol for the ORBIT experimental framework has been extended to allow access to geographically diverse nodes. A proxy service handles communication with remote wired nodes using GRE tunnels on behalf of the experiment controller, thereby enabling distant nodes to be transparently included as a part of the wireless experiment.

In the second approach, we consider wired users who require an abstraction of a wireless edge to be included in the experiment. The second model would provide a PlanetLab-ORBIT gateway as a node that users of a PL slice can access whenever they want an emulated wireless edge network. This gateway will provide abstractions for setup, control, and measurement on a specified wireless topology.

Proof of concept experiments conducted over this integrated network testbed is described in Mahindra et al.[48] This unified design can serve as a practical foundation for wired/wireless integration in heterogeneous testbeds in the bigger GENI concept.

10.4.1.5 Spiral 2 "Meso-Scale" Deployments

As part of Spiral 2, the GENI GPO launched two related campus build-outs based on OpenFlow[50] and WiMAX technologies, with significant commonality in the technologies employed and overlaps in the campuses chosen. In an effort to create a large-scale testbed, the GPO solicited resources from both Internet2 (I2) and National LambdaRail (NLR) to provide both bandwidth and operational support to connect the campus deployments into a larger, national research networks. The main objectives of the current "meso-scale" deployments are as follows:

- create a large-scale experimentation infrastructure that will enable new forms of network science and engineering;
- support broad community participation and enable "opt in" by early users across 13 major campuses;
- ensure participation of equipment manufacturers, such as HP, Juniper, NEC, Arista, Cisco, and Nicira, to GENI-enable commercial equipment.

10.4.1.6 OpenFlow and Enterprise GENI

The hardware platform of choice for the network build-out is based on the OpenFlow architecture[50] enabling researchers to run experimental protocols in the networks they already use every day. OpenFlow is based on a standardized interface to control the forwarding table of an Ethernet switch. More specifically, OpenFlow provides control over individual flows at the individual device level. This provides the basis for multilayer network slicing, creation of virtual network slices that can be defined across a combination of physical layer, link layer, network layer, and/or transport layer flow rules.

As shown in Figure 10.11, the integration of OpenFlow into the GENI architecture consists of two components: a FlowVisor and an Aggregate Manager. The

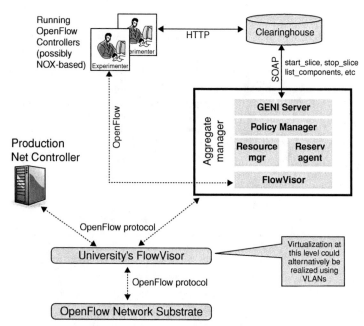

Figure 10.11. Enterprise GENI architecture. (Picture source: http://www.openflowswitch .org/wk/index.php/E-GENI)

FlowVisor partitions a physical OpenFlow switch into multiple logical (slice) switches and acts as a proxy for multiple OpenFlow controllers. FlowVisor is also responsible for guaranteeing isolation between multiple experiments running on the same switching fabric. The Aggregate Manage is a GENI-compliant OpenFlow controller that can control a subset of switch (network) resources allocated for experimentation. It enables GENI experimenters' access to Open-Flow environments along with campus/enterprise access to GENI experimental network infrastructure.

10.4.1.7 WiMAX

Extending the work on WiMAX virtualization undertaken Spiral 1,[42] the WiMAX meso-scale project in Spiral 2 will create an open, programmable, GENI-enabled "cellular-like" infrastructure across eight major research university campuses as shown in Figure 10.12. This project leverages a commercial 802.16e base station from NEC, replacing the standard WiMAX controller with an open GENI software implementation that supports virtualization and layer 2/3 programmability.

Open WiMAX base stations provide network researchers with wide-area coverage and the ability to support both mobile and fixed end-users. This will open up a path for direct "opt in" of student users in these campuses into GENI research experiments, via WiMAX modems and, as they become available, WiMAX handsets.

The core of this Spiral 2 project is the "GENI WiMAX base station kit" that consists of a NEC 802.16e base station (BS) with indoor and outdoor unit

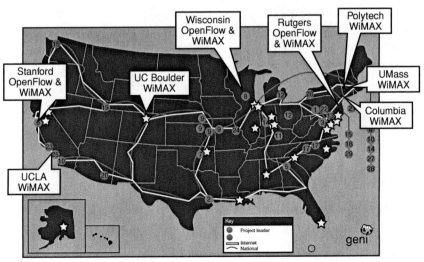

Figure 10.12. Proposed WiMAX deployments for GENI. (With permission from: http://groups.geni.net)

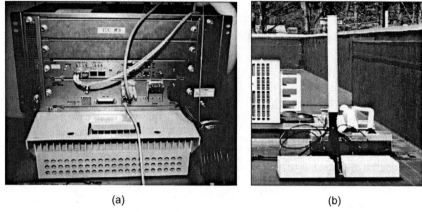

(a)   (b)

Figure 10.13. WiMAX base station.

(Figure 10.13a), a BS controller, antenna (Figure 10.13b), and a number of WiMAX clients. The BS is a 5U rack-based system that can be populated with up to three channel cards (sectors). It operates in the 2.5 GHz or the 3.5 GHz bands and can be tuned to use 5, 7, or 10 MHz wide channels. The WiMAX base station has an external Linux-based PC controller that runs the open API GENI control software offering programmable wireless networking capabilities and provides support for multiple GENI slices (virtualization). Each controller PC provides each experimenter's slice with its own virtual machine and the ability to control the respective radio resources allocated to each associated client. In the GENI context, each slice can be viewed as a dedicated BS (vBS) supporting multiple clients and multiple traffic types per client.

The WiMAX base stations will be integrated into OMF (ORBIT management framework), which is one of the supported GENI control frameworks. An experimenter will be able to access and create their own virtual WiMAX network through the portal and use available OMF tools for experiment control, management, and measurement to conduct their experiments. Further details can be found at www.Geni.net.[51]

10.5 Concluding Remarks

The large-scale wide area networking testbed with heterogeneous networking elements including wired and wireless devices can thus create infrastructure for entirely new forms of experimentation at a much larger scale than has previously been available. The new generation of a networking testbed described in this chapter is capable of supporting multiple future Internet prototypes in parallel, and can thus serve as an important tool not only for research-stage validations but also for early service deployments with real-world applications and end-users.

Worldwide deployment and eventual federation of these testbeds should lower the barrier for large-scale release of new protocols and services, thus improving the prospects for clean-slate networking ideas discussed in other chapters of this book.

References

[1] Kotz, D., Newport, C., Gray, R. S., Liu, J., Yuan, Y., and Elliott, C. 2004. Experimental Evaluation of Wireless Simulation Assumptions. *MSWiM 2004*.

[2] Pawlikowski, K., Jeong, H.-D. J., and Lee, J.-S. R. 2002. On Credibility of Simulation Studies of Telecommunication Networks. *IEEE Communications Magazine*, 40(1): 132–139.

[3] NS Simulator, http://www.isi.edu/nsnam/ns/

[4] Wireless and Mobility extensions to ns, http://www.monarch.cs.rice.edu/cmu-ns.html

[5] Zeng, X., Bagrodia, R., and Gerla, M. 1998. GloMoSim: a Library for Parallel Simulation of Large-scale Wireless Networks. *Proceedings of the 12th Workshop on Parallel and Distributed Simulations – PADS '98*.

[6] OPNET, http:/www.opnet.com

[7] Qualnet, http://www.scalable-networks.com/

[8] Lundgren, H., Lundberg, D., Nielsen, J., Nordström, E., and Tschudin, C. 2002. A Large-scale Testbed for Reproducible Ad hoc Protocol Evaluations. *Proceedings of IEEE Wireless Communications and Networking Conference*.

[9] Zhang, Y., and Li, W. 2002. An Integrated Environment for Testing Mobile Ad-Hoc Networks. *Proceedings of the Third ACM International Symposium on Mobile Ad Hoc Networking and Computing (MobiHoc'02)*.

[10] Kiess, W., and Mauve, M. 2007. A Survey on Real-World Implementations of Mobile Ad-hoc Networks. *Ad Hoc Networks*, 5(3), 324–339.

[11] MIT Roofnet. http://www.pdos.lcs.mit.edu/roofnet/

[12] Peterson, L., Shenker, S., and Turner, J. 2004. Overcoming the Internet Impasse Through Virtualization. *Third Workshop on Hot Topics in Networking (HotNets-III)*.

[13] Huang, M. *VNET: Planetlab Virtualized Network Access*. http://www.planet-lab.org/PDN/PDN-05–029

[14] Rao, A., and Stoica, I. 2005. An Overlay MAC Layer for 802.11 networks. *Proceedings of ACM Mobisys*.

[15] Neufeld, M., Fifield, J., Doerr, C., Sheth, A., and Grunwald, D. 2005. SoftMAC – Flexible Wireless Research Platform. *Proceedings of the Fourth Workshop on Hot Topics in Networks (HotNets)*.

[16] VANU. http://www.vanu.com

[17] GNURadio. http://www.gnu.org/software/gnuradio

[18] Joint Tactical Radio System. http://jtrs.army.mil

[19] Minden, G. *KU Agile Radio Technology*. http://www.ittc.ku.edu/techreview2005/presentations/ Minden_Agile Radios.ppt

[20] Ackland, B., Bushnell, M., Rose, C., Seskar, I., Raychaudhuri, D., Tentzeris, M., Papapolymerou, J., Laskar, J., and Sizer, T. 2004. NeTS-ProWiN: High Performance Cognitive Radio Platform with Integrated Physical and Network Layer Capabilities. *National Science Foundation*.

[21] Wireless Open Access Research Platform, http://warp.rice.edu/trac

[22] Crawdad, A. *Community Resource for Archiving Wireless Data at Dartmouth*. http://crawdad.cs.dartmouth.edu/index.php

[23] Raychaudhuri, D., Seskar, I., Ott, M., Ganu, S., Ramachandran, K., Kremo, H., Siracusa, R., Liu, H., and Singh, M. 2005. Overview of the ORBIT Radio Grid

Testbed for Evaluation of Next-Generation Wireless Network Protocols. *Proceedings of the IEEE Wireless Communications and Networking Conference.*

[24] Ganu, S., Seskar, I., Ott, M, Raychaudhuri, D., and Paul, S. 2006. Architecture and Framework for Supporting Open-Access Multi-User Wireless Experimentation. *First International Conference on Communication Systems Software and Middleware.*

[25] Ott, M., Seskar, I., Siracusa, R., and Singh, M. 2005. ORBIT Testbed Software Architecture: Supporting Experiments as a Service. *Proceedings of IEEE Tridentcom 2005.*

[26] Singh, M., Ott, M., Seskar, I., and Kamat, P. 2005. ORBIT Measurement Framework and Library (OML): Motivations, Design, Implementation and Features. *Proceedings of IEEE Tridentcom 2005.*

[27] Lei, J., Yates, R., Greenstein, L., and Liu, H. 2005. Wireless Link SNR Mapping onto an Indoor Testbed. *IEEE International Conference of Testbeds and Research Infrastructures for the Development of Networks and Communities (TRIDENTCOM).*

[28] Rakotoarivelo, T., Ott. M., Jourjon, G., and Seskar, I. 2009. OMF: A Control and Management Framework for Networking Testbeds. *SOSP Workshop On Real Overlays and Distributed Systems.*

[29] White, J., Jourjon, G., Rakatoarivelo, T., and Ott, M. 2010. Measurement Architectures for Network Experiments with Disconnected Mobile Nodes. *Proceedings of IEEE Tridentcom 2010.*

[30] The Kansei Sensor Testbed. http://ceti.cse.ohio-state.edu/kansei/

[31] Stargate platform, http://platformx.sourceforge.net/Links/resource.html

[32] Murty, R., Mainland, G., Rose, I., Chowdhury, A. R., Gosain, A., Bers, J., and Welsh, M. 2008. CitySense: A Vision for an Urban-Scale Wireless Networking Testbed,. *Proceedings of the 2008 IEEE International Conference on Technologies for Homeland Security.*

[33] Banerjee, N., Corner, M. D., Towsley, D., and Levine, B. 2008. Relays, Meshes, Base Stations: Enhancing Mobile Networks with Infrastructure. *Proceedings of ACM Mobicom.*

[34] UMass DieselNet, http://prisms.cs.umass.edu/dome/umassdieselnet

[35] EMULAB, http://www.emulab.net/

[36] Universal Software Radio Peripheral, http://www.ettus.com

[37] http://users.emulab.net/trac/emulab/wiki/wireless

[38] http://users.emulab.net/trac/emulab/wiki/gnuradio

[39] VT-CORNET, http://wireless.vt.edu/coreareas/cognitive.html

[40] Software Communications Architecture, http://sca.jpeojtrs.mil

[41] Global Environment for Network Innovations (GENI) System Overview Document ID: GENI-SE-SY-SO-02.0 September 29, 2008.

[42] GENI Spiral One: http://groups.geni.net/geni/wiki/SpiralOne

[43] GENI Spiral Two: http://groups.geni.net/geni/wiki/SpiralTwo

[44] Devine, S., Bugnion, E., and Rosenblum, M. 1998. Virtualization System Including a Virtual Machine Monitor for a Computer with a Segmented Architecture. *US Patent 6397242.*

[45] Barham, P., Dragovic, B., Fraser, K., Hand, S., Harris, T., Ho, A., Neugebauer, R., Pratt, I., and Warfield, A. 2003. Xen and the Art of Virtualization. *SOSP.*

[46] Dike, J. 2001. User-mode Linux. *Proceedings of the 5th Annual Linux Showcase and Conference.*

[47] Paul, S., Ganu, S., Kamat, P., and Royer, E. B. *Requirements for Wireless GENI Experiment Control and Management, GDD-07–43.* http://www.geni.net/GDD/GDD-07–43.pdf

[48] Mahindra, R., Bhanage, G., Hadjichristofi, G., Ganu, S., Seskar, I., and Raychaudhuri, D. 2008. Integration of Heterogeneous Networking Testbeds. *Proceedings of 4th International Conference on Testbeds and Research Infrastructures for the Development of Networks & Communities.*

[49] Smith, G., Chaturvedi, A., Arunesh Mishra, A., and Suman Banerjee, S. 2007. Wireless Virtualization on Commodity 802.11 Hardware. *ACM WinTECH Workshop (co-located with ACM Mobicom)*.

[50] McKeown, N., Anderson, T., Balakrishnan, H., Parulkar, G., Peterson, L., Rexford, J., Shenker, S., and Turner, J. 2008. OpenFlow: Enabling Innovation in Campus Networks. *ACM SIGCOMM Computer Communication Review*, 38(2).

[51] http://groups.geni.net/geni/attachment/wiki/WiMAX/GENI_WiMax_System_Engg_v0.1.pdf

11

Concluding Remarks

Dipankar Raychaudhuri and Mario Gerla

In the previous chapters of this book, we have covered a broad range of networking requirements for emerging wireless scenarios along with the protocol features needed to support them. Clearly, not all of these requirements will be reflected in the general purpose architecture of the Internet, but it may be expected that many of the core capabilities will gradually migrate into mainstream networking protocols that will be in use ten to twenty years into the future. In this concluding chapter, we provide a brief discussion of the roadmap for network evolution or revolution in response to the changes in usage and technology that have been identified in this book.

Although it is impossible to predict exactly how the future Internet of the year 2025 will be realized, we can still enumerate a few alternative scenarios by which the Internet might evolve to meet the many challenges of cellular convergence and mobility. These are:

(1) *Incremental evolution of IP features*: This scenario assumes that the IP standardization process (e.g., IETF and ITU) will anticipate a reasonable set of future requirements and incorporate them into next-generation standards. This would be similar in spirit to IPv6, which improved on IPv4 by providing key features for addressing, mobility, and security. As discussed in Chapter 2, standards processes are already responding to emerging wireless technologies (such as IP-based cellular networks) and usage scenarios (such as multihop wireless access). However, evolution of a widely deployed protocol like IP is quite complicated in its own right, because the degree of design freedom is severely limited by backward compatibility requirements. Also, as new requirements and standards subcommittees are added, it becomes more and more difficult to manage complexity of the resulting protocol standard. Nevertheless, this gradual evolution scenario is perhaps

the most likely outcome given the inertia of changing any large system as big as the Internet.

(2) *Special-purpose wireless access networks with Internet backhaul*: The second evolution scenario is based on the belief that the core Internet protocol will remain relatively stable and will continue to be used for backbone connectivity. In this concept, the special requirements of wireless access networks for cellular, vehicular, ad hoc, mesh, sensor, and the like will require a specialized access network with its own unique protocol features. This scenario has the disadvantage of requiring protocol gateways between networks, making it difficult for an Internet host across the network to access the features of the wireless access network. This is how cellular networks are currently integrated with the Internet, but there are serious concerns about scalability and the lack of a seamless protocol framework.

(3) *Overlay network architectures for mobility services*: An alternative to the preceding scenario with special-purpose wireless access networks connecting to the general-purpose Internet is to create multiple customized "overlay networks" for different mobility services such as content delivery or video broadcasting. The concept of an overlay network is to deploy routers with new protocols and features on top of IP tunnels used strictly for link-layer connectivity. Overlay networks have proven to be a very flexible strategy for deployment of new networking capabilities while continuing to utilize prior investments in IP as the foundation. The main drawback of overlay networks is the fact that they lack the universality of a single Internet protocol and thus tend to have niche application developer communities. The second problem with overlays is that packets must be processed by two sets of networking protocols, thus increasing latency and processing workload at routers handling large amounts of traffic.

(4) *New clean-slate Internet protocol for emerging mobility services*: In this approach, anticipated requirements for wireless/mobile systems will be integrated into a single clean-slate protocol. Clearly, this is a complex design problem due to the relatively large number of requirements for dealing with user/device mobility, wireless link properties, geographic location, context, and so on. Any new design would have to be validated extensively on large-scale testbeds such as those discussed in Chapter 10, and would subsequently need to be released for more general use with real end-users and applications. The initial bootstrapping of such a new network service is a major challenge, though the availability of programmable networks with virtualized routers and wireless access points will gradually make it easier to try out new protocols at scale. The clean-slate scenario is considered relatively impractical because of the difficulty of competing with the large worldwide base of IP networks, but there are certainly examples of new protocols being

adopted in a short period – IP and the Web are themselves good examples of this.

(5) *Multiprotocol pluralistic architecture based on network virtualization*: Rather than attempting to combine qualitatively different requirements into a single protocol, it is possible to use network virtualization technology as a foundation capable of supporting a set of otherwise incompatible protocols that have been individually optimized for very different transport services. This so-called *pluralistic network architecture*, which first gained attention in context of experimental networks such as PlanetLab and GENI (see Chapter 10), uses network virtualization and programmability to accommodate multiple independent protocols within the same physical infrastructure. End-user devices attaching to the network will use one or more of the available services by opting in to one or more of the virtual networks. Such a pluralistic architecture also offers the important features of legacy support and graceful migration because existing protocols such as IPv4 can be supported on one of the virtual networks, while speculative new services can be added in parallel by appropriately provisioning additional network slices.

At the time of writing of this book, a number of future Internet research and prototyping activities have been initiated in the United States, Japan, South Korea, and Europe. These activities include the Future Internet Design (FIND) and Global Environment for Network Innovation (GENI) programs started under the auspices of the U.S. National Science Foundation (NSF). The FIND program in particular has supported a large number of clean-slate protocol research ideas during the first phase (2006–2009). These projects span a wide range of topics including network security, economic incentives, content delivery, network virtualization, and the mobile/wireless scenarios considered in this book. The European Union's FP7 program has similar coverage of various future Internet research topics, and includes several projects with a particular focus on mobility aspects. The Japanese future Internet research program (called the "new generation network project") is also a significant activity with coverage of enabling core technologies as well as new protocol architectures.

The NSF FIND program in the United States is currently moving to a second phase (called the Future Internet Architecture [FIA] initiative) aimed at converging early-stage protocol ideas into a small number of competing Internet architectures to be validated and deployed at large scale. It is still too early to assess the impact of these future Internet research initiatives, but it may be expected that the results from these programs will gradually influence the Internet mainstream in terms of design requirements and new protocol components. There is also a slim but nontrivial probability that one or more of the new architectures under evaluation will gain growing acceptance first in research trials and later in real-world deployments. Regardless of the eventual outcome, the journey from today's network to the future Internet with ubiquitous mobility will be an interesting one and will result in many technology and service innovations.